Mine Ventilation

Carlos Sierra

Mine Ventilation

A Concise Guide for Students

 Springer

Carlos Sierra
Department of Mining, Topography
and Structure Technology
University of Leon
León, Spain

ISBN 978-3-030-49805-4   ISBN 978-3-030-49803-0   (eBook)
https://doi.org/10.1007/978-3-030-49803-0

This Springer imprint is published by the registered company Springer Nature Switzerland AG
The registered company address is: Gewerbestrasse 11, 6330 Cham, Switzerland

Preface

This document was initially conceived to cover the topic Mine Ventilation included as part of the module Technology of Exploitation of Mining Resources in the Master's Degree in Mining Engineering from the University of Cantabria. It has now been expanded and updated to meet my new teaching obligations covering the same topic in the module Advanced Mining Techniques in the Master's Degree in Mining Engineering and Energy Resources at the University of León.

The volume includes all the concepts necessary to succeed in any general course on the ventilation of mines and underground works. It also contains a significant number of questions and exercises which aim to strengthen the understanding of this theoretical base. When writing it, I took into account the chronic shortage of time that students in technical schools have, so I sacrificed rhetoric in favour of synthesis.

My aim was to produce a text that would serve the student as a link between the disciplines of fluid mechanics and mine ventilation. This volume also incorporates an updated compilation of regulations as well as a large number of worked-out examples of realistic problems, which, I hope, will make it a useful reference for professionals in the field of underground ventilation systems.

I am indebted to Mr. Emilio Andrea Blanco, a mining engineer and retired adjunct professor at the University of Cantabria for his invaluable help and wise advice with the calculations for ventilation networks, his many useful comments and patient review of the first manuscript. Despite our careful scrutiny, I am aware that there may be errors remaining in the document. I assume all responsibility for these and invite the readers, should they spot any error, to contact me so that they can be corrected for future editions.

Contact through: csief@unileon.es.

León, Spain   Carlos Sierra, Ph.D.
June 2020   M.Sc. Civil Engineering, M.Sc. Process Engineering
                                                      Mining Engineer and Geological Engineer
                                          https://www.researchgate.net/profile/Carlos_Sierra18

The original version of the book was revised: Incorporation of the Introduction section and correction of figure placement and typographical errors have been carried out. The correction to the book is available at https://doi.org/10.1007/978-3-030-49803-0_9

Contents

Chapter 1
Fundamental Concepts of Fluid Mechanics for Mine Ventilation

1.1 Introduction

This chapter offers a very concise review of the main concepts of fluid mechanics, with a particular focus on aerodynamics, the mastery of which is essential for understanding the rest of the text. Although many students may be familiar with this subject, they should revisit it to refresh concepts and units.

1.2 Fluid Statics

1.2.1 Pressure Scales

Two fundamental pressure scales are (Fig. 1.1):

- *Absolute pressure* (P_{abs}), which is measured in relation to an absolute vacuum. It is always positive.
- *Gauge pressure* (P_g), which is measured in relation to the atmospheric pressure (P_{atm}) at any time. It can be positive or negative.

The relationship between the two is expressed in Eq. 1.1.

$$P_{abs} = P_{atm} + P_g \tag{1.1}$$

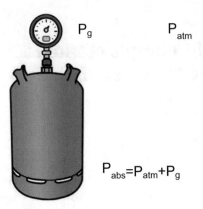

$$P_g \qquad\qquad P_{atm}$$

$$P_{abs}=P_{atm}+P_g$$

Fig. 1.1 Atmospheric (P_{atm}), gauge (P_g) and absolute (P_{abs}) pressure in a container

1.2.2   Laws of Perfect Gases

An ideal gas is one in which the following theoretical conditions are met:

- The volume occupied by gas molecules is negligible compared to the total volume of the gas.
- There are no forces of attraction or repulsion between the gas molecules and their container.
- There is no loss of internal energy due to collisions between gas molecules.

In reality, these conditions are usually met when the density is low, that is, when the pressure is low and the temperature high. Under these simplifications, the following laws, called the laws of perfect gases in theoretical physics, are applicable.

Boyle's Law[1]

"For a given quantity of an ideal gas that remains at a fixed temperature (T), the pressure (P_i) and the volume (V_i) are inversely proportional" (Fig. 1.2a). That is to say (Eq. 1.2):

$$P_1 V_1 = P_2 V_2 \tag{1.2}$$

Charles' Law

"For a given quantity of an ideal gas that remains at a fixed pressure (P), the volume (V_i) and the temperature (T_i) are directly proportional" (Fig. 1.2b). Then, mathematically (Eq. 1.3):

---

[1] Also known as Boyle-Mariotte's law.

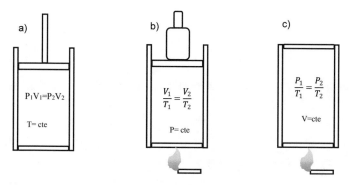

Fig. 1.2 Laws of ideal gases. **a** Boyle's law; **b** Charles's law; **c** Gay-Lussac's law

$$\frac{V_1}{T_1} = \frac{V_2}{T_2} \qquad (1.3)$$

Gay–Lussac's Law

"For a fixed volume of an ideal gas (V), the pressure (P_i) and temperature (T_i) are directly proportional" (Fig. 1.2c). So (Eq. 1.4):

$$\frac{P_1}{T_1} = \frac{P_2}{T_2} \qquad (1.4)$$

General Gas Equation (Ideal Gas Law)

The *general gas equation* is stated as follows: "For a gas in which pressure and temperature are kept constant, the volume it occupies is directly proportional to the number of molecules (n)".

Deduction:

According to Boyle's law, if T and n are constant, then:

$$V \alpha \frac{1}{P}$$

According to Charles' law, if P and n are constant, then:

$$V \alpha T$$

Bearing in mind Avogadro's law[2]:

$$V \alpha n$$

---

[2] Avogadro's law (1811): "Equal volumes of different gaseous substances, measured under the same conditions of pressure and temperature, contain the same number of molecules."

Therefore:

$$V \alpha \frac{nT}{P}$$

The concept of proportionality is resolved by introducing a constant, then:

$$V = R\frac{nT}{P}$$

Or in its most common form (Eq. 1.5):

$$PV = nRT \tag{1.5}$$

where R is the universal constant of the molar, universal, or ideal gas constant. To obtain its value, we can start from the previous expression. Since a mole of any gas at 0 °C and one atmosphere pressure occupies 22.414 l, it can be deduced that:

$$R = \frac{PV}{nT} = \frac{1 \, \text{atm} \cdot 22.414 \, \text{l}}{1 \, \text{mol} \cdot 273 \, \text{K}} = 0.082 \frac{\text{atm} \, \text{l}}{\text{K} \, \text{mol}}$$

Alternatively:

$$R = 0.0831 \frac{\text{bar} \, \text{l}}{\text{K} \, \text{mol}} = 8.314 \frac{\text{kPa} \, \text{l}}{\text{K} \, \text{mol}} = 8.314 \frac{\text{J}}{\text{K} \, \text{mol}} = 1.99 \frac{\text{cal}}{\text{K} \, \text{mol}}$$

1.2.3 Real Gases: van der Waals Equation

To better model the behaviour of real gases, van der Waals established a modification of the previous equation. Specifically, it corrects the pressure (P) and volume (V) terms (Eq. 1.6):

$$\left(P + \frac{an^2}{V^2} \right)(V - nb) = nRT \tag{1.6}$$

where

- a: Coefficient to account for intermolecular attractions, and
- b: Coefficient to account for the finite volume of the molecules.

1.2.4 Hydrostatic Pressure

This type of pressure is due to a column of fluid on an object. The equation by which it is governed can be easily obtained from the weight (W) and density (ρ) of a mass (m) of air contained in a column of height (h) over its base area[3] (A) (Fig. 1.3).

This relationship is expressed mathematically as:

$$P = \frac{W}{A} = \frac{mg}{A} = \frac{\rho V g}{A} = \frac{\rho A h g}{A}$$

Then, finally (Eq. 1.7):

$$P = \rho g h \tag{1.7}$$

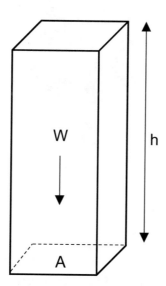

Fig. 1.3 Force exerted by a fluid prism on its base

---

[3]Note that, although surface (S) denotes the topological object and area (A) its measurement, we do not make any distinctions in this text for simplicity reasons.

1.3   Fluid Dynamics

1.3.1   Types of Pressure Present in a Moving Fluid

A moving fluid has the following types of pressure:

Static Pressure (P_s)

Static pressure is a concept comparable to atmospheric pressure or the pressure of gas-filled vessels (Fig. 1.4). This type of pressure is the one that a fluid has regardless of its speed. In addition, it is the same in all directions and can be positive or negative; that is, it can help the flow or put up resistance.

Fig. 1.4 Static pressure acting inside an air balloon

Dynamic Pressure (P_v)

The *dynamic pressure* is a measure of the kinetic energy per unit volume possessed by a fluid particle in its path along a pipe (Fig. 1.5). In other words, it is the type of pressure caused by the inertia of fluid particles striking a surface perpendicular to their movement. Mathematically, it can be expressed as (Eq. 1.8):

$$P_v = \frac{1}{2}\rho v^2 \tag{1.8}$$

Its relationship with kinetic energy derives from the fact that density (ρ) represents the mass and v represents the velocity of the fluid.

Total Pressure (P_T)

The *total pressure* is the sum of the two previous ones (Fig. 1.5), that is, (Eq. 1.9):

$$P_T = P_s + P_v \qquad (1.9)$$

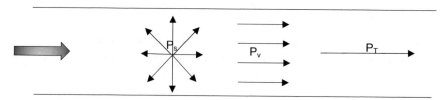

Fig. 1.5 Types of fluid pressure in forced conduction, static pressure (P_s), dynamic pressure (P_v) and total pressure (P_T)

1.3.2 Pressure Units

The pascal (Pa) is the International System (SI) pressure unit. It is the pressure exerted by a force of 1 N on a 1 m^2 surface normal to it. Other units exist. Their conversion factors are listed in Table 1.1.

From the definition of pressure as force (F) per unit area (A):

$$P = \frac{F}{A} \qquad (1.10)$$

Multiplying up and down by the displacement (d), it becomes:

$$P = \frac{F\,d}{A\,d}$$

This relationship means that pressure can be expressed in units of energy or work (W) per units volume (V) (Eq. 1.11):

Table 1.1 Pressure units most frequently used in engineering

	bar	Pa	atm	kgf $\times$ cm^{-2}	mmHg	mmwc
1 bar	1	1×10^5	0.987	1.02	7.5×10^2	1.02×10^4
1 Pa	1×10^{-5}	1	9.87×10^{-6}	1.02×10^{-5}	7.5×10^{-3}	0.102
1 atm	1.013	1.013×10^5	1	1.033	7.6×10^2	1.033×10^4
1 kgf $\times$ cm^{-2}	0.981	9.807×10^4	0.968	1	7.356×10^2	10^4
1 mmHg	1.333×10^{-3}	1.333×10^2	1.316×10^{-3}	1.36×10^{-3}	1	13.59
1 mmwc	9.807×10^{-5}	9.807	9.677×10^{-5}	10^{-4}	7.354×10^{-2}	1

$$P = \frac{W}{V}\left[\frac{N}{m^2}, \frac{J}{m^3}\right] \tag{1.11}$$

In this way, it is possible to indicate that the ambient pressure is one atmosphere, 101,300 Pa or 101,300 J m^{-3}.

Exercise 1.1 A tank contains 1 m^3 of air at 11 atm (absolute pressure). Approximate, according to what you have learned, the energy stored in it.

Solution

Normally pressure vessels are calculated with gauge pressure (P_g) as the forces generated by the atmospheric pressure inside the tank (P_{atm}) are cancelled out by the forces generated by the atmospheric pressure outside the tank (P_{atm}). This concept is depicted in the figure:

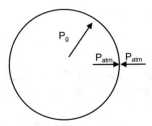

The student would tend to use $W = P\,V$ straightaway. However, the energy stored in the vessel depends on thermodynamics.

Thus, if we consider that the air expands adiabatically, its maximum expansion work is (Coleman et al. 1988):

$$W_{A\rightarrow B} = -k(11-1) \cdot 101,300\,\frac{J}{m^3} = -k \cdot 1,013,000\,J = -k \cdot 1.01 \times 10^6\,J.$$

where $k = \frac{1}{\gamma-1}$, and γ is the ratio of the specific heats.

Similarly, if we consider an isothermal process—as in Compressed-Air Energy Storage (CAES) system—we have (Coleman et al. 1988):

$$W_{A\rightarrow B} = P_B V_B \ln \frac{P_A}{P_B} + (P_B - P_A)V_B =$$

$$= 1.01 \times 10^6\,Pa \cdot 1\,m^3 \cdot \ln\frac{101,300\,Pa}{1.01 \times 10^6\,Pa} + \left(1.01 \times 10^6 - 101,300\right)Pa \cdot 1\,m^3 =$$

$$= -1.41 \times 10^6\,J$$

1.3.3 Energy Conservation: Bernoulli's Equation

It is assumed that a control volume of an incompressible fluid[4] travels through forced conduction. The sum of the kinetic (K), potential (U) and due to pressure (W) energies, must be the same for any point of the fluid under the principle of energy conservation, which is expressed as:

$$K_1 + U_1 + W_1 = K_2 + U_2 + W_2$$

If a fluid control volume is assumed in a pipeline, the particles to the left of control volume (boundary 1) exert positive work, i.e. they act in the direction of favouring circulation. However, those on the right of control volume (boundary 2) perform negative work, opposing the movement (Fig. 1.6).

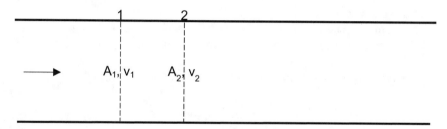

Fig. 1.6 Boundaries 1 and 2 of a control volume

The work exerted by these particles can be defined as:

$$W = Fd = PAd = PV$$

If then we apply the expressions of kinetic energy $\left(\frac{1}{2}mv^2\right)$, potential energy ($mgh$) and pressure energy ($W$), we have:

$$\frac{1}{2}m_1v_1^2 + m_1gh_1 + P_1V_1 = \frac{1}{2}m_2v_2^2 + m_2gh_2 + P_2V_2$$

For each mass (m): m $= \rho V$, where ρ is the density and V the volume, therefore:

$$\frac{1}{2}\rho V_1v_1^2 + \rho V_1gh_1 + P_1V_1 = \frac{1}{2}\rho V_2v_2^2 + \rho V_2gh_2 + P_2V_2$$

---

[4]A fluid is incompressible (*isochoric*) if the effects of pressure on its density are negligible at a certain speed. Normally it is valid to consider air as incompressible for ventilation calculations, given its moderate circulation speed and the reduced compression ratio provided by fans. This simplification cannot be done, for example, in the case of mine compressed air networks.

As the fluid is incompressible, the control volume crossing 1 and 2 is the same size, and therefore has the same mass. Taking out the common factor, one arrives at:

$$\frac{1}{2}\rho V\left(v_2^2 - v_1^2\right) + \rho V g(h_2 - h_1) = (P_2 - P_1)V$$

Which simplifies to:

$$\frac{1}{2}\rho\left(v_2^2 - v_1^2\right) + \rho g(h_2 - h_1) = (P_2 - P_1)$$

Then, regrouping terms leads to Bernoulli's equation (Eq. 1.12):

$$\frac{1}{2}\rho v_1^2 + \rho g h_1 + P_1 = \frac{1}{2}\rho v_2^2 + \rho g h_2 + P_2 \qquad (1.12)$$

This expression is stated as:

"The sum of kinetic per unit volume $\left(\frac{1}{2}\rho v_i^2\right)$, potential per unit volume $\left(\rho g h_i\right)$ and static energy per unit volume (P_i) remains constant for any two points along a streamline".

Sometimes the Bernoulli equation is regrouped as follows (Eq. 1.13):

$$\frac{P_i}{\rho g} + \frac{v_i^2}{2g} + z_i = H_i \qquad (1.13)$$

where

- $\frac{P_i}{\rho g}$: Pressure head or static head (m),
- $\frac{v_i^2}{2g}$: Velocity head (m),
- z_i : Elevation head (m), and
- H_i : Total head (m).

Keep in mind that the units are meters of the circulating fluid and not the fluid present in the gauge, which will generally be different from the circulating fluid.

Exercise 1.2 Demonstrate that the units in the terms $\frac{P}{\rho}g$ and $\frac{v^2}{2g}$ of Bernoulli's equation are meters.

Solution

$$\frac{P}{\rho g}\left[\frac{\frac{N}{m^2}}{\frac{Kg}{m^3}\frac{N}{Kg}} = m\right]$$

$$\frac{v^2}{2g}\left[\frac{\left(\frac{m}{s}\right)^2}{\frac{m}{s^2}} = m\right]$$

1.3.4 Continuity Equation and Bernoulli Effect

The continuity equation is nothing more than the law of conservation of mass applied to the flow of a fluid through a duct. Therefore, you have that (Eq. 1.14):

$$\rho A_1 v_1 = \rho A_2 v_2 \tag{1.14}$$

where ρ is the density of the fluid, A is the cross-sectional area of the duct and V is the velocity in the section considered.

The above expression finds its most characteristic example in the Venturi effect. This is the reduction in static pressure when the fluid velocity increases in the constricted section of a duct.

Thus, for the continuity equation to be fulfilled, if area A_2 is smaller than area A_1, velocity v_2 will be greater than v_1 and P_{v_2} will be higher than P_{v_1} (Fig. 1.7). In addition, according to Bernoulli's expression, an increase in the dynamic pressure in one of the sections implies a decrease in static pressure, which translates into static pressure P_2 being less than P_1. Similar reasoning can be made for area cross section A_3 of Fig. 1.7 with respect to area cross sections A_1 and A_2.

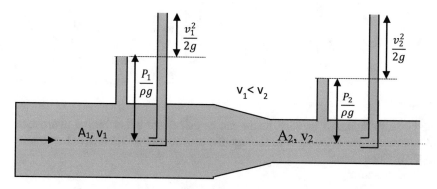

Fig. 1.7 Reduction of the static pressure experienced in the contraction of a pipe

1.3.5 Measurements

It is important to note that the measuring elements located perpendicular to the surfaces of the ducts through which a fluid circulates provide static pressures (Fig. 1.8a). On the other hand, measurements made in the direction of flow indicate total pressures (Fig. 1.8c). If part of the total pressure is compensated by the static pressure, then the resulting measurement is of dynamic pressure (Fig. 1.8b).

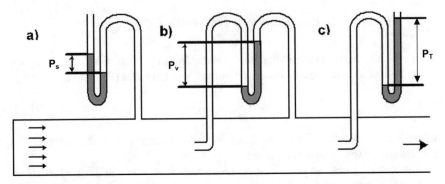

Fig. 1.8 Various arrangements for measuring the three types of pressures: static (P_s), dynamic (P_v) and total (P_T)

Exercise 1.3 Indicate the types of pressures measured in the figure. Bearing in mind that the heights are in cm of water column, transform these units to Pa.

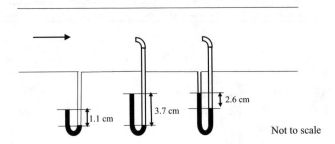

Solution

The pressure indicated by each metre can be calculated either by the hydrostatic column expression ($P = \rho g h$) or through the equivalence of Table 1.1, 1 mmwc = 9.807 Pa. Therefore:

Static pressure:

$$1.1 \cdot 10 \, \text{mmwc} \cdot \frac{9.807 \, \text{Pa}}{1 \, \text{mmwc}} = 107.88 \, \text{Pa}$$

Total pressure:

$$3.7 \cdot 10 \, \text{mmwc} \cdot \frac{9.807 \, \text{Pa}}{1 \, \text{mmwc}} = 362.86 \, \text{Pa}$$

Dynamic pressure:

$$2.6 \cdot 10 \, \text{mmwc} \cdot \frac{9.807 \, \text{Pa}}{1 \, \text{mmwc}} = 254.98 \, \text{Pa}$$

Exercise 1.4 The diagram represents a Pitot tube inserted into a duct through which air circulates. The difference in height of mercury measured on the Pitot was 10 mm. You are asked to determine the speed at which air moves inside the duct. Data: Air density: 1.2 kg m^{-3}, mercury density: 13.60 g cm^{-3}.

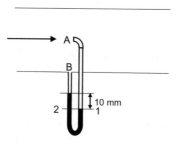

Solution

Considering the arrangement at the inlet A, both static and dynamic pressure enter the Pitot tube. On the other hand, at inlet B, only static pressure enters the tube. The pressure in points at the same high (1 and 2) must be the same in the Pitot tube. Thus:

$$P + \rho_{Hg}gh = \frac{1}{2}\rho_a v^2 + P$$

You finally get that:

$$v = \sqrt{2gh\frac{\rho_{Hg}}{\rho_a}} = \sqrt{2 \cdot 9.81\frac{m}{s^2}0.01\,m\frac{13{,}600}{1.2}} = 47.2\frac{m}{s}$$

1.3.6 Conservation of Linear Momentum

To immobilize a tennis ball travelling at 200 km h^{-1} in a time t, is much simpler than to stop, at that same time, a vehicle that moves at the same speed. To evaluate the effects of motion, it is, therefore, necessary to know not only the velocity of bodies but also their mass. This can be done using the kinetic energy equation; however, what cannot be considered is the time it takes for the body to stop. It is, therefore, necessary to have an expression to explain this phenomenon.

Thus, for two interacting bodies, if F_1 is the force exerted by m_A in m_B and F_2 by m_B in m_A we have by Newton's second law that:

$$F_1 = m_B\frac{dv_B}{t} = m_B\frac{v'_B - v_B}{t}$$

Wherein the superscript ($'$) represents the final velocities (v).
Also, by Newton's third law: $F_2 = -F_1$, so:

$$-F_1 = m_A \frac{dv_A}{t} = m_A \frac{v'_A - v_A}{t}$$

Then:

$$m_B \frac{v'_B - v_B}{t} = -m_A \frac{v'_A - v_A}{t}$$

And therefore we get Eq. 1.15:

$$m_B v_B + m_A v_A = m_B v'_B + m_A v'_A \qquad (1.15)$$

With what one finally obtains the law of conservation of the linear moment for the system (Eq. 1.16):

$$p_{\text{before}} = p_{\text{after}} \qquad (1.16)$$

Exercise 1.5 The figure represents a regulator with a 0.75 m diameter orifice, which is located in a 3 m diameter circular gallery in which the air circulates at 8 m s^{-1}. Determine the force (F) acting on the walls of the regulator if the pressure difference on both sides is 200 Pa. Assume that there is no *vena contracta*.

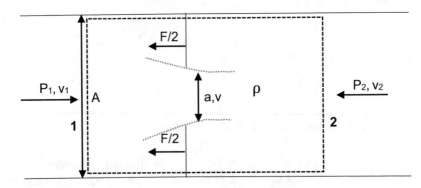

Resolve the conservation of the linear momentum for the indicated control volume. Assume the air is incompressible and do not consider its friction on the walls of the system.

Note: The indication of $\frac{F}{2}$ is by symmetry in the figure.

Solution

By the continuity equation, the mass flow ($\dot{m}$) is the same on both sides of the control volume, so since the cross section (A) is the same, the speed (v_i) also coincides. Applying then the momentum conservation equation, one obtains:

$$\sum F_x = P_1 A - F - P_2 A = \dot{m}_2 v_2 - \dot{m}_1 v_1 = \dot{m}v - \dot{m}v = 0$$

Solving for F yields

$$F = P_1 A - P_2 A = (P_1 - P_2)A$$

Finally, substituting values gives:

$$F = 200 \, \text{Pa} \, \pi \frac{3^2}{4} \text{m}^2 = 450 \, \text{N}$$

Note that in the solution of this exercise the formation of the *vena contracta* has not been taken into account.

See Sect. 1.3.10 for further details on the *vena contracta* concept.

1.3.7 Viscosity

Viscosity is a characteristic of fluids that represents their resistance to flow. Conceptually, it corresponds to the internal frictional force between sheets in contact with a moving fluid. Formally, it is the relationship between the shear force and the speed gradient. It can be expressed in two forms, (a) *Dynamic viscosity* (μ) in direct application of the definition and whose units are Pa s, N s m^{-2} or kg m^{-1} s^{-1}; or (b) *kinematic viscosity* (ν), which is the dynamic viscosity divided by the density of the fluid. In which case the units are m^2 s^{-1}.

According to the above, kinematic viscosity can be written as (Eq. 1.17):

$$\nu = \frac{\mu}{\rho} \tag{1.17}$$

1.3.8 Laminar and Turbulent Flow: Reynolds Number

There are three fundamental fluid flow regimes: *laminar, turbulent* and *transition*.[5] The velocity distribution in the cross section of a duct is different in each of them. Thus, the profile will be parabolic if the flow is laminar (Fig. 1.9a), and in the turbulent regime the parabola will tend to flatten out as the turbulence gets higher (Fig. 1.9b).

---

[5]This transition regime occurs, for example, in dense media mineral separation and in certain froth flotation cases.

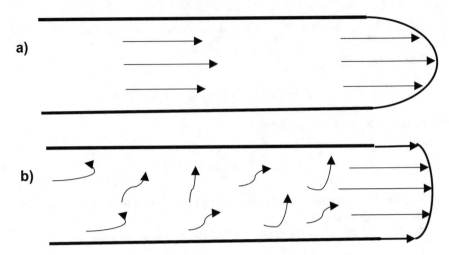

a)

b)

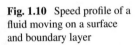

Fig. 1.9 Velocity profile of the fluid circulating through a duct: **a** Laminar, **b** turbulent

Fig. 1.10 Speed profile of a
fluid moving on a surface
and boundary layer

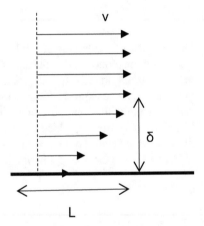

In a like manner, when a fluid flows on a plate, minimum velocity is reached in contact with the plate. Then, the speed increases with distance from the plate up to a point at which it remains constant (Fig. 1.10).

The *Reynolds number* is a dimensionless number, whose square root is inversely proportional to the distance from a surface at which the liquid velocity remains constant (δ) (boundary layer) (Fig. 1.10). This number arises from the relationship between inertial (destabilizing) and viscous (stabilizing) forces acting on the fluid[6] and is, therefore, dimensionless. Thus:

---

[6]Dimensionless numbers have proven very useful in scaling-up operations. That is to say, in quantifying the expected results on a real scale from those derived from tests on a smaller scale.

$$Re = \frac{\text{Inertial forces}}{\text{Viscous forces}} = \frac{m\,a}{\mu\,\frac{v}{L}\,A} = \frac{\rho L^3 \frac{v}{t}}{\mu \frac{v}{L} L^2} = \frac{\rho L^3 \frac{1}{t}}{\mu \frac{1}{L} L^2} = \frac{\rho L^2 \frac{1}{t}}{\mu}$$

Or (Eq. 1.18):

$$Re = \frac{\rho v L}{\mu} \tag{1.18}$$

where

- m: Mass (kg),
- a: Acceleration (m s^{-1}),
- A: Area (m^2),
- ρ: Fluid density (kg m^{-3}),
- v: Maximum fluid velocity (m s^{-1}),
- μ: Dynamic viscosity of the fluid (Pa s, N s m^{-2}, or kg m^{-1} s^{-1}), and
- L: Characteristic length of the system or diameter (D) of the pipe through which the fluid flows (m). This parameter will often be replaced by the hydraulic diameter[7] (Eq. 1.19):

$$D_h = \frac{4A}{O} \tag{1.19}$$

- A: Cross-sectional area (m^2) of the duct, and
- O: Perimeter (m).

Making use of Eq. 1.17, the Reynolds number can be defined as (Eq. 1.20):

$$Re = \frac{v\,D_h}{\nu} \tag{1.20}$$

For air, which has a kinematic viscosity between 1.42×10^{-5} m^2 s^{-1} (10 °C) and 1.60×10^{-5} m^2 s^{-1} (30 °C), it can be written that:

$$Re = 67{,}280\,D_h\,v$$

The Reynolds number delimits the change between flow regimes. The limits vary depending on closed-conduit flows or open flows. Thus, for a circular pipe, if Re $>$ 10^3 the regime will be turbulent, whereas for fluid flowing over a flat plate this same regime is reached at Re $> 3.5 \times 10^5$ to 1×10^6. In mine galleries, it is frequent to

---

[7]The hydraulic diameter is the quotient between the cross-sectional area and its perimeter. This parameter can be used to characterize an irregular surface using a single parameter. For this reason, its use is extensible to other fields that have nothing to do with fluid mechanics, such as, for example, in dimensioning the pillars in room and pillars mining when the pillar cross section has to be characterized.

indicate that if Re < 2000 the flow is in the laminar regime and if Re > 4000 the flow is turbulent. Between 2000 and 4000 is the transition regime. Novitzky (1962) notes that the change of regime from laminar to turbulent in mining stopes occurs at Re around 1000 to 1500.

It will be seen below that most principal ventilation circuits operate in a turbulent regime. This is more a consequence of the large diameters of the galleries than the high speeds at which the air moves in them. The turbulent regime, characterized by the formation of internal eddies, is desirable in mine ventilation, since it improves the mixture of clean gases with toxic gases, favouring their evacuation.

Exercise 1.6 The volumetric flow rate of air circulating through a vertical shaft with a 6 m diameter is 250 m^3 s^{-1}. Determine whether the regime is laminar or turbulent. Assume:

- Average temperature: 20 °C,
- Air density: 1.2 kg m^{-3}, and
- $\mu = (17.0 + 0.045T[°C]) \times 10^{-6} [N \frac{s}{m^2}]$.

Solution

If you start from the continuity equation:

$$v = \frac{Q}{A} = \frac{Q}{\frac{\pi D^2}{4}} = \frac{250 \frac{m^3}{s}}{\frac{\pi 6^2}{4} m^2} = 8.84 \frac{m}{s}$$

Bearing in mind that viscosity can be obtained from:

$$\mu = (17.0 + 0.045 \, T[°C]) \times 10^{-6} = (17.0 + 0.045 \cdot 20) \times 10^{-6}$$

$$\mu = 1.79 \times 10^{-5} \, N \frac{s}{m^2}$$

Then the Reynolds' is:

$$Re = \frac{\rho Dv}{\mu} = \frac{1.2 \frac{Kg}{m^3} \cdot 6\,m \cdot 8.84 \frac{m}{s}}{1.79 \times 10^{-5} \, N \frac{s}{m^2}} = 3.56 \times 10^6$$

If Re < 2000 the flow is laminar, if Re > 4000 the flow is turbulent. Thus, in the present case, the air flows in a turbulent regime.

1.3.9 Friction Losses

As air flows through a duct or gallery, it loses energy due to friction between fluid layers, as well as friction between air and walls. In the particular case of mining, they account for 70–90% of total mine losses. Figure 1.11 depicts the static pressure losses suffered by a water pipe due to friction when circulating through a horizontal pipe.

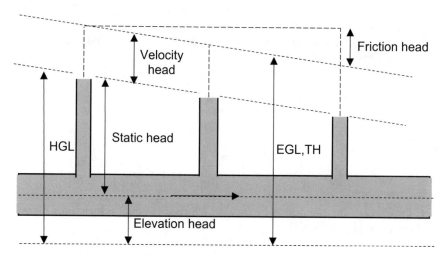

Fig. 1.11 Energy gradients in a flow. HGL, Hydraulic Gradient Line; EGL, Energy Gradient Line or Total Head (TH)

The total pressure losses by friction[8] (ΔP_f) of a fluid with the walls of the duct can be calculated using the Darcy–Weisbach equation (Eq. 1.21):

$$\Delta P_f = f \frac{L}{D_h} \rho \frac{V^2}{2} \tag{1.21}$$

where

- ΔP_f: Pressure loss due to friction (Pa),
- f: Darcy friction factor (dimensionless),
- L: Length of the duct (m),
- D_h: Hydraulic diameter of the pipe (m),
- v: Fluid velocity (m s^{-1}), and
- ρ: Density of the fluid (kg m^{-3}).

---

[8]As a consequence of the continuity equation, if the cross section in a duct is the same, so is the velocity. Thus, any pressure loss is total pressure loss if the duct has a constant cross section. Therefore, as in this case there is no variation in the dynamic pressure, the total pressure loss coincides with the static pressure loss.

Or as a head loss (H_f), in meters of fluid column (Eq. 1.22):

$$H_f = f \frac{L}{D_h} \frac{V^2}{2g} \tag{1.22}$$

where

- g: Standard acceleration due to gravity (m s^{-2}).

From the above expressions, it can be deduced that the larger the diameter of a pipe, the lower the friction losses.

If friction losses are to be considered, the term H_f must be added to account for them in the Bernoulli expression:

$$\frac{P_1}{\rho g} + \frac{v_1^2}{2g} + z_1 = \frac{P_2}{\rho g} + \frac{v_2^2}{2g} + z_2 + H_f$$

Thus:

$$\frac{P_1}{\gamma} + \frac{v_1^2}{2g} + z_1 = \frac{P_2}{\gamma} + \frac{v_2^2}{2g} + z_2 + f \frac{L}{D_h} \frac{v_1^2}{2g}$$

If the duct has the same cross section, which implies equal velocity, but there is a difference in height, then $v_1 = v_2$, so:

$$\frac{P_1 - P_2}{\rho g} = z_2 - z_1 + f \frac{L}{D_h} \frac{v^2}{2g}$$

The coefficient of friction is usually calculated using the *Moody diagram* or multiple expressions. If the regime is laminar and the duct has a circular cross section,[9] the *Hagen–Poiseuille equation* (1840) can be used (Eq. 1.23):

$$f = \frac{64}{\mathrm{Re}} \tag{1.23}$$

If the regime is turbulent various expressions are better adapted to reality according to the type of duct. Thus, if the flow is turbulent ($\mathrm{Re} < 10^5$) and the duct is smooth, the coefficient of friction depends more on the Reynolds number than on the roughness. In these cases, the *Blasius equation* (1913) can be used (Eq. 1.24):

$$f = \frac{0.316}{\mathrm{Re}^{0.25}} \tag{1.24}$$

---

[9]The "f" used in this text corresponds to the friction factor of Darcy's law sometimes denoted as f_D. Although this is the most widespread, it is possible to find sources that use Fanning's friction factor (f_F). The relationship between the two is $f_D = 4f_F$.

Karmann–Prandtl (e.g. Prandtl 1935) obtained an implicit function[10] for smooth ducts with wider applicability within the turbulent regime (Re: $4000–10^8$) (Eq. 1.25):

$$\frac{1}{\sqrt{f}} = -2.0 \log\left(\frac{2.51}{Re\sqrt{f}}\right) \tag{1.25}$$

If the flow is turbulent and the duct very rough, the roughness has much more influence than the Reynolds number, which has almost no influence. In this case, the *Nikuradse formula* applies (1932, 1933) (Eq. 1.26):

$$\frac{1}{\sqrt{f}} = -2.0 \log\left(\frac{\frac{\varepsilon}{D_h}}{3.71}\right) \tag{1.26}$$

where

- ε: Absolute conduction roughness (m),
- D_h: Hydraulic diameter (m), and
- r: Relative roughness $\left(\frac{\varepsilon}{D_h}\right)$.

If the flow is turbulent and the roughness conditions are intermediate ($\varepsilon/D_h = 0 - 0.05$), one of the most used relationships is the *Colebrook–White equation* (1937) (Eq. 1.27):

$$\frac{1}{\sqrt{f}} = -2.0 \log\left(\frac{\frac{\varepsilon}{D_h}}{3.71} + \frac{2.51}{Re\sqrt{f}}\right) \tag{1.27}$$

Although the *USBM equation* (Smith 1956), modified it for such conditions by proposing (Eq. 1.28):

$$\frac{1}{\sqrt{f}} = -2.0 \log\left(\frac{\frac{\varepsilon}{D_h}}{3.71} + \frac{2.825}{Re\sqrt{f}}\right) \tag{1.28}$$

The last two expressions are implicit for f, so approximate relationships such as *Haaland's equation* (1983) have been proposed (Eq. 1.29):

$$\frac{1}{\sqrt{f}} = -1.8 \log\left[\frac{6.9}{Re} + \left(\frac{\frac{\varepsilon}{D}}{3.71}\right)^{1.11}\right] \tag{1.29}$$

A practical alternative for determining the friction factor is the Moody diagram. This chart depicts the Darcy friction factor as a function of both the Reynolds number and the relative roughness. There are four zones in the diagram, namely (Fig. 1.12):

---

[10]An equation that establishes a relationship between two or more variables but cannot be written as $y = f(x)$.

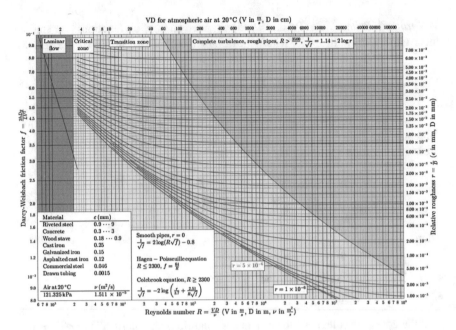

Fig. 1.12 Moody diagram for atmospheric air at 20 °C

1. *Laminar flow zone*, where f is a linear function of Re.
2. *Critical zone* (Re: 2000–4000), where the flow is unstable and might have either laminar or turbulent characteristics.
3. *Transition zone*, where f depends on the values of both Re and the relative roughness.
4. *Fully developed turbulence zone*, where f is just a function of the relative roughness and thus independent of Re (horizontal asymptote).

Exercise 1.7 Air is flowing downstream through a 500 mm diameter cast iron ventilation duct with a circular cross section. The flow rate is 0.9 m³ s⁻¹ and the duct is inclined 15° to the horizontal. Determine the static pressure loss that takes place along 100 m of the duct.

Data: Roughness of cast iron duct (ε): 0.0015 mm, air density: 1.2 kg m⁻³, dynamic viscosity of air (μ): 1.86×10^{-5} N$\frac{s}{m^2}$.

Solution

Given that:

$$v = \frac{Q}{A} = \frac{Q}{\frac{\pi D^2}{4}} = \frac{0.9 \frac{m^3}{s}}{\pi \frac{0.5^2}{4} m^2} = 4.584 \frac{m}{s}$$

Reynolds' number is:

$$Re = \frac{\rho D v}{\mu} = \frac{1.2 \frac{Kg}{m^3} \cdot 0.5 \cdot 4.584 \frac{m}{s}}{1.86 \times 10^{-5} N \frac{s}{m^2}} = 147{,}870.96$$

With Re calculated, it can be verified that the conditions to apply the Haaland expression are fulfilled:

$$Re = 4000{-}10^8$$

$$\frac{\varepsilon}{D} = \frac{0.0015 \times 10^{-3}}{500 \times 10^{-3}} = 3 \times 10^{-6}$$

$\frac{\varepsilon}{D}$ belongs to the interval : $(0 - 0.05)$

Now, you can estimate f from the Haaland equation:

$$\frac{1}{\sqrt{f}} = -1.8 \log\left[\frac{6.9}{Re} + \left(\frac{\frac{\varepsilon}{D}}{3.71}\right)^{1.11}\right] = -1.8 \log\left[\frac{6.9}{147{,}870.96} + \left(\frac{3 \times 10^{-6}}{3.71}\right)^{1.11}\right]$$

$$f = 0.016466$$

If the duct has the same cross section, which implies the same speed, but there is a difference in height, then:

$$\frac{P_1 - P_2}{\rho g} = z_2 - z_1 + f \frac{L}{D_h} \frac{v^2}{2g}$$

Taking into account the definition of flow, then $v = \frac{Q}{S}$, so substituting:

$$\frac{P_1 - P_2}{\rho g} = 0 - z_1 + f L \frac{\left(\frac{Q}{\frac{\pi D^2}{4}}\right)^2}{D 2g}$$

Simplifying:

$$\frac{\Delta P}{\rho g} = -z_1 + fL\frac{16Q^2}{2\pi^2 g D^5}$$

Since:

$$\sin 15° = \frac{z_1}{L}$$

Therefore:

$$\Delta P = \left(f\frac{z_1}{\sin 15°}\frac{16Q^2}{2\pi^2 g D^5} - z_1\right)\rho g = 265.54\,\text{Pa}$$

Note: Note that if the fluid was water instead of air, the difference in pressure due to height would become more important due to its greater specific weight.

Exercise 1.8 The attached table provides the values corresponding to the flow of an air stream through a duct. Calculate the value of the coefficient of friction using the expressions of von Karman–Prandtl, Nikuradse, Colebrook, USBM and Haaland.

Variable	Symbol	Units	Value
Diameter	D	m	0.5
Flow	Q	$m^3\,s^{-1}$	0.9
Absolute roughness	ε	m	0.0015
Air density	ρ	$kg\,m^{-3}$	1.2
Dynamic viscosity	μ	$N\,s\,m^{-2}$	0.0000186
Mean velocity	v	$m\,s^{-1}$	4.58366
Reynolds number	Re	–	147,860

Solution

Solutions search by iterative calculations are simplified by making $\frac{1}{\sqrt{f}} = A$ and searching for the value of that makes $f(x) = A - f\,(A, \text{parameters}) = 0$. Therefore:
Using von Karman–Prandtl's formula (Eq. 1.25) (turbulent flow and smooth duct) we have:

$$\frac{1}{\sqrt{f}} = -2.0\,\log\left(\frac{2.51}{Re\sqrt{f}}\right)$$

$A = 7.76057; f(x) = -9.765 \times 10^{-7}$
$f = 0.01660$

Nikuradse formula (Eq. 1.26) (turbulent flow and rough duct) gives:

$$\frac{1}{\sqrt{f}} = -2.0 \log\left(\frac{\frac{\varepsilon}{D_h}}{3.71}\right)$$

$A = 6.18451$
$f = 0.02615$

Colebrook–White equation (Eq. 1.27) (turbulent flow and intermediate roughness) gives:

$$\frac{1}{\sqrt{f}} = -2.0 \log\left(\frac{\frac{\varepsilon}{D_h}}{3.71} + \frac{2.51}{\mathrm{Re}\sqrt{f}}\right)$$

$A = 6.08016$
$f = 0.02707$

USBM approximate expression (Eq. 1.28) (turbulent flow and intermediate roughness) provides:

$$\frac{1}{\sqrt{f}} = -2.0 \log\left(\frac{\frac{\varepsilon}{D_h}}{3.71} + \frac{2.825}{\mathrm{Re}\sqrt{f}}\right)$$

$A = 6.06813; f(x) = 8.3489 \times 10^{-14}$
$f = 0.02716$

Haaland's approximate formula (Eq. 1.29) (turbulent flow and intermediate roughness):

$$\frac{1}{\sqrt{f}} = -1.8 \log\left[\frac{6.9}{\mathrm{Re}} + \left(\frac{\frac{\varepsilon}{D}}{3.71}\right)^{1.11}\right]$$

$A = 6.08534$
$f = 0.02700$

The indicated conditions correspond to turbulent flow and medium roughness pipeline, for this reason, von Karman–Prandtl's expression is not indicated. Moreover, Nikuradse's expression, despite being indicated for rough pipelines, approaches Colebrook–White's expression well enough. Finally, Haaland and USBM expressions for friction factor, which approximate the Colebrook–White's expression offer also significantly accurate results.

1.3.10 Shock Losses

Any variation from a straight duct of constant cross section generates *shock losses*.[11]
These losses, from a physical point of view, are due to variations in the momentum,
as a consequence of changes in direction, speed and acceleration in fluid molecules.
They are usually divided into:

(a) Those due to changes of direction such as elbows, joints, or splits; and
(b) Those resulting from variations in the cross section such as obstructions,
     expansions/contractions, entrances and exits of the fluid in the duct.

A theoretically demonstrable example of this type of loss is what happens when
fluid circulates inside a pipe and suddenly an orifice is found, in which case one of
the most notable effects is its contraction in the form of a *vena contracta* (Fig. 1.13).
Its study also serves to introduce the concept of equivalent orifice and regulator that
will be discussed later.

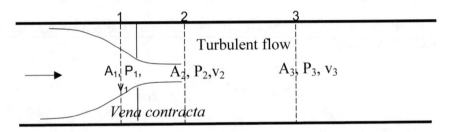

Fig. 1.13 *Vena contracta* formation in an orifice plate

Therefore, starting from the Bernoulli Equation, assuming that there are no viscous
losses and that the flow is laminar, we have that:

$$\frac{P_1}{\rho} + \frac{v_1^2}{2} = \frac{P_2}{\rho} + \frac{v_2^2}{2}$$

where regrouping:

$$P_1 - P_2 = \frac{\rho v_1^2}{2}\left(\frac{v_2^2}{v_1^2} - 1\right)$$

If the continuity equation is also considered:

$$v_1 A_1 = v_2 A_2$$

---

[11] Also termed local losses, dynamic losses or minor losses.

Then:

$$P_1 - P_2 = \frac{\rho v_1^2}{2}\left[\left(\frac{A_1}{A_2}\right)^2 - 1\right]$$

This shows that the losses $P_1 - P_2$ depend on the initial dynamic pressure $\left(\frac{\rho v_1^2}{2}\right)$ and a constant function of the geometry of the obstacle. If a control volume in the turbulent regime zone is assumed and the equations of the conservation of momentum applied to it, we have:

$$A_3 P_2 - A_3 P_3 = \dot{m}v_3 - \dot{m}v_2$$

where

$$\dot{m} = \rho A v$$

With what:

$$P_1 - P_3 = \frac{\dot{m}}{A_3}(v_3 - v_2) = \rho v_3(v_3 - v_2) = \rho v_3^2\left(1 - \frac{v_2}{v_3}\right)$$

In addition, by the conservation of mass, $v_3 = v_1$, with which[12]:

$$P_1 - P_3 = \frac{1}{2}\rho v_1^2\left(1 - \frac{v_2}{v_1}\right)^2 = \frac{1}{2}\rho v_1^2 X$$

Therefore, losses $(P_1 - P_3)$, can be expressed in international system units as Eq. 1.30:

$$P_1 - P_3 = P_{shock} = X P_v \tag{1.30}$$

Or in pressure head, as Eq. 1.31:

$$H_{shock} = \frac{P_1 - P_2}{\rho g} = X\frac{v^2}{2g} \tag{1.31}$$

There are several coefficients, depending on the nature of the obstacle, which we will deal with in later chapters. The Schauenburg house supplies the following impact loss factors for its ducts (Table 1.2).

There are two types of local losses: *permanent* and *recoverable* (Fig. 1.14). The above expression and the coefficients are used to calculate the first of them.

---

[12]A loss of energy in a pipeline is always a loss of total pressure. What happens is that, in points that have the same speed (as in the case of 1 and 3 that have the same cross section), the losses of static and total pressure coincide. Actually, the equation should look like:

$$P_{shock} = \left(P_1 + \frac{v_1^2}{2}\rho\right) - \left(P_3 + \frac{v_2^2}{2}\rho\right)$$

Table 1.2 Coefficients of total permanent losses for some system elements. Note that X_1 and X_2 represent loss coefficients referring to A_1 and A_2, respectively. Modified from Schauenburg Ltd

Element		Angle ($\theta°$)	Loss coefficients (X, X_1, X_2)
Gradual contraction			X_2
		30	0.02
		45	0.04
		60	0.07
Equal area transformations			X
		≤ 14	0.15
Sharp-edged (flush) entrance			X
		–	0.34
Plain (open) entrance			X
		–	0.85
Bell-mouth (formed) entrance			X
		–	0.03
Gradual expansion			X_1
		5	0.17
		7	0.22
		10	0.28
		20	0.45
		30	0.59
		40	0.73
Sharp-edged discharge			X
			1

Fig. 1.14 Static (P_s), total (P_T) and permanent (P_p) pressure losses in a sharp-edged entrance

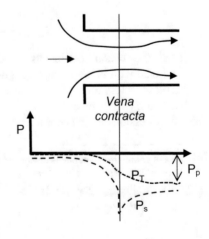

Exercise 1.9 A fan 1.5 m in diameter moves a flow rate of 100 m³ s⁻¹. Determine the pressure losses that occur in the inlet, if:

(a) It has a bell-mouth shape ($X = 0.05$).
(b) Its flank is plain ($X = 0.9$).

Air density: 1.2 kg m⁻³.

Solution

First, the velocity is calculated:

$$v = \frac{Q}{A} = \frac{100\frac{m^3}{s}}{\pi\,(0.75\,\text{m})^2} = 56.59\frac{m}{s}$$

Then the dynamic pressure is:

$$P_v = \frac{1}{2}\rho v^2$$

$$P_v = \frac{1}{2}\cdot 1.2\frac{kg}{m^3}\cdot\left(56.59\frac{m}{s}\right)^2 = 1921.46\,\text{Pa}$$

Taking into account the expression for calculating shock losses:

$$P_{shock} = X\,P_v$$

(a) In the case of a flared inlet:

$$P_{shock\,(flared)} = 0.05\cdot 1921.46\,\text{Pa} = 96.07\,\text{Pa}$$

(b) In the case of a plain inlet:

$$P_{shock\,(flat)} = 0.9\cdot 1921.46\,\text{Pa} = 1729.31\,\text{Pa}.$$

References

Blasius, P. H. (1913). Das Aehnlichkeitsgesetz bei Reibungsvorgängen in Flüssigkeiten. *VDI Mitt. Forschungsarb, 131,* 39–46.

Colebrook, C. F., & White, C. M. (1937). Experiments with fluid friction in roughened pipes. *Proceedings of the Royal Society of London Series A, 161*(904), 367–381.

Coleman, M., Cain, M., Danna, R., Harley, C., & Sharp, D. (1988). *A review of energy release processes from the failure of pneumatic pressure vessels.* Eastern Space & Missile Center.

Haaland, S. E. (1983). Simple and explicit formulas for the friction factor in turbulent pipe flow. *Journal of Fluids Engineering, 105*(1), 89–90.

Hagen, G. H. L. (1839). Ueber die Bewegung des Wassers in engen cylindrischen Rohren. *Poggendorfs Annalen der Physik und Chemie* (2), 46, 423.

Nikuradse, J. (1932). Gesetzmässigkeit der turbulenten Strömung in glatten Rohren. *VDI Forschungs Heft,* 356.

Nikuradse, J. (1933). Strömungsgesetze in rauhen Rohren. VDI Forschungs Heft, 361.

Novitzky, A. (1962). *Ventilación de minas* (p.155). Buenos Aires. Argentina: Yunque.

Poiseuille, J. L. (1840). Recherches experimentales sur le mouvement des liqnides dans les tubes de tres petits diametres. *Compte Rendus, 11,* 961.

Prandtl, L. (1935). The mechanics of viscous fluids. *Aerodynamic Theory, 3,* 155–162.

Smith, R. (1956). Flow of natural gas through experimental pipelines and transmission lines, *Monogram 9, US Bureau of Mines.* American Gas Association.

Bibliography

Boon, J. P., Yip, S., Burgers, J. M., Van de Hulst, H. C., Chandrasekhar, S., Chen, F. F., et al. (1988). *An introduction to fluid mechanics.* Cambridge, UK: Cambridge University Press.

Elger, D. F., & Roberson, J. A. (2016). *Engineering fluid mechanics* (pp. 170–185). Hoboken, NJ: Wiley.

Kreith, F., & Bohn, M. S. (1993). *Principles of heat transfer* (5th ed.). West Publishing.

Nakayama, Y. (2018). *Introduction to fluid mechanics.* Butterworth-Heinemann.

Shaughnessy, E. J., Katz, I. M., & Schaffer, J. P. (2005). *Introduction to fluid mechanics* (Vol. 8). NY: Oxford University Press.

Smits, A. J. (2000). *A physical introduction to fluid mechanics.* Wiley.

Young, D. F., Munson, B. R., Okiishi, T. H., Huebsch, W. W. (2010). *A brief introduction to fluid mechanics.* Wiley.

Wang, S. K., & Wang, S. K. (2000). *Handbook of air conditioning and refrigeration.* NY: McGraw-Hill.

Chapter 2
Environmental Conditions in the Mine

2.1 Introduction

It has been known since ancient times that various toxic gases are present in mines. *Carbon monoxide*[1] (*CO*) is the perhaps best known of them. Unfortunately, the consequences of this gas are frequently observed in domestic accidents, as a result of combustion at low oxygen concentration. Another gas that humankind has known about since early times is *methane*[2] (CH_4), due to its explosiveness and the catastrophic accidents caused in the coal-mining industry.

A product of combustion processes, *carbon dioxide*[3] (CO_2), is also a very well-known gas, especially in the area of climate change. However, this gas is less obviously perceivable. One of its key characteristics is its displacement of oxygen, which causes asphyxiation. *Hydrogen sulphide*[4] (H_2S) is primarily identifiable owing to its characteristic odour of rotten eggs. What is less widely known about it, however, is that it is extremely toxic. All these gases are present in mining environments, but so are others—particularly those released during blasting. The most important of these are *nitrogen monoxide* (*NO*) and *nitrogen dioxide* (NO_2), both of which can have fatal consequences. In addition to the gases mentioned above, large volumes of powder are produced during mining activity. Dust is the cause of many occupational diseases that engineers must be aware of, for example, silicosis. Another parameter that must be considered in underground environments is heat and, here, ventilation will play a major role in its dissipation.

All these parameters, together with the high humidity that results from the presence of water in the ground, considerably worsen working conditions in mines. In this section, we discuss the origins, regulatory values, prevention and control measures relating to them.

---

[1] In mining engineering terminology, this gas is termed *white damp*.

[2] *Fire damp.*

[3] *Black damp.*

[4] *Stink damp.*

C. Sierra, *Mine Ventilation*, https://doi.org/10.1007/978-3-030-49803-0_2

2.2 Gases and Hygiene

2.2.1 Air and Oxygen

Air is the homogeneous mixture of gases that covers the Earth's surface. It consists mainly of N_2, O_2, Ar and CO_2 (Table 2.1).

Within the human breathing process, there are compositional differences between inhaled air and exhaled air. These differences mainly relate to oxygen and other gases (Table 2.2).

Oxygen is considered the main air component for breathing. If the air's oxygen concentration falls below its normal values, the normal functioning of the human body is affected. Such alterations produce a condition known as *Confined Spaces Hypoxic Syndrome* (*CSHS*).

The effects of the reduction of air's oxygen concentration on humans are (MSA 2007):

- 21–19.5%: Normal breathing, any activity,
- 19.5–16%: Simple breathing, and
- 16–12%: Faster and deeper respiratory rhythm, accelerated pulse, attention and coordination problems.

Table 2.1 Composition of "fresh" air at sea level by volume (v/v) and mass (m/m) percentage (Mackenzie and Mackenzie 1998)

Gas	Concentration	
	% $_{v/v}$	% $_{m/m}$
N_2	78.08	75.72
O_2	20.95	23.14
Ar	0.934	1.29
CO_2	0.037	0.05
Ne	1.82×10^{-3}	1.27×10^{-3}
He	5.24×10^{-4}	7.24×10^{-5}
CH_4	1.7×10^{-4}	9.4×10^{-5}
Kr	1.14×10^{-4}	3.3×10^{-4}

Table 2.2 Approximated concentrations of gases in inhaled and exhaled air (Španěl and Smith 2001)

Gas	Concentration dry air (% $_{v/v}$)	
	Inhaled air	Exhaled air
O_2	21	~16
Ar	1	1
CO_2	0.04	~4
N_2	78	78
Water vapour	Low content	High content

- 14–10%: Sensation of light-headedness, blurred vision, headache, accelerated pulse;
- 10–6%: Nausea and vomiting, loss of movement capacity, fainting and even unconsciousness; and
- 6%: Convulsions, respiratory arrest, cardiac arrest.

The most common cause of a reduction in air oxygen concentration is oxygen being displaced by another gas or consumed during combustion. *Altitude sickness* is also a very common disorder that occurs in some mines in Latin America, and particularly in Bolivia. The cause of this syndrome is decreased oxygen availability at high altitude. Although the mole fraction of oxygen in air (X_{O_2}) remains approximately constant within the troposphere, the total pressure (P_T) decreases with altitude. Therefore, oxygen partial pressure (P_{O_2}) falls, in accordance with $P_{O_2} = X_{O_2} \cdot P_T$ (Peacock 1998). This partial pressure is what drives gas exchange in the alveoli. Its decrease generates a cascade effect whereby heart and breathing rates increase in an attempt to make up the oxygen shortfall.

Question 2.1 Helium (He) is released inside an unsealed chamber until the concentration of He in the air of the chamber is 15%$_{v/v}$. Indicate the variation of the concentrations of O_2 and N_2.

Answer

If the concentration of He is 15%, the sum of $O_2 + N_2$ is approximately 85%. O_2 comprises approximately 21% of that 85%. Therefore, the total concentration of O_2 in the air is 17.85%. N_2 comprises 78% of the 85%, meaning the total N_2 concentration of the air is 66.3%.

2.2.2 Mine Gases

A diverse group of gases can be found in underground environments. The main sources of these gases are shown in Table 2.3.

When their concentration is excessive, workers experience the symptoms indicated in Table 2.4. When you go through this table, keep in mind that the longer the exposure time, the more accentuated the health effects will be.

The main organoleptic characteristics of mine gases are shown in Table 2.5. Knowledge of these can be very useful, as they can be used as warning signs when there is no specific detector available.

Table 2.3 Sources of the
various mine gases

Gas	Sources
CO	Incomplete combustion Emanations Explosions Blasting
CO_2	Complete combustion Fermentation Breathing Chemical reactions between acidic water and carbonates Blasting
CH_4	Decomposition of organic matter in the presence of water Carbon blasting (methane in pores)
SO_2	Diesel engines Emanations from rocks Explosions Blasting
N_2	Blasting
NO_x	Diesel engines Explosions Blasting
NH_3	Reaction of ANFO with cement
H_2S	Rarely present Bacterial decomposition of organic matter Stagnant waters Diesel engines (sulphur in fuels)
H_2	Electric motors Electric batteries Fires Reactions between strong acids and metals
Benzene	Diesel motors

2.2.3 Occupational Exposure Limits

Occupational Exposure Limits (*OELs*) are the upper limits concerning exposure to certain hazardous substances in workplace settings. Values recorded in the workplace are compared to these limits to determine whether workplace conditions pose any risks to human health (Deveau et al. 2015). Limits of this type were first introduced in South Africa and the United States in the first decade of the twentieth century. In the 1920s, the United States Bureau of Mines (USBM) became the first body to produce extensive regulations on exposure limits. From the early 1940s, the American Conference of Governmental Industrial Hygienists (ACGIH) began to establish *Threshold Limit Values* (*TLV*). These describe the conditions under which

Table 2.4 Effects of mine gases on health based on their concentration. Modified from MSA (2007) and Simonton and Spears (2007)

Gas	General effect	Concentration	Effect at that concentration
CO	CO has a greater affinity for haemoglobin than oxygen does	0.005%	Maximum daily limit of 8 h
		0.02%	Mild headache within 2–3 h
		0.04% 0.16%	Death within 3 h Death within 1 h
CO_2	O_2 displacement	0.5%	Deeper and faster breathing
		3%	Sweats, rapid pulse
		5%	The body can tolerate only a few minutes of exposure to this level Respiratory rhythm becomes three times higher than normal
		7%	Death
CH_4	O_2 displacement		May cause suffocation
SO_2	Formation of sulphuric acid, effect similar to that of NO_2	10 ppm	SO_2 attacks the mucous membrane
		500 ppm	Life-threatening level
N_2	O_2 displacement		N_2 can cause suffocation if it displaces oxygen
NO	Irritates the respiratory system's moist surfaces because nitrous and nitric acids form when NO comes into contact with water	25 ppm 0–50 ppm 60–150 ppm >200 ppm	Mild irritation of the eyes and respiratory tract Mild odour Intense irritation accompanied by coughing NO can be fatal at this level even if exposure lasts only a short time. It is extremely toxic. The effects can be hidden for as long as 72 h
NO_2	Formation of nitric acid in lungs and blood plasma, which can cause death	5–10 ppm 20 ppm	Sore throat Irritation of the eyes
		100 ppm	Cough. Danger within half an hour
		150 ppm	Dangerous even for short periods of exposure
		250 ppm	Fatal even for short periods of exposure
NH_3	Formation of nitric acid on the respiratory tract's moist surfaces[a]	24–50 ppm	Irritation of the nose and throat within 10 min
		72–134 ppm	Irritation of the nose and throat within 5 min
		700 ppm	Severe and immediate irritation of the respiratory system
		5000 ppm	Respiratory spasms. Rapid suffocation

(continued)

Table 2.4 (continued)

Gas	General effect	Concentration	Effect at that concentration
		>10,000 ppm	Pulmonary oedema. Fatal accumulation of fluid in the lungs. Death
H_2S	Paralysis of the respiratory system	0–10 ppm	Irritation of the eyes, nose and throat; headaches; sensation of light-headedness; miscarriages
		10–50 ppm	Irritation of the eyes and respiratory tract. Nausea and vomiting, coughing, shortness of breath
		50–200 ppm	Conjunctivitis, seizures, coma, death
H_2	Displacement of O_2		May cause suffocation

[a]This gas is key in the formation of peroxyacetyl nitrate, nitrosamines and nitro-polycyclic aromatic hydrocarbons (nitro-PAHs)

Table 2.5 Gases to be considered in the context of ventilation of mines

Gas	Smell	Taste	Colour
CO	Odourless	Tasteless	Colourless
CO_2	Odourless	Acidic if the concentration is too high	Colourless
CH_4	Odourless	Tasteless	Colourless
SO_2	Sulphur, harsh odour	Acidic, bitter	Colourless
N_2	Odourless	Tasteless	Colourless
NH_3	Harsh odour	Soapy, rotten	Colourless
NO	Odourless Sharp sweet smell	Tasteless	Colourless, although in high concentrations tends to be orange
NO_2	Harsh odour, similar to chloride	Flavourless. Causes a burning sensation in the nose	Orange
H_2S	Rotten eggs	Sweet	Colourless
H_2	odourless	Tasteless	Colourless

most workers can function without suffering negative consequences to their health. This system establishes three levels:

- *Time-Weighted Average* (*TLV-TWA*): Average concentration of pollutant, for eight hours a day or forty hours a week, to which the majority of workers can be exposed without suffering from adverse effects.
- *Ceiling* (*TLV-C*): Concentration of the contaminant that must not be exceeded at any time during the working day.

- *Short-Term Exposure Level (TLV-STEL)*: Exposure at or above this level should not last more than fifteen minutes; such exposure should not be repeated more than four times per day, and there should be a break of at least an hour between such exposures.

Since the 1970s, the Occupational Safety and Health Administration (OSHA) of the United States has set *Permissible Exposure Limits (PEL)*; there is a legal requirement for workplaces to comply with these. The PEL values are divided into two categories:

(a) *Ceiling values (PEL-C), and*
(b) *Average values for eight-hour workday (PEL-TWA).*

These standards are contained in Sect. 1910.1000 of the Code of Federal Regulations (CFR). The OSHA recommends the most restrictive values set by the ACGIH, though this is not considered to be a legal requirement. Another important reference is the *Immediately Dangerous for Life or Health (IDLH)* values offered by the National Institute for Occupational Safety and Health (NIOSH).

The European Union's equivalents to OSHA's mandatory values are the Binding *Occupational Exposure Limit Value (BOELV)*, which must be adopted directly or assumed in a more restrictive manner by EU Member States (Council Directive 98/24/EC). In addition, EU Directive 2017/164 establishes *Indicative Occupational Exposure Limit Values* (IOELV). Member States are obliged to establish national OELs, and in doing so, they must take into account those set by the EU. Table 2.6 shows a selection of the aforementioned values.

Table 2.6 Summary table with various Occupational Exposure Limits for mine gases

Gas	ACGIH		BOELV/IOELV		OSHA	NIOSH
	TWA (ppm)	STEL (ppm)	TWA (ppm)	STEL (ppm)	PEL (ppm)	IDLH (ppm)
CO	25	–	20	100	50	1200
CO_2	5000	30,000	5000	15,000	5000	40,000
NH_3	25	35	20	50	50	300
NO	25	–	$25^a \to 2$	35	25	100
NO_2	3	5	$3^a \to 0.5$	$5^a \to 1$	5	20
H_2S	10	15	5	10	20	100
SO_2	2	5	0.5	1	5	100

[a]In the case of NO_x, the indicated limits are valid for the mining and tunnelling sectors until 2023. They will then become more restrictive. Values that will change are followed by an arrow

Individuals states also have specific mining regulations—for example, those set by Safe Work Australia (2013) or those set by Spain's ITC 04.7.02, though the latter is based on EU regulations. Both Australia's and Spain's regulations are summarized in Table 2.7.

Table 2.7 Exposure limits in Australian (Safe Work Australia 2013) and Spanish mining regulations (RGNBSM 1985)

Gas	Safe Work Australia		RGNBSM	
	TWA (ppm)	STEL (ppm)	TWA (ppm)	Ceiling (ppm)
CO	30	–	50	100
CO_2	5000 (mines other than coal mines) 12,500 (coal mines)	30,000	5000 (0.5%)	12,500 (1.25%)
NO/NO_2	25	–	10	25
H_2S	10	15	10	50
SO_2	2	5	5	10
H_2	–	–	1000 (0.1%)	10,000 (1.0%)
CH_4	1000	–	15,000 (1.5%)	25,000 (2.5%)

Question 2.2 In relation to CH_4, CO, H_2S, CO_2, N_2, O_2, NO, NO_2 and SO_2, indicate:

(a) Which can be detected by a characteristic smell.
(b) Which can be explosive.

Answer

(a) The smell of H_2S becomes detectable from 0.13 ppm. At concentrations lower than 50 ppm, NO has a moderately sweet odour. Between 0.2 and 1 ppm, NO_2 has an intense and harsh odour. From 3 ppm, the sulphurous odour of SO_2 can be detected.
(b) CH_4, CO and H_2S.

Question 2.3 (a) Describe two hazards associated with CO; (b) indicate the types of mining activities in which CO is present; and (c) convert a concentration of 400 mg m^{-3} of CO into ppm.

Answer

(a) CO is a combustible and explosive gas. Compared to O_2, it has a greater affinity with haemoglobin. This can lead to death caused by oxygen deficiency in the bloodstream.
(b) This gas is present in areas where incomplete combustion has occurred—for example, combustion engines, fires and explosions. Due to its density slightly smaller than that of air, is usually found at the top of galleries and stopes. The forced circulation of air in mines makes their air composition homogeneous. That is why these considerations must be taken carefully.
(C) The simplified expression for the conversion of ppm of any gas (1 atm and 25 °C) into mg m^{-3} is:

$$C\,[ppm] = 24.45 \frac{C\,[mg\ m^{-3}]}{molecular\ weight}$$

Therefore: 349.161 ppm of CO.

Radioactive Gases: Radon

Rn is a colourless, odourless and tasteless gas that is found naturally in rocks and soils. Of the 37 known isotopes of radon, ^{195}Rn to ^{231}Rn, the commonest is ^{222}Rn. Nowadays, this gas is considered as one of the most significant causes of lung cancer (ICRP 2011; WHO 2009). The literature provides different reference values for its effects on the health of miners (e.g. Navaranjan et al. 2016; Kreuzer et al. 2017). A first occupational reference value can be obtained by measuring radioactivity levels. An annual average equivalent to more than 300 radioactive decays per m^3 (300 Bq m^{-3}) indicates the need to restrict gas exposure (Shahrokhi et al. 2017). Other references (e.g. DHHS 1987) suggest values above 0.11–0.25 WL.[5] In terms of gas doses absorbed by living matter, the recommended parameters range from 1 to 20 mS year^{-1} (ICRP 2011). Among the protective measures that have been considered to reduce concentrations of this gas are (ICRP 2011): (a) adequate selection of the exploitation method (such as mining by remote control); (b) intense ventilation; (c) general air filtering;[6] (d) individual protection equipment for dust filtering; and (e) work rotation to reduce exposure time.

2.2.4   Weight of Gases and Stratification

In mine-ventilation calculations, the weight of air is usually assumed to be 1.2 kg m^{-3}, based on 1 atm of pressure, a temperature of 15 °C and a 60% humidity.

The diffusion of gases consists of a net movement of gas molecules from zones of higher gas concentration to zones of lower concentration. A gas's speed of diffusion is inversely proportional to the square root of its molar mass (*Graham's law*). Gases' diffusivity increases as temperature rises (*Chapman–Enskog theory*). The effect of pressure on diffusivity is insignificant except at very high pressures (>20 atmospheres). Thus, the diffusivity is considered as a constant in the context of mine ventilation.

Gases that undergo a diffusion or mixing process do not usually become stratified according to density. This is because of the predominance of random thermal movements of molecules, which produce forces that are much more significant than those of gravity. This phenomenon is very similar to that which can be observed in the case of "fresh air" gases, which do not become separated due to density difference (Badino 2009). Gas-stratification occurs when gases do not have enough time to inter-diffuse. Thus, if a gas leak has occurred only recently, the mixture may not

---

[5] A WL (*Working Level*) equals 2.08 x 10^{-5} J m^{-3} of potential alpha energy concentration of Rn progeny. One WLM^{-1} (*Working Level Month*) equals to 1WL · 173 h or 0.5 WL ·340 h, as 173 h of work a month are established.

[6] Although it is difficult to eliminate Rn because of its gaseous nature, its decay products are solid and usually adhere to dust particles. They can, therefore, be filtered and precipitated electrostatically.

Table 2.8 Relative specific density of different gases. The specific density data are based on standard conditions

Gas	Molecular weight (g mol^{-1})	Specific gravity (relative density)
H_2	2	$2/29 = 0.07$
CH_4	16	$16/29 = 0.55$
NH_3	17	$17/29 = 0.59$
CO	28	$28/29 = 0.97$
Air	29	**1**
NO	30	$30/29 = 1.03$
H_2S	35	$35/29 = 1.21$
CO_2	44	$44/29 = 1.53$
NO_2	46	$46/29 = 1.59$
SO_2	64	$64/29 = 2.26$

yet have taken place, and so some stratification may be observed, although this will disappear over time. Graham's law states that the diffusion rate of a gas is inversely proportional to its density. Accordingly, less dense gases will have a higher diffusion rate and therefore, tend less to stratify (Table 2.8).

Thus, gases tend to appear in galleries and stopes in a similar way to that shown in Fig. 2.1. According to the graph, the light gas will move to occupy the upper part

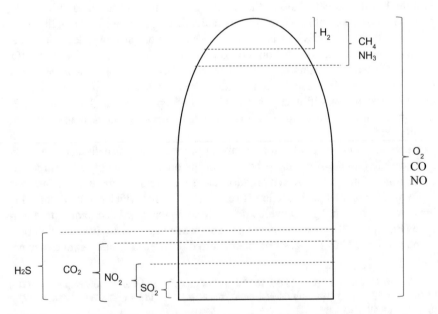

Fig. 2.1 Strict theoretical stratification of gases in a mining gallery. Figure does not take into account either the molecular movement or the effects of forced ventilation

of the gallery, unless it is occupied by another gas, in which case there will be top-to-bottom stratification in order of increasing density. CO is lighter than air, so it will direct to the roof of the airway. However, because it is not much lighter, assuming that it can be found at any height represents a conservative approach. Regarding nitrogen oxides, it should be taken into account that NO will rapidly oxidize to become NO_2.

2.3 Gas Detection

The way in which gases become stratified implies that sensors must be located not in the centre of the gallery[7] but in an arrangement similar to that indicated in Fig. 2.2.

The main recommendations relating to sensor location are as follows (MSA 2007):

- Since H_2 and CH_4 are lighter than air, their detectors should be located on the ceiling of the gallery.
- As SO_2, NO_2, CO_2, H_2S and some hydrocarbon gases are denser than air, their detectors must be placed in the lower part of the gallery.
- CO and O_2 are located approximately at human head height. NO is slightly denser than air, so it tends to descend a little more.
- Sensors should not be located directly in zones with direct fresh air supply because the fresh air will produce a dilution effect.
- Instead, sensors must be located in the areas where these gases tend to accumulate.

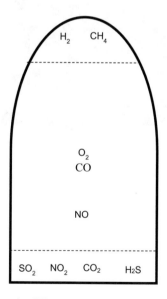

Fig. 2.2 Location of the sensors for different gases in a mine gallery

---

[7]Forced ventilation will act varying considerably gases stratification.

- In certain cases, the real risk is the displacement of O_2 by another gas such as N_2, rather than the presence of a specific gas.
- Toxic gases should always be monitored at human head height due to the effects of stratification.
- The exception to the previous rule is the case of flammable gases; it is more logical to monitor these gases in areas of predominant concentration.

There are four basic options for determining the concentration of gases inside mines: reactive tubes, specific detectors, multigas detectors and chemical analysis.

2.3.1 Reactive Tubes

This method of gas detection is not common nowadays. A *reactive tube* is a glass ampule inside which there is a reagent that reacts in the presence of a particular gas. The procedure consists of breaking the ampoule at both ends, performing from 3 to 10 strokes with the bellows (Fig. 2.3) and finally reading the concentration in the coloured area of the ampoule after a couple of minutes. There are tubes of this type for most of the gases in the mine.

Fig. 2.3 Reactive tube for gas measurement

2.3.2 Electrochemical Sensors

This is the most common type of sensor (MSA 2007). *Electrochemical sensors* are actually small reactors in which, when certain gases are present, a chemical reaction that releases electrons takes place (Kumar et al. 2013). To be more specific, the gases usually penetrate a hydrophobic membrane and come into contact with an electrode embedded in an electrolyte. This causes oxidation and reduction reactions that discharge electrons (Fig. 2.4). The electrons that are generated are measured using electronic equipment that establishes the relationship between the current generated and the concentration.

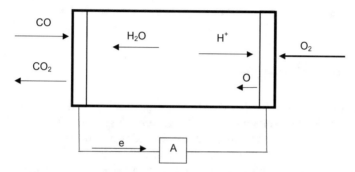

Fig. 2.4 Scheme of a CO meter

2.3.3 Catalytic Sensors

Catalytic sensors are based on oxidizing flammable gases or vapours and measuring the heat released from the reaction. The most common sensors consist of two platinum pellistors that have an electrical connection between them in the form of a Wheatstone bridge (Fig. 2.5). The coil of one of the pellistors is treated with a catalyst for the reaction, and the other is treated with a substance that inhibits the reaction (Kumar et al. 2013).[8] An electric current passes this structure to promote the oxidation reaction of the gas that is being measured at approximately 450 °C. When the catalytic reaction takes place, the temperature of this coil rises, increasing its resistance and unbalancing the bridge (MSA 2007). The bridge current is approximately proportional to the gas concentration at the Lower Explosive Limit (LEL).

---

[8]Pellistors are miniature calorimeters used to measure the heat released during the burning of a gas or combustible vapour.

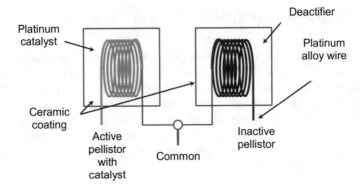

Fig. 2.5 Scheme of a catalytic sensor used in gas measurement. Modified from Equipco Services

2.3.4 Infrared Sensors

Infrared sensors sensors are based on the principle that gas molecules are excited by infrared radiation of a certain wavelength, generating vibrations that absorb energy from it. A typical infrared photometer consists of two infrared rays. One of these passes through a container in which the gas to be analysed is located, and the other through a container which does not contain the gas, and is used as a reference (Fig. 2.6) (MSA 2007). Concentration can be deduced based on the difference in the absorbed radiation between the two tubes (Kumar et al. 2013).

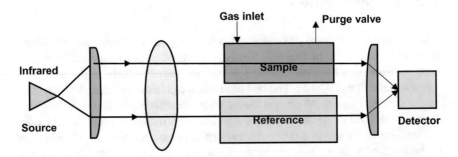

Fig. 2.6 Scheme of an infrared (IR) sensor

2.3.5 Atmospheric Monitoring Systems

An *Atmospheric Monitoring System* (*AMS*) consists of a network of sensors (which mainly detect gases) located inside the mine. These sensors send data continuosly to a device on the surface, which processes the data in real-time (e.g. Rowland et al. 2018). An AMS makes it possible to determine if an area is safe or not. It also actuates the mine's alarm systems and interacts with the ventilation system to configure it based on what the mining operations require.

The AMS was originally used in coal mines to monitor CH_4, but the use of them has been extended to all types of underground mining, where they also control possible fires in electrical installations.

2.4 Basic Psychrometry

Psychrometry is the branch of thermodynamics and climatology that studies mixtures of gases and vapours. It relies on two essential parameters: temperature and humidity. The former is important because the ventilation rate needs to increase as the temperature rises. Normally, recommended maximum temperatures underground are: 30–32 °C (dry-bulb) and 27–28 °C (wet-bulb) (Hartman et al. 2012). The other parameter that determines the needs of ventilation (or air conditioning) is an increase in humidity. The next section describes these and other psychrometric parameters.

Dry-bulb Temperature (T_d)

An ordinary thermometer measures *dry-bulb temperature*. This type of measurement is completely independent of air humidity and airvelocity, as well as from the heat radiation of nearby objects.

Wet-bulb Temperature (T_w)

Wet bulb-temperature measurement involves air humidity content, which is determined by using a *wet-bulb thermometer*. This is a common thermometer that is covered with a cotton bulb moistened with water, meaning that the temperature is affected by the cooling effect of the water evaporating in the air. Therefore:

- If the air is drier, the bulb will dry faster and the temperature will be lower.
- Air currents will accelerate the water evaporation from the bulb, dropping the temperature as a result of the latent heat for water passing from a liquid to a gas state.

Dry- and wet- bulb temperatures coincide when the air is saturated.

Equivalent Temperature (T_e)

This measurement involves environmental temperature, relative humidity and air velocity. It offers an approximation of the thermal sensation perceived by a human being. The temperature of a saturated atmosphere without air movement is what causes the sensation of heat in most people. This temperature cannot be measured directly. Instead, it must be obtained through graphs or mathematical equations. The equations that are most commonly used in mining are Eqs. 2.1 and 2.2:

$$T_e = 0.9\, T_w + 0.1\, T_d \tag{2.1}$$

or

$$T_e = 0.7\, T_w + 0.3\, T_d - v \tag{2.2}$$

where

- T_e: Equivalent temperature (°C),
- T_w: Wet-bulb temperature (°C),
- T_d: Dry-bulb temperature (°C), and
- v: Air velocity (m s^{-1}).

Absolute Humidity (AH)

Humidity is the mass of water vapour per air mass or volumen unit. If the air is humid, then we speak about absolute humidity (kg of water vapour/m^3 of humid air) or *specific humidity* (kg of water vapour/kg of humid air) depending on whether mass or volume units are used for humid air. If the data refer to dry air (kg of water vapour/kg of dry air), it is termed *mixing ratio* or *humidity ratio*.[9]

Maximum Humidity (Saturation Humidity) (SH)

Saturation humidity is the maximum amount of water vapour that a volume of air can contain at a given temperature. It is usually measured in g m^{-3}.

Relative Humidity (RH)

Relative humidity is the ratio between the mass of water vapour in the air and the maximum possible mass of water vapour in the air at that temperature. In other words, it is the relationship between the water vapour pressure in the air (P_v) and the saturation pressure (maximum water vapour pressure) at the same temperature and pressure (P_s) (Eq. 2.3):

$$RH(\%) = \frac{P_v}{P_s} 100 \tag{2.3}$$

It can also be expressed in terms of mole fractions as the ratio between the mole fraction of water vapour in the air with certain humidity and the mole fraction of water vapour that saturates the sample at the same pressure and temperature.

---

[9]There is a lot of confusion regarding specific humidity and mixing ratio because they offer similar results, which has often encouraged people to use them interchangeably for many purpuses.

Exercise 2.1 Determine the relative humidity of an air mass if the partial pressure of the water in the air at 15 °C is 872 Pa and the pressure of water vapour at the same temperature is 1705 Pa.

Solution

$$\text{RH}(\%) = \frac{P_v}{P_s}100 = \frac{1705\,\text{Pa}}{872\,\text{Pa}}100 = 51.15\%$$

Saturated Vapour Pressure ($P_{s(w)}$)

The *saturated vapour pressure* is the pressure of a vapour in equilibrium with its liquid phase. It is connected to the solid's or liquid's capacity to pass into a gaseous state—that is, their volatility. This parameter only depends on temperature. As the temperature increases, so does it. The most frequently used correlations are those from Magnus in 1884, as in Eq. 2.4:

$$P_{s(w)} = 0.6105\,e^{\left(\frac{17.27t_w}{t_w+237.3}\right)} \tag{2.4}$$

where

- $P_{s(w)}$: Saturated vapour pressure (kPa), and
- t_w: Wet-bulb temperature (°C).

Note that this expression changing t_w for t_d would generate $P_{s(d)}$. An extensive review can be found in Alduchov and Eskridge (1996).

Vapour Pressure (P_v)

A key parameter in psychrometric analysis, which can be used to calculate air humidity, is *vapour pressure*. This is the pressure of a solid or liquid system when it is in equilibrium with its vapour. The following is an update of the original correlation by Sprung (1888) (Eq. 2.5):

$$P_v = P_s - 0.000644\,P_b(T_d - T_w) \tag{2.5}$$

with

- P_v: Vapour pressure (kPa),
- T_d: Dry-bulb temperature (°C), and
- T_w: Wet-bulb temperature (°C).

Specific Volume (V_e, v)

Specific volume is the total volume of air plus water vapour per unit mass of dry air. It should not be confused with density, which is the total ratio of gas mass plus vapour to the total volume of gas plus vapour. It is calculated, therefore, according to Eq. 2.6:

$$\nu = \frac{V}{M_{\text{dry}}} \tag{2.6}$$

where

- ν: Specific volume (m^3 kg^{-1}),
- V: Total volume of the mixture (m^3), and
- M_{dry}: Dry-air mass (kg).

Question 2.4 Work out whether dry air has a higher or lower density than humid air. Base your assumptions on the molecular weight of the substances that compose air.

Answer

Air is composed mainly of N_2 and O_2. The molecular weight of N_2 is 28 g mol^{-1}, and that of O_2 is 32 g mol^{-1}. Water vapour has a molecular weight of 18 g mol^{-1}, thus the more the water vapour displaces other gases, the lighter the gas mixture will be. Therefore, dry air is heavier than wet air.

Dew Temperature (DT)

Dew temperature is the temperature to which air must be cooled for it to become saturated with water vapour. When air cools below the saturation temperature, water begins to condense and become dew. In the SI, it is expressed in K.

Enthalpy (H)

Variation in *enthalpy* expresses the amount of energy absorbed or transferred by the air. In the SI, it is expressed in J kg^{-1}.

Sensible Heat and Latent Heat

The total amount of energy in the air is composed of *sensible heat* and *latent heat*. Sensible heat is the heat obtained or transferred during changes in air temperature. Latent heat is the heat exchanged during liquid–vapour phase changes. In the SI, it is expressed in J kg^{-1}.

Psychometric Chart

The psychometric chart represents the curves of the state equations for the air–vapour mixture. The atmospheric pressure is fixed and the dry temperature (abscissas) and the humidity ratio (ordinates) are left as variables. For any point within the chart, the parameters of the air can be determined according to the arrows in Fig. 2.7.

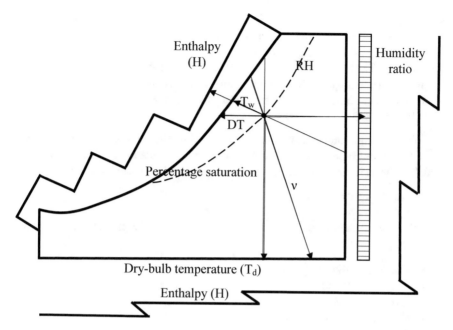

Fig. 2.7 Psychrometric chart

2.5 Aerosols in Mining Environments

Dust is the main type of *aerosol* present in mine air. Dust particles usually have a size between 1 and 100 μm. In addition to dust, fumes (<1 μm) are common. These are usually the result of combustion processes—mainly those of diesel machines (*Diesel Particulate Matter, DPM*)—or fires.

In addition to dispersions of solids, it is also common for liquids to be present in the air in the form of *fog* (denser) and *mist* (lighter). The presence of liquids is a consequence of the high temperature and humidity of the mining environment (Fig. 2.8).

The possible formation of *secondary particles* must also be taken into account. For example, studies have described the formation of NH_4NO_3 particles as a result of a reaction between HNO_3 ($NO_2 + H_2O$) and NH_3. Similarly, the presence of $(NH_4)_2SO_4$ following a reaction between H_2SO_4 ($SO_3 + H_2O$) and NH_3 has been described (e.g. Ge et al. 2017; Saarikoski et al. 2018).[10]

According to the manner in which it interacts with the respiratory system, *airborne dust* can be divided into the following types (CEN 1993):[11]

---

[10]A small part (<2%) of the SO_2 emitted by diesel combustion oxidizes to become SO_3 (Bugarski et al. 2012).

[11]Note that the above intervals seek to represent by means of a parameter what is actually a function of size distribution. In addition, they are dependent on individual physiognomies and breathing

- *Inhalable*: Particles that can enter the respiratory system through the nose and mouth (30–100 μm) and that may continue deeper into the respiratory tract, precipitate or leave it (e.g. pollen).
- *Thoracic*: These are particles that reach beyond the larynx getting into the thorax (trachea, bronchi, bronchioles) (<30 μm) (e.g. coal dust).
- *Respirable*: These particles penetrate to the unciliated airways (alveoli) (<10 μm) (e.g. bacteria and cigarette smoke).

The most common occupational diseases suffered by miners is *pneumoconiosis*. This is a chronic lung condition caused by the accumulation of dust in the lung, which causes a tissue reaction. There are several types of pneumoconiosis:

- *Silicosis*: Crystalline silica;
- *Coal workers' pneumoconiosis*: Carbon, graphite;
- *Silicatosis*: If the minerals contain other elements besides silica (that is, they are mixed powders):

  - *Kaolinosis*: Kaolin,
  - *Talcosis*: Talcum,
  - *Asbestosis*: Asbestos,
  - *Berylliosis*: Beryllium, and
  - *Siderosis (welders' pneumoconiosis)*: Iron.

Among the different types of pneumoconiosis, silicosis is the main one that affects miners' respiratory systems. The main polymorphs of silica (which in regulations is referred to as *free silica*) are *quartz*, *tridymite* and *cristobalite*. Of the three, quartz is

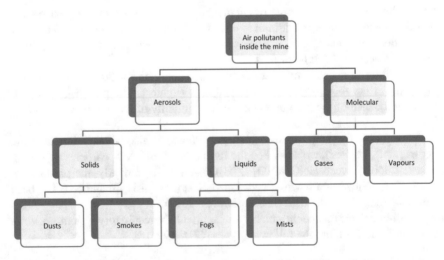

Fig. 2.8 Types of air pollutants in mines: aerosols vs. gases and vapours

patterns, therefore, they are approximations that may vary according to local regulations and over time.

the most common. When it is heated to 867 °C, it is transformed into tridymite and into cristobalite at 1470 °C. Effects on health depend on the type of polymorph.

Regarding silica, OSHA recommends a PEL-TWA of 0.05 mg m^{-3} for *Respirable Crystalline Silica* (*RCS*), while the ACGIH recommends a TLV-TWA of 0.025 mg m^{-3}. EU Directive 2017/2398 indicates a BOELV of 0.1 mg m^{-3} TWA. Silica has been classified as carcinogenic to humans by the International Agency for Research on Cancer (IARC).

In Spain, the daily exposure limits (Spanish: *Valores Límites Ambientales de Exposición Diaria, VLA-ED*) in the breathable fraction are 0.05 mg m^{-3} for quartz and 0.05 mg m^{-3} for cristobalite (INSHT 2018). In addition, the free silica contained in the breathable dust fraction must not be higher than 0.1 mg m^{-3}, and the respirable dust fraction must not exceed 3 mg m^{-3} (ITC2585/2007).

2.5.1 Measures Against Dust

Although in the past it was common to use *dry dust collection systems*, the apparatuses used were often affected by the hardness of the mining environment. Moreover, they required considerable maintenance, and they were also impractical due to their large dimensions. Nowadays, airborne dust prevention and suppression usually involve wet spray systems and wet drilling (Fig. 2.9).

Fig. 2.9 Cutting tool with external spray for dust control

Various *Emission Factors* (*EFs*) that aim to approximate the amount of dust produced in mining operations have been developed. Dust emission factors for drilling vary from 0.1 kg to 0.59 kg per drill (USEPA 1998; Roy et al. 2010). The NPI (2012) recommends using 0.59 kg per drill for the total suspended particles and 0.31 kg for the PM10. More complex EFs have been proposed, such as that recommended by NPI (2012) for total particle emissions after blasting[12] (Eq. 2.7):

---

[12]Drill length is required to be <21 m.

$$FE_v = 0.00022\,A^{1.5} \tag{2.7}$$

where

- FE_v: Blast emission (kg blast^{-1}), and
- A: Horizontal area blasted (m^2).

PM10 can be estimated by multiplying Eq. 2.7 by 0.52 (NPI 2012).
PM2.5 can be estimated by multiplying Eq. 2.7 by 0.03 (USEPA 1998).

These factors are a very large simplification of a complex physical reality that is affected by rock characteristics, blasting methods and explosive types. They have changed over the years; they vary in different sources, and they refer to surface mining (USEPA 2005). For all these reasons, they should be taken as mere approximations.

2.5.2 Radioactive Dust

Miners may also be exposed to inhalation of *radioactive dust*. Emissions of such dust mainly take place in uranium mines. Although it cannot be said that uranium minerals are highly radioactive, the decay of their natural isotopes (^{238}U [relative abundance of 99%], ^{235}U and ^{234}U), leads to the formation of others with higher radioactivity such as ^{234}U, ^{230}Th, ^{226}Ra and ^{210}Po (Macfarlane and Miller 2007). In fact, isotopes in the ^{235}U radioactive series (0.72% relative abundance) and those in the ^{232}Th series (Macfarlane and Miller 2007) should also be considered. One limit that is widely found in international regulations is an effective dose of 20 mSv/year (average of 5 years) (Valentin 2007). This figure requires workers to be extensively monitored. Among the techniques used to protect against this type of dust are those which we have explained previously in the section regarding Rn and the products of its decay. In general, a high level of precaution is necessary. Emissions should be limited by wetting the face or using *High-Efficiency Particulate Air* (*HEPA*) or *Ultra-Low Particulate Air* (*ULPA*) filters, and by establishing, where necessary, individual respiratory protection.

2.5.3 Diesel Particulate Matter

Emissions from diesel engines (*Diesel Exhaust, DE*) include gases (e.g. carbon, sulphur and nitrogen oxides), vapours (e.g. aromatic hydrocarbons), and solids (e.g. burned and unburned hydrocarbons, sulphates, fragments of metals and their oxides and salts). Solid emissions are known as *Diesel Particulate Matter* (*DPM*). These are particles that individually vary in size between 5 and 50 nm and form clusters with sizes in the sub-micrometre range. The majority of these particles are <100 nm in size and are termed *nano Diesel Particulate Matter* (nDPM). They are the most dangerous as they can pass directly from the lungs into the blood thence diffusing throughout the body even as far as the central nervous system (Black and Mullins 2019).

Occupational exposure to DPM causes irritation of the throat and eyes, dizziness and headaches in the short term. In the long term, exposure increases the risk of cardiopulmonary and lung cancer (Steiner et al. 2016; Silverman 2017). Mining is the industrial sector in which workers are most affected by this type of emissions (Monforton 2006). The Code of Federal Regulations of the United States (30 CFR 57.5060) indicates a PEL of 160 μg m^{-3} TC[13] (Mischler and Colinet 2009) but, a lot of criticism has appeared in the scientific literature concerning carbon measurements as a proxy for DPM health effects. This criticism is mainly based on the fact that carbon measurement is more closely correlated with the mass of DPM rather than the number, specific surface or chemical composition of the exhaust (Black and Mullins 2019).

Measures to prevent DPM formation and protect miners include (OCRC 2017):

- Replacement of old engines by new, lower-emission ones.
- Use of particle filters and catalytic converters (Diesel Oxidation Catalyst, DOC).
- Use of alternative fuels such as biodiesel and emulsified fuels (Water Emulsion Fuels, WEF).
- Adequate maintenance programmes that allow the correct functioning of diesel equipment.
- Use of sealed operator cabins (environmental cabs).

2.5.4 Measurement of Aerosols

Diesel Particulate Matter

The sample is obtained by generating a suction current by means of a pump. The current is then filtered, and the collected material is subjected to chemical analysis. These devices are always attached to the worker at all times so as to monitor working conditions as closely as possible. The two most popular variants are:

(a) *Respirable Combustible Dust (RCD)*, consisting of (Vermeulen et al. 2010):

- Filtration of air in a 0.8 μm silver membrane filter.
- Determination of the mass retained in the filter.
- The filter is transferred to a metal tray.
- The set is placed in a muffle furnace at 400 °C for 90 min.
- The particulate material reacts with the oxygen, assisted by the catalytic action of the filter's silver.
- Mass loss is considered the RCD.

---

[13]Total Carbon.

This is a simple and economical method in which the quartz collected in the same filter can also be analysed. However, it does not offer adjusted results at low concentrations (<0.6 mg m^{-3}) and is not selective, since carbonaceous or sulphurous mineral powders, oil mists and cigarettes can interfere with the measurement (Ayers 2017).

(b) *Thermal–Optical Analysis (TOA)* (NIOSH 5040):

This method uses *Organic Carbon (OC)* and *Elemental Carbon (EC)* as a proxy for determining DPM. The procedure involves the following previous steps:

- The thickest particles (>1 μm) are separated from the airstream by means of cycloning.
- The undersized material (<1 μm) passes through a quartz fibre filter.
- Particles <0.9 μm are retained in the filter.
- A portion of the filter is separated and placed in a special oven.
- Finally, OC and EC are measured from the evolved gas by means of TOA.

Mineral Dust

Samples are obtained with devices that are similar to those described in the previous section. These are mainly of relevance to the monitoring of respirable dust, for which standard Dorr-Oliver cyclones are generally used. The procedure is as follows (Verpaele and Jouret 2012):

- The air is aspirated at a speed of 1.0 l s^{-1}.
- The weight of the powder mass is obtained through weighing by difference.
- The concentration of dust in the air (mg m^{-3}) is obtained from the ratio between 1000 times the weight gains in the filter (mg) and the product of the airflow (l min^{-1}) times the monitoring time (min).
- Silica dust can be studied via X-Ray Diffraction (XRD) (NIOSH 7500). A sufficient amount of sample must be collected to do so (Stacey et al. 2014).

2.6   Heat Inside Mines

2.6.1   *Geothermal Gradient*

Geothermal gradient is the main reason why the temperature increases as one goes deeper into a mine. This increase is approximately 1 °C every 70–110 m in the case of deep metal mines, and 1 °C every 20–50 m in the case of European coal mines (Houberechts 1962). The heat that is present is usually the residual heat from planetary accretion from Earth's earlier eras and the decay of radioactive isotopes. Both of these things cause the terrestrial nucleus to be around 5700 °C and its heat flow outwards to be an average of 0.07 W m^{-2}. This heat penetrates the mine. The thermal conductivity of local materials influences the entry of heat into the mine.

This phenomenon is modelled through *Fourier's law*, which allows calculation of the heat flow for an isothermal surface, and the *Laplace's equation*, which describes the temperature field to offer an approximation of heat flow to the air (e.g. Goch and Patterson 1940).

2.6.2 Outdoor Climate

External climate has a direct impact on mines' internal temperatures. This affects the ground temperature and the variation in the air temperature that is introduced from the outside. Seasonal temperature variations generate a similar effect. In general, below a depth of 25 m, the temperature inside mines is fairly independent of external conditions. At this depth, which is called the *neutral zone*, the recorded temperature coincides with the average annual temperature of the outside air.

2.6.3 Air Self-compression

Air self-compression is the process by which a column of descending air compresses and its temperature increases. If one assumes that the humidity remains constant, and there is no friction with the walls, there is no heat exchange, so the process can be considered adiabatic. At the downcast shaft, this phenomenon is the opposite of that which takes place at the upcast shaft. At the latter, as the air rises, a decrease in temperature and pressure is observed.

The temperature increases at the downcast shaft, usually at a dry-bulb temperature of about 1 °C per 100 m of descent. Brake (2008) considers the real values of increase to be 0.6 °C and 0.4 °C per 100 m when the wet-bulb temperatures outside are 6 °C and 25 °C, respectively.

Starting from the equation of hydrostatic pressure in its differential form:

$$dP = \rho g dH$$

Solving for dH and operating (Eq. 2.8):

$$dH = \frac{dP}{\rho g} = \frac{dP}{\gamma} = V_e dP \qquad (2.8)$$

If the process is adiabatic, without heat exchange with the outside, then $k = C_p/C_v$ where C_p and C_v are the specific heats at constant volume and pressure, respectively. The relationship is (Eq. 2.9):

$$P V_e^k = constant \qquad (2.9)$$

By differentiating and simplifying, Eq. 2.10 is obtained:[14]

$$V_e dP + k P dV_e = 0 \qquad (2.10)$$

Considering the Clapeyron equation:

$$P V_e = R_w T$$

Differentiating the function (Eq. 2.11):

$$P dV_e = R_w dT - V_e dP \qquad (2.11)$$

If in Eq. 2.10, we substitute the corresponding terms for those of Eqs. 2.8 and 2.11, we obtain Eq. 2.12 below:

$$dH + k(R_w dT - dH) = 0 \qquad (2.12)$$

Working the expression out:

$$dH + k R_w dT - k dH = 0$$
$$dH - k dH + k R_w dT = 0$$
$$(1 - k) dH + k R_w dT = 0$$

Integrating:

$$(1 - k) \int dH + k R_w \int dT = 0$$
$$(1 - k) H + k R_w T + C = 0$$

We obtain:

$$T = \frac{(k - 1) H}{k R_w} - C$$

If we establish initial values of $H_0 = 0$ and $T = T_0$, we find that $C = -T_0$. If an adiabatic process is considered, the equation will be (Eq. 2.13):

$$T = \frac{(k - 1) H}{k R_w} + T_0 \qquad (2.13)$$

As for dry air $R_w = 29.29$ kpm (kg K)$^{-1}$ and $k = 1.41$, we obtain (Eq. 2.14):

$$T = T_0 + 0.0098\, H \qquad (2.14)$$

---

[14] $d(P\, V^k) = dP\, V^k + P k\, V^{k-1} dV = dP\, V\, V^{k-1} + P k\, V^{k-1} dV = 0 \rightarrow V\, dP + P k\, dV = 0.$

Navarro and Singh (2011) propose an increase of 0.98 °C per 100 m of descent, a figure that also takes into account the possible inclination of the shaft by means of Eq. 2.15:

$$\Delta t_a = T_2 - T_1 = 0.0098\, L\, sen(\alpha) \tag{2.15}$$

where

- Δt_a: Temperature increase with depth (°C),
- α: Angle between the shaft and the horizontal (sexagesimal degrees),
- L: Shaft depth (m),
- T_1: Surface temperature (°C), and
- T_2: Temperature at the bottom of the shaft (°C).

2.6.4 Electromechanical Equipment

Several electromechanical machines are commonly used in mining—for example, conveyor belts, armoured conveyors, pumps, locomotives and fans. These machines only convert some of the energy that they receive into useful work; most of it is dissipated as heat.

In the case of diesel equipment, heat losses cluster around the exhaust gas system (about 1/3), the engine's cooling system (about 1/3), and friction losses (5–10%). The rest (<1/3) is useful work (Mollenhauer and Tschöke 2010). Some authors—e.g., Ganesan (2012)—give a figure of 80% for heat losses meaning that only 20% of energy is available for useful work.

However, this useful work will ultimately degrade to heat unless it is used to increase the potential energy of a body. Therefore, if a machine's activity does not transform some of its input energy into mass elevation, all of it ends up being dissipated as heat (e.g. horizontal drilling). Thus, lifting machinery such as pumps or conveyor belts do not dissipate all their input energy in the form of heat, as a part of it will be stored as potential energy. In addition, combustion not only generates gases but also water vapour, meaning that only part of the heat generated will be felt as a temperature change in the air (*sensible heat*), while another part will be stored as *latent heat*. This latent heat is recovered when the water vapor condenses in the cooling towers of the mine.[15]

Heat release can be estimated based on fuel consumption and its calorific value (Eq. 2.16) (Calizaya and Marks 2011):

$$q = C\, PC\, E \tag{2.16}$$

---

[15]The *Lower Calorific Value* (*LCV*) of a fuel does not include the latent heat used in water vaporization. This heat is considered in the calculation of *Higher Calorific Value*.

where

- q: Heat released per unit time (kJ s^{-1}).
- C: Fuel consumption (l s^{-1}).
- PC: Diesel fuel calorific value. An approximation may be 34,000–38,000 kJ l^{-1} of fuel.
- E: Efficiency of combustion. In this case, the heat released is less than the theoretical heat due to dissociation and lack of oxygen. Combustion efficiency is around 90–97%.

Some authors suggest that the heat released by a diesel engine is between 2.8 and 3 kW per kW of the equipment's effective power (Banerjee 2003; Calizaya and Marks 2011). These values are based on a consumption of 0.24 kg of fuel per kW of power, with a calorific value of 44 MJ kg^{-1} (Vutukuri and Lama 1986).

The quantity of heat released and the increase in air temperature caused are not directly correlated. This is because not all this heat will pass into the air, nor will all of it be released at the same time. In fact, dissipation of the heat can be very slow. This is why some authors fix this value at 0.9 kW (kW$_{power}$)$^{-1}$. The method proposed by McPherson (1993a) to estimate air temperature increase due to working diesel engines (based on hours as the time unit) is:

1. Approximate the total rate of heat flow (Eq. 2.17):

$$q_T = \frac{C\,PC}{3600} \qquad (2.17)$$

where

- q_T: Total heat released per unit time (kW),
- C: Fuel consumption (l h^{-1}), and
- PC: Diesel calorific value (kJ l^{-1}).

2. Subtract from the previous rate of heat flow, the rate of heat flow used to increase the potential energy of bodies and fluids (Q_p) (Eq. 2.18):

$$q_T' = q_T - q_p \qquad (2.18)$$

3. Estimate the rate of latent heat flow by means of Eq. 2.19:

$$q_L = \frac{V_{H_2O}\,L_{H_2O}}{3600} \qquad (2.19)$$

where

- q_L: Latent heat per unit time (kW),
- V_{H_2O}: Water volume produced per unit time (l h^{-1}), and

- L_{H_2O}: Latent heat of water vaporization (kJ kg^{-1}).

Between 3 and 10 l of water per l of fuel consumed are produced.

4. Calculate the sensible heat (Eq. 2.20):

$$q_s = q'_T - q_L \tag{2.20}$$

5. Determine the temperature decrease considering the airflow (Eq. 2.21):

$$\Delta T = \frac{q_s}{\dot{m}_{air} C_{air}} \tag{2.21}$$

where

- ΔT: Increase in air temperature (K),
- $\dot{m}_{air}$: Mass flow rate of air (kg s^{-1}), and
- C_{air}: Specific heat of air (kJ kg^{-1} K^{-1}).

In a similar manner to the previous case, all the energy generated in the electrical equipment is transformed into either potential energy (e.g. pumps, conveyor belts) or heat. This can be shown by means of Eq. 2.22 (Calizaya and Marks 2011):

$$q_{em} = P_{abs} - \frac{\dot{m} g h}{1000} \tag{2.22}$$

In which

- q_{em}: Heat generated per unit time (kW),
- P_{abs}: Electric power absorbed (kW),
- $\dot{m}$: Mass flow rate (kg s^{-1}),
- g: Acceleration of gravity (N kg^{-1}), and
- h: Height difference (m).

With regard to the absorbed power, no efficiency is quoted because all the energy that is not transformed into potential energy, useful or otherwise, becomes heat.

In the case of electric machines, whose efficiency is much higher than that of internal combustion engines, the total efficiency can be approximated to 75–85%. Regarding lighting, an incandescent bulb has a light emission efficiency close to 5%;[16] that of fluorescent bulbs is 20% and that of LED bulbs is 95% (potentially 100%). The remaining energy is immediately transformed into heat. Moreover, fans can also increase air temperature and should also be considered during heat emission calculations. A good approximation here, is to consider that they increase air temperature at a rate of 0.25 K per kPa.

---

[16]Only 5% of the total electrical power is transformed into visible optical power and 95% is wasted in the form of thermal power (mostly in the infrared).

2.6.5 Explosives

Heat released from explosives can be approximated from the *explosion heat at constant volume*. This ranges from 3 MJ kg^{-1} for explosive emulsions and heavy ANFO, and 3.9 MJ kg^{-1} for the ANFO, to 4.9 MJ kg^{-1} for the aluminized ANFO [17]. Of this released energy, approximately 50–95% eventually becomes heat. However, not all of this heat will go directly to the air masses, because a large part first passes into the rock, delaying the process. A smaller part is used as energy in the rock fragmentation.

2.6.6 Heat of the Rock Mass

Rocks can experience temperature variations due to: (a) exposure to a mining atmosphere whose temperature is falling; (b) transfer of the rocks from one area inside the mine to another one at a lower temperature; (c) evacuation of stored heat after blasting.

If the temperature variation, the mass of exposed or displaced material and the rock's specific heat are known, it is possible to estimate the heat emitted by means of Eq. 2.23:

$$q = \dot{m} C_p \left(T_i - T_f \right) \qquad (2.23)$$

where

- q: Rate of heat flow (kW) or heat (J).
- $\dot{m}$: Mass flow rate (kg s^{-1}) or mass (kg).
- C_p: Specific heat (kJ kg^{-1} °C^{-1}). Frequent values for different types of rocks can be found in Eppelbaum, Kutasov and Pilchin (2014).
- T_i: Initial temperature (°C).
- T_f: Final temperature (°C).

Exercise 2.2 A blasting of 30 m^3 of rock with a specific explosive consumption of 5 kg m^{-3} is being carried out in a development end. It is estimated that 75% of all the heat generated by blasting remains in the rock an hour after the blasting takes place. Determine:

(a) The heat released by blasting, and
(b) The temperature increase experienced by the rock.

Data: Explosive heat at a constant volume: 3.5 MJ kg^{-1}; specific heat of the rock: 0.22 kcal kg^{-1} °C^{-1}.

---

[17]Data extracted from the commercial catalogue of Maxam Ltd.

Solution

(a) $q_T = 30\,\text{m}^3 \dfrac{5\,\text{kg explosive}}{1\,\text{m}^3\text{rock}} \dfrac{3.5 \times 10^6\text{J}}{1\,\text{kg explosive}} = 5.25 \times 10^8\,\text{J}$

(b) $q_r = 5.25 \times 10^8 \cdot 0.75\,\text{J} = 3.9375 \times 10^8\,\text{J}$

A density for the rock must be considered. An acceptable value is 2.5 t m^{-3}, therefore 75,000 kg. Therefore:

$$\Delta t = \frac{q_r}{C_p\, m} = \frac{3.9375 \times 10^8\,\text{J}}{916.67\frac{\text{J}}{\text{kg}\,°\text{C}}75,000\,\text{kg}} = 5.73\,°\text{C}$$

2.6.7 Human Metabolism

Workers generate *metabolic heat* (M) which is then dissipated from the organism according to Eq. 1.17:

$$M = RE + R + C + E + A + C_d \tag{2.24}$$

where

- RE: *Rate of heat flow due to respiratory exchange* (W),
- R: *Rate of heat flow due to radiation* (W),
- C: *Rate of heat flow due to convection* (W),
- E: *Rate of heat flow due to evaporation* (W),
- A: *Heat accumulation rate* (W), and
- C_d: *Rate of heat flow due to conduction* (W).

The temperature raise due to metabolic heat is not usually an important factor in mining, although it could be a key consideration in specific zones such as small stopes. For an office worker can be assumed to emit 100 W (equivalent to an incandescent bulb), while for a manual worker, the figure is 174–622 W, depending on the degree of activity (McPherson 1993b). Houberechts (1962) mentions values of 382 W for shovellers, 250 W for drillers, and 318 W for stope miners. For extremely heavy works, the figure can reach up to 400 W. An accepted approach in the case of mine ventilation is to assume around 250 W per miner.

2.6.8 Mine Waters

Water that enters the mine through cracks has a temperature corresponding to the temperature of the rock prior to the work. In addition, so called *mine water*, is that which is either already present or is generated within the mine. This category includes the water that has to be brought to the surface for mine activities to take place and, for example, the water from ventilation-related processes such as refrigeration. The temperature of mine water may be higher than that of water from other sources. Depending on the water's origin, if one knows the temperature drop that it undergoes, the rate of heat flow can be estimated by adapting Eq. 2.23 to water cooling. To reduce the release of heat into the air, measures can be taken such as covering and rapidly channelling any water present in the mine.

2.6.9 Other Sources

Oxidation reactions involving sulphides, wood and coal, as well as the hydration of cement and particular salts such as kieserite or carnallite are exothermic, and so pose a potential fire risk. The heat produced from these sources is, however, very difficult to estimate and although there are some exceptions—for example, in certain coal mines and potash mines—it is generally of moderate importance.

References

Alduchov, O. A., & Eskridge, R. E. (1996). Improved Magnus form approximation of saturation vapor pressure. *Journal of Applied Meteorology, 35*(4), 601–609.

Ayers, D. M. (2017). Comparing the NIOSH method 5040 to a diesel particulate matter meter for elemental carbon. Theses and Dissertations (All). *1573.*

Badino, G. (2009). The legend of carbon dioxide heaviness. *Journal of Cave and Karst Studies, 71*(1), 100–107.

Banerjee, S. P. (2003). *Mine ventilation.* Dhanbad, India: Lovely Prakashan.

Black, S., & Mullins, B. (2019). A study of nano diesel particulate matter (nDPM) Behaviour and physico-chemical changes in underground hard rock mines of Western Australia.

Brake, D. J. (2008). *Psychrometry, mine heat loads, mine climate and cooling.* Brisbane, Australia: Mine Ventilation Australia.

Bugarski, A. D., Janisko, S. J., Cauda, E. G., Noll, J. D., & Mischler, S. E. (2012). Controlling exposure to diesel emissions in underground mines (p. 37). NIOSH.

Calizaya, F., & Marks, J. (2011). Heat, humidity and air conditioning. In *SME mining engineering handbook.* Society for Mining, Metallurgy and Exploration.

CEN. European Committee for Standardization. (1993). DIN EN 481, Workplace atmospheres–Size fraction definitions for measurement of airborne particles.

Department of Health and Human Services (DHHS) (1987). *A recommended standard for occupational exposure to radon progeny in underground mines.* Publication No. 88-101.

Deveau, M., Chen, C. P., Johanson, G., Krewski, D., Maier, A., Niven, K. J., ... & Zalk, D. M. (2015). The global landscape of occupational exposure limits—Implementation of harmonization principles to guide limit selection. *Journal of Occupational and Environmental Hygiene, 12*(sup1), S127–S144.

Eppelbaum, L. V., Kutasov, I., & Pilchin, A. (2014). Thermal properties of rocks and density of fluids. In *Applied Geothermics*. Lecture Notes in Earth System Sciences. Berlin: Springer.

European Commission: Council Directive 98/24/EC of 7 April 1998 on the protection of the health and safety of workers from the risks related to chemical agents at work. https://osha.europa.eu/en/legislation/directives/75.

Ganesan, V. (2012). *Internal combustion engines*. McGraw Hill Education (India) Pvt Ltd.

Ge, X., He, Y., Sun, Y., Xu, J., Wang, J., Shen, Y., et al. (2017). Characteristics and formation mechanisms of fine particulate nitrate in typical urban areas in China. *Atmosphere, 8*(3), 62.

Goch, D. C., & Patterson, H. S. (1940). The heat flow into tunnels. *Journal of the Chemical, Metallurgical and Mining Society of South Africa, 41.*

Hartman, H. L., Mutmansky, J. M., Ramani, R. V., & Wang, Y. J. (2012). *Mine ventilation and air conditioning* (3rd ed.) (p. 611). Wiley.

Houberechts, A. (1962). Température-Humidité-Climat.In Aérage. Document SIM N1. *Revue de l'Industrie Minérale.*

ICRP. (2011). Radiological protection against radon exposure. ICRP ref 4829-9671-6554. Draft Report for Consultation. Atlanta, USA: Elsevier.

INSHT. (2018). *Límites de exposición profesional para agentes químicos*. Madrid: Instituto Nacional de Seguridad e Higiene en el Trabajo.

Kreuzer, M., Sobotzki, C., Schnelzer, M., & Fenske, N. (2017). Factors modifying the radon-related lung cancer risk at low exposures and exposure rates among German uranium miners. *Radiation Research.*

Kumar, A., Kingson, T. M. G., Verma, R. P., Mandal, R., Dutta, S., Chaulya, S. K., et al. (2013). Application of gas monitoring sensors in underground coal mines and hazardous areas. *International Journal of Computer Technology and Electronics Engineering, 3*(3), 9–23.

Macfarlane, A. M., & Miller, M. (2007). Nuclear energy and uranium resources. *Elements, 3*(3), 185–192.

Mackenzie, F. T., & Mackenzie, J. A. (1998). *Our changing planet: An introduction to earth system science and global environmental change* (No. 504.3/. 7 MAC). Upper Saddle River, NJ: Prentice Hall.

McPherson, M. J. (1993a). *Subsurface ventilation and environmental engineering* (p. 560). Springer Science & Business Media.

McPherson, M. J. (1993b). *Subsurface ventilation and environmental engineering* (p. 608). Springer Science & Business Media.

Mischler, S. E., & Colinet, J. F. (2009, November). Controlling and monitoring diesel emissions in underground mines in the United States. In *Mine Ventilation: Proceedings of the Ninth International Mine Ventilation Congress*, New Delhi, India (Vol. 2, pp. 879–888).

Mollenhauer, K., & Tschöke, H. (Eds.). (2010). *Handbook of diesel engines*. Springer Science & Business Media.

Monforton, C. (2006). Weight of the evidence or wait for the evidence? Protecting underground miners from diesel particulate matter. *American Journal of Public Health, 96*(2), 271–276.

MSA. (2007). *Gas detection handbook*. Mine Safety Appliances Company.

Navaranjan, G., Berriault, C., Do, M., Villeneuve, P. J., & Demers, P. A. (2016). Cancer incidence and mortality from exposure to radon progeny among Ontario uranium miners. *Occupational and Environmental Medicine. 73*(12), 838–845

Navarro, V. F., & Singh, R. N. (2011). Thermal state and human comfort in underground mining. In *Developments in heat transfer*. InTech.

NPI. (2012). Emission estimation technique manual for mining. National Pollutant Inventory. Department of Sustainability, Environment, Water, Population and Communities. Canberra, Australia.

OCRC. Occupational Cancer Research Centre. (2017). *Diesel particulate matter control strategies in mining*. Ontario (Canada).

Peacock, A. J. (1998). ABC of oxygen: oxygen at high altitude. *BMJ: British Medical Journal, 317*(7165), 1063.

RGNBSM. (1985). Reglamento General de Normas Básicas de Seguridad Minera. *Real Decreto,* 863.

Rowland, J. H., Harteis, S. P., & Yuan, L. (2018). A survey of atmospheric monitoring systems in US underground coal mines. *Mining Engineering, 70*(2), 37.

Roy, S., Adhikari, G. R., & Singh, T. N. (2010). Development of emission factors for quantification of blasting dust at surface coal mines. *Journal of Environmental Protection, 1*(4), 346.

Saarikoski, S., Teinilä, K., Timonen, H., Aurela, M., Laaksovirta, T., Reyes, F., ... & Junttila, S. (2018). Particulate matter characteristics, dynamics, and sources in an underground mine. *Aerosol Science and Technology, 52*(1), 114–122.

Safe Work Australia. (2013). *Workplace exposure standards for airborne contaminants.* Australian Government-Safe Work Australia.

Shahrokhi, A., Vigh, T., Németh, C., Csordás, A., & Kovács, T. (2017). Radon measurements and dose estimate of workers in a manganese ore mine. *Applied Radiation and Isotopes, 124,* 32–37.

Silverman, D. T. (2017). Diesel exhaust causes lung cancer: Now what? *Occupational and Environmental Medicine, 74*(4), 233–234.

Simonton, D. S., & Spears, M. (2007). Human health effects from exposure to low-level concentrations of hydrogen sulfide. *Occupational Health & Safety, 76*(10), 102–104.

Španěl, P., & Smith, D. (2001). On-line measurement of the absolute humidity of air, breath and liquid headspace samples by selected ion flow tube mass spectrometry. *Rapid Communications in Mass Spectrometry, 15,* 563–569.

Sprung, A. (1888). Über die Bestimmung der Luftfeuchtigkeit mit Hilfe des Assmannschen Aspirationspsychrometers. *Das Wetter, 5,* 105–108.

Stacey, P., Lee, T., Thorpe, A., Roberts, P., Frost, G., & Harper, M. (2014). Collection efficiencies of high flow rate personal respirable samplers when measuring Arizona road dust and analysis of quartz by x-ray diffraction. *Annals of Occupational Hygiene, 58*(4), 512–523.

Steiner, S., Bisig, C., Petri-Fink, A., & Rothen-Rutishauser, B. (2016). Diesel exhaust: Current knowledge of adverse effects and underlying cellular mechanisms. *Archives of Toxicology, 90*(7), 1541–1553.

USEPA. (1998). *Revision of emission factors for AP-42 section 11.9: Western surface coal mines.* EPA Contract 68-D2-0159 WA 4-02, Research Triangle Park NC.

USEPA. (2005). *Compilation of air pollutant emission factors, AP-42* (6th Edn.). Research Triangle Park, NC: Office of Air Quality Planning and Standards, Emission Inventory Branch.

Valentin, J. (2007). *The 2007 recommendations of the international commission on radiological protection* (pp. 1–333). Oxford: Elsevier.

Vermeulen, R., Coble, J. B., Yereb, D., Lubin, J. H., Blair, A., Portengen, L., ... & Silverman, D. T. (2010). The diesel exhaust in miners study: III. Interrelations between respirable elemental carbon and gaseous and particulate components of diesel exhaust derived from area sampling in underground non-metal mining facilities. *Annals of Occupational Hygiene, 54*(7), 762–773.

Verpaele, S., & Jouret, J. (2012). A comparison of the performance of samplers for respirable dust in workplaces and laboratory analysis for respirable quartz. *Annals of Occupational Hygiene, 57*(1), 54–62.

Vutukuri, V. S., & Lama, R. D. (1986). *Environmental engineering in mines* (p. 518). Cambridge University Press: Great Britain. ISBN 0521157390.

WHO. World Health Organization. (2008). *Hazard prevention and control in the work environment: airborne dust.* WHO Press.

WHO. World Health Organization. (2009). *WHO handbook on indoor radon, a public health perspective.* Geneva: WHO Press.

Chapter 3
Flow Rates and Pressure Measurements

3.1 Introduction

In the previous chapter, an overview of environmental conditions in mining was offered from the viewpoint of occupational health. In this chapter, that summary is fleshed out by considering measurement and control of the main physical variables in a ventilation system. In general, these variables are pressures, airflow speeds and, indirectly, flow rates. These measurements, which are crucial in guaranteeing an airflow for miners and for the equipment used, are among the main inputs for mathematical models of ventilation. This section covers the main techniques and devices used in determining them.

3.2 Measurement of Static Pressure

The instruments most commonly used in measuring air pressure are *mercury barometers*[1] and *aneroid barometers*. The first of these (Fig. 3.1a) comprises a tube filled with mercury with a closed top in which a vacuum is made and an open bottom that is immersed in a vessel containing this metal. The rises and falls in the level of mercury inside the tube indicate increasing and decreasing air pressure. In aneroid barometers (Fig. 3.1b), measurements are based on the expansion and contraction of a flexible-walled cylinder in which a partial vacuum has been produced. These movements are amplified by means of a mechanical linkage and finally are indicated on a dial.

Barometers can indicate obstructions at points along the airflow circuit. Hence, an increase in circuit resistance usually implies a growth in static pressure if the air speed remains constant.

---

[1]The European Commission restricts the use of measuring devices containing mercury under Commission Regulation (EU) No 847/2012.

C. Sierra, *Mine Ventilation*, https://doi.org/10.1007/978-3-030-49803-0_3

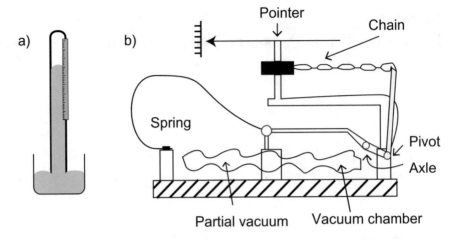

Fig. 3.1 a Mercury barometer and **b** aneroid barometer

Exercise 3.1 The atmospheric pressure at the topographic surface level measured by a barometer is 765 mm Hg. The static pressure generated by a fan is 60,000 Pa. Estimate the reading of this same barometer at the bottom of a 750 m shaft, ventilated from the surface by the fan. Omit friction losses in your calculations.

Solution

First, units are converted:

$$765 \, \text{mm Hg} = \frac{1 \, \text{atm}}{760 \, \text{mm Hg}} \frac{101,300 \, \text{Pa}}{1 \, \text{atm}} = 101,966 \, \text{Pa}$$

Next, the pressure of the static column of air in the shaft is calculated:

$$P = \rho g h = 1.2 \, \frac{\text{kg}}{\text{m}^3} \, 9.8 \, \frac{\text{N}}{\text{kg}} \, 750 \, \text{m} = 8820 \, \text{Pa}$$

By adding the pressure generated by the fan, the absolute pressure at the bottom of the shaft is obtained, thus:

$$101,966 \, \text{Pa} + 8820 \, \text{Pa} + 60,000 \, \text{Pa} = 170,786 \, \text{Pa}$$

3.3 Measurement of Dynamic Pressure

There are two basic types of dynamic pressure gauge: Lind and Prandtl. The Lind (Fig. 3.2) consists of a U-shaped tube with open ends filled with a liquid, usually water. One of the ends forms a horizontal elbow to receive the airflow. Both dynamic and static pressure enter this end, while only static pressure enters the other. Thus, static pressure cancels out at the two ends and only dynamic pressure is measured.

The functioning of a Prandtl anemometer is similar to the Lind, but its layout is much more compact. Static pressure enters through points a (Fig. 3.3), while total pressure enters through b. The static pressure entering through a cancels out with that entering through b, so that only dynamic pressure is measured.

Once the dynamic pressure has been measured, air speed can be determined by means of Eq. 3.1:

$$v = \sqrt{\frac{2P_v}{\rho}} \tag{3.1}$$

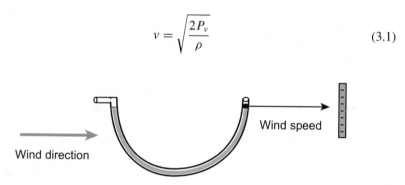

Fig. 3.2 Lind anemometer

Fig. 3.3 Prandtl anemometer

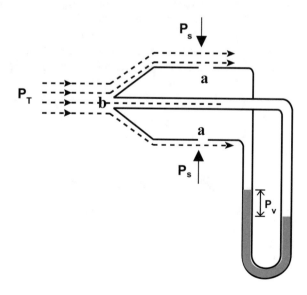

where

- v: Air speed (m s^{-1}),
- P_v: Dynamic pressure (Pa), and
- ρ: Air density (kg m^{-3}).

3.4 Pressure Drop Surveys

As air flows through an airway a pressure drop occurs. This pressure loss can be determined by two procedures, namely (Prosser and Loomis 2004): (a) Direct measurement of the differential pressure (*gauge-and-tube method* or *trailing hose*); (b) Indirect determination by the absolute pressure difference between two points (*barometer* or *altimeter method*).

In the gauge-and-tube method, the total pressure difference between two locations connected by a rubber tube is directly read by a manometer (Fig. 3.4). The system will be displaced along the wind direction from the intake path to the return path. This method is time-consuming and there is a limitation on the distance between measuring points given by the length of the hose. However, it is the most accurate method, so it is preferred when small pressure differences are to be measured.

The indirect method is carried out by simultaneously measuring the absolute static pressure at two stations with the aid of a barometer. The method can be performed even at long distances between measurement points but requires the existence of large-pressure differences to reduce measurement errors. To conduct a differential pressure survey by the indirect method, there are two alternatives, namely single-base method and leapfrogging.

In the *single-base method*, one barometer is located underground while the other one is located at a fixed base which can be either outside or inside the mine. The base must be located at a point with minimal temperature and pressure fluctuations that might be caused by sunlight, fans, regulators, hoisting equipment, etc. Measurements are recorded or logged at regular time intervals (approx. 5 min). This method requires correction for atmospheric pressure and air speed changes, as well as elevation.

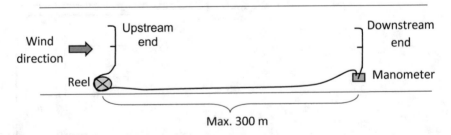

Fig. 3.4 Differential pressure measurement by the gauge-and-tube method. Modified from Prosser and Loomis (2004)

In the *leapfrogging* method, both instruments are located underground and readings are performed simultaneously, so correction for atmospheric pressure changes does not need to be performed. First, a number of measuring stations are fixed, and then, the two barometers are moved along the stations such that the downstream barometer used for one pair of measurements becomes the upstream barometer of the next pair.

Barometer surveys can be approximated employing Bernoulli equation between the two measuring points as if air were incompressible and velocity pressure changes negligible. The frictionless pressure drop ($\Delta P_{2_{calc}}$) can be approximated according to Hall (1981) and Prosser and Loomis (2004) as:

$$P_{2_{calc}} = P_2 \frac{2P_1 + Dg\rho_1}{2P_2 - Dg\rho_2} \tag{3.2}$$

- P_1: Barometric pressure at station 1 (Pa),
- P_2: Barometric pressure at station 2 (Pa),
- ρ_1: Air density at station 1 (kg m^{-3}),
- ρ_2: Air density at station 2 (kg m^{-3}), and
- D: Depth below datum (m).

Then, the frictional pressure drop (ΔP_{12}) is given by:

$$\Delta P_{12} = P_{2_{calc}} - P_2$$

Exercise 3.2 A barometric pressure survey is being conducted within two points of an interior incline traversed by an upflow of 20 m^3 s^{-1}. Station A is located at the −300 m level and station B at the −200 m level. Barometric pressure data, as well as air density calculations are presented in the table.

Station	Barometric pressure (Pa)	Air density (kg m^{-3})
A	103,690	1.245
B	104,712	1.250

You are asked to determine:

(a) The frictional pressure drop between both stations.
(b) The approximate value of the frictional airflow resistance between both stations.

Solution

(a) We start by applying Eq. 3.2:

$$P_{B_{calc}} = P_B \frac{2P_A + Dg\rho_A}{2P_B + Dg\rho_B}$$

Then, we can calculate D, as:

$D = 300 \, \text{m} - 200 \, \text{m} = 100 \, \text{m}$

Thus:

$$P_{B_{calc}} = 104{,}712 \, \frac{\text{N}}{\text{m}^2} \left(\frac{2 \cdot 103{,}690 \, \frac{\text{N}}{\text{m}^2} + 100 \, \frac{\text{N}}{\text{m}^2} \cdot 9.81 \, \frac{\text{N}}{\text{kg}} \cdot 1.245 \, \frac{\text{kg}}{\text{m}^3}}{2 \cdot 104{,}712 \, \frac{\text{N}}{\text{m}^2} - 100 \, \frac{\text{N}}{\text{m}^2} \cdot 9.81 \, \frac{\text{N}}{\text{kg}} \cdot 1.250 \, \frac{\text{kg}}{\text{m}^3}} \right)$$

$$= 104{,}915 \, \text{Pa}$$

Therefore, the frictional pressure drop is:

$$\Delta P_{AB} = P_{B_{calc}} - P_B = 104{,}915 \, \text{Pa} - 104{,}712 \, \text{Pa} = 203 \, \text{Pa}$$

(b)   Given that the average airflow rate in the airway ($20 \, \text{m}^3 \, \text{s}^{-1}$) is provided, the frictional resistance (R) can be approximated as:

$$R = \frac{\Delta P_{AB}}{Q^2} = \frac{203 \, \frac{\text{N}}{\text{m}^2}}{\left(20 \, \frac{\text{m}^3}{\text{s}} \right)^2} = 10.15 \, \frac{\text{N s}^2}{\text{m}^8}$$

3.5   Estimating the Section of a Gallery

The area of a cross section of a mine gallery can be found very precisely by laser scanning. If this technique is not available, the area can be estimated according to its approximate geometry, as shown in Fig. 3.5 (Luque Cabal 1988).

Hence, for the case of a steel straight walls semicircular arch (Fig. 3.5a), Eq. 3.3 would apply:

$$A = 0.83 \, B \, H \tag{3.3}$$

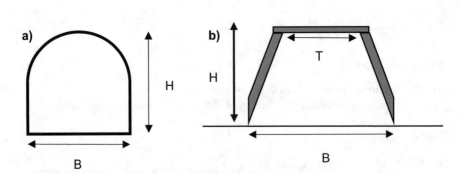

Fig. 3.5 Geometry of a section with: **a** a steel semicircular arch and **b** timber support

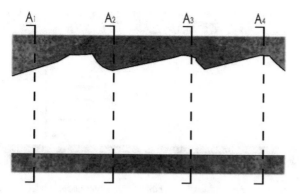

Fig. 3.6 Variations in cross section along a mine gallery

where

- B: Breadth at floor level (m), and
- H: Height of the support (m).

In the case of rigid timber supports (Fig. 3.5b) Eq. 3.4 can be applied:

$$A = \frac{(B + T)}{2} H \qquad (3.4)$$

where

- T: Width of the cap (m).

When making area measurements, it must be kept in mind that the area of the cross section of the gallery may vary along its length (Fig. 3.6). The mean cross section (A_m) for basic ventilation calculations can be obtained as the average of individual cross sections (Eq. 3.5):

$$A_m = \frac{\sum_{i=1}^{n} A_i}{n} \qquad (3.5)$$

where

- A_i: Cross section under consideration (m^2), and
- n: Number of cross sections under consideration.

3.6 Air Speed Measurement

Air, like any other fluid, presents a speed profile inside a mine gallery or in an air duct. This speed is at its greatest in the centre and at its lowest near the walls (Fig. 3.7). To measure air speeds, both *anemometers* and dynamic pressure meters are used. There

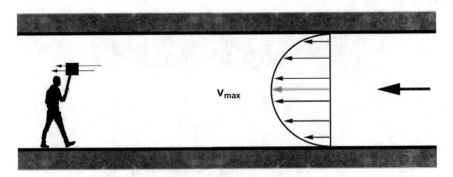

Fig. 3.7 Air speed profiling in a mine gallery. Note that the operator is located behind the apparatus so as to avoid distortions in the air current

are several types of anemometer, classified in accordance with the technology they use: *rotating-vane*, *hot-wire* and *ultrasonic*.

3.6.1   Rotating-Vane Anemometers

The *Robinson cup* is the type of anemometer usually found in meteorological stations. This equipment uses the force exerted by the wind on the concave and convex surfaces of hemispherical cups turning around a vertical axis. However, for mining purposes, it is common practice to use radial vanes rather than hemispheres (Fig. 3.8). The flow rate is proportional to the rotation velocity of this axis.

Fig. 3.8 Rotating-vane anemometer

3.6.2 Hot-Wire Anemometer

The *hot wire anemometer* is a thermo-electric device based on placing a hot wire in the midst of an airflow. The wire's temperature drops as a consequence of the flow of air. The temperature decrease in the wire is proportional to the air speed. Variations in the temperature of the wire produce changes in its resistance. These are measured by means of a Wheatstone bridge (Fig. 3.9).

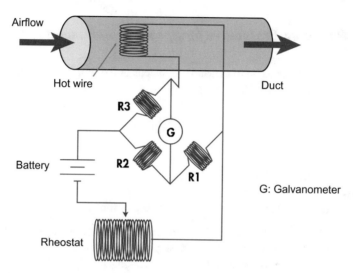

Fig. 3.9 Hot-wire anemometer (constant current system)

3.6.3 Ultrasonic Flowmeter

The *ultrasonic flowmeter* uses ultrasounds to measure the speed and direction of airflow. Sound is a wave that spreads through the air in all directions, forming a sphere-like shape. Two transducers are arranged so that one emits ultrasounds at one end of the gallery cross section and the other reads them at the other end. Normally, the transducers are positioned at an angle to the transverse section (Fig. 3.10). This means that the time the ultrasound takes to reach the other transducer is lower if it travels along with the airflow and larger if travels in countercurrent. The working principle is grounded on sending two signals—one with the flow and another opposing it—and measuring the time difference in the reception of the two signals. This time difference is directly proportional to air speed. This method provides very accurate and fast readings, valid even at very low air speeds.

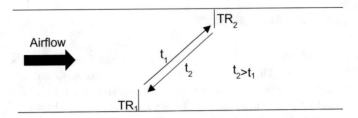

Fig. 3.10   Air speed measurement with an ultrasonic flowmeter within a mine gallery

3.6.4   Operating Procedures for Speed Measurements

The first approach would be to measure air speeds at fixed points. Equation 3.6 can be used as a first approximation in a gallery:

$$v_m = K \, v_D \tag{3.6}$$

where

- K: Coefficient, ranging from 0.75 to 0.8,
- v_m: Average speed (m s^{-1}), and
- v_D: Maximum air speed measured at the centre of the gallery (m s^{-1}).

There are more accurate procedures for calculating the average air speed in a traverse section (Simode 1962). In all these procedures, the arithmetic mean of all measurements is taken as the actual air speed[2] (Eq. 3.7):

$$v_m = \frac{\sum_{i=1}^{n} v_i}{n} \tag{3.7}$$

where

- v_m: Average speed (m s^{-1}),
- v_i: Speed in each of the grid sections (m s^{-1}), and
- n: Number of squares in the grid.

Figure 3.11a depicts a grid of approximately equal area squares[3]; Fig. 3.11b calculates the average air speed through 4 sampling points located in a homothetic curve (average value, in this case, should be then affected a coefficient, namely 0.98); Fig. 3.11c indicates the geometry for the direct determination of average speed through 5 local measurements.

---

[2]To calculate the average speed of an air current, the speed should be directly averaged. That is to say, if dynamic pressures are obtained, for example by means of a Pitot tube, they should first be transformed into speeds, and then averaged. This is the case because dynamic pressure is proportional to the square of air speed.

[3]Some quadrilaterals formed in the adjacent to the walls are discarded for practical reasons.

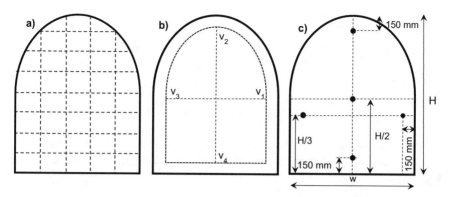

Fig. 3.11 Breaking down of a mine gallery to calculate average air speed. **a** Breaking down into squares of approximately the same area. Measurements are taken in the middle point of each square; **b** Measurements with a single homothetic curve; **c** Determination based on local measurements

Determinations that are more precise can be obtained by means of the *polar method* of the homothetic curves, as depicted in Fig. 3.12 (Simode 1962).

According to this procedure points *A*, *B*, *C* and *D* represent a distance to the wall of:

- *A*: 1/10 half the width of the gallery,
- *B*: 1/4 of half the width of the gallery,
- *C*: 1/2 of half the width of the gallery, and
- *D*: Centre of the gallery.

To calculate the speed in accordance with the method above Eq. 3.8 is employed:

Fig. 3.12 Determination of average airflow speed in a mine gallery by the polar method of the homothetic curves

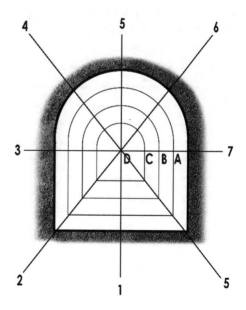

$$v_m = 0.083\, v_D + 0.313\, v_C + 0.286\, v_B + 0.282\, v_A \qquad (3.8)$$

The configuration above is complex to perform in practice, for this reason homothetic curves can be equally spaced (e.g. 60 cm), and the approximation of Eq. 3.9 can be used:

$$v_m = 0.07 v_D + 0.3(v_A + v_B + v_C) \qquad (3.9)$$

If a speed-integrating device is in use, the method recommended by the manufacturer should be followed. Traverses will usually be similar to those shown in Fig. 3.13.

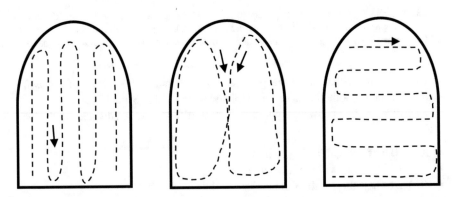

Fig. 3.13 Examples of traversing movements made in determining airflow speed in a mine gallery

Exercise 3.3 The table shows air speeds in a grid arrangement similar to what is shown in Fig. 3.10. Calculate the average airflow speed in the gallery. Data are quoted in metres per second.

N	A	B	C	D
0	–	–	–	4.5
1	3.4	3.5	4.1	–
2	3.5	3.6	4.2	–
3	3.3	3.6	4.3	–
4	3.4	3.8	4.2	–
5	3.4	3.7	4.1	–
6	3.4	3.6	4.2	–

(continued)

(continued)

N	A	B	C	D
7	3.4	3.7	4	–
8	3.4	3.7	4.1	–
D	–	–	–	4.7

Solution

N	A	B	C	D
0	–	–	–	4.5
1	3.4	3.5	4.1	–
2	3.5	3.6	4.2	–
3	3.3	3.6	4.3	–
4	3.4	3.8	4.2	–
5	3.4	3.7	4.1	–
6	3.4	3.6	4.2	–
7	3.4	3.7	4	–
8	3.4	3.7	4.1	–
D	–	–	–	4.7
$\sum_{i=1}^{n} v_i$	27.1	29.2	33.2	
$\frac{\sum_{i=1}^{n} v_i}{n}$	3.4	3.65	4.18	4.6

Exact formula:

$$v_m = 0.083 \, (4.6 + 0.313) \, (4.18 + 0.286) \, (3.65 + 0.282) \, 3.4 = 3.693$$

Approximate formula:

$$v_m = 0.07 \, (4.6 + 0.3) \, (3.4 + 3.65 + 4.18) = 3.691$$

Simplified formula:

$$v_m = 0.8 \cdot 4.6 = 3.68$$

3.6.5 Smoke-Cloud Generator (Smoke Tube)

Smoke tubes are one of the devices traditionally used to measure slow air speeds in mine galleries. They consist of a glass tube that can be broken at one end so that the substances it contains can be expelled by squeezing a flexible bulb (*smoke-cloud generator*). These substances react with the humidity in the air, the reaction generating a cloud of smoke that moves with the airflow, filling the entire section of the gallery.

The speed of the current can be determined by measuring the time it takes the smoke to move between any two points. The average cross section is usually calculated as an arithmetic mean of sections at a separation distance of 2 m.

Exercise 3.4 The air moves in a westerly to easterly direction along a mine gallery. A smoke tube is broken in section K, and is moved around quadrants, i, ii, iii, iv. The sections A_1, A_2 and A_3 were measured as being 9.8, 8.9 and 9.5 m^2, respectively. The average time the smoke took to get from each quadrant in K to its counterpart in S, a distance of 7.5 m away, is given in the table. Determine the best possible approximation of the ventilation airflow rate in this gallery.

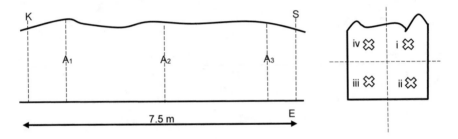

Quadrant	Time from K to S in each repetition (s)			
	1	2	3	4
I	10	9	9	10
II	11	11	10	10
III	9	9	10	9
IV	10	11	11	10

Solution

Quadrant	Time from K to S for each repetition (s)				
	1	2	3	4	Average
I	10	9	9	10	9.5
II	11	11	10	10	10.5
III	9	9	10	9	9.25
IV	10	11	11	10	10.5
Overall average					**9.94**

$$v = \frac{s}{t} = \frac{7.5 \, \text{m}}{9.94 \, \text{s}} = 0.754 \, \frac{\text{m}}{\text{s}}$$

$$A_m = 9.4 \, \text{m}^2$$

$$Q = A_m v = 0.754 \, \frac{\text{m}}{\text{s}} \cdot 9.4 \, \text{m}^2 = 7.1 \, \frac{\text{m}^3}{\text{s}}$$

Note that the speed measurements obtained by this method are frequently increased by 10%, so 0.9v should be considered as the approximate speed value.

3.6.6 Tracers

The use of *tracers* is a procedure similar to the preceding method. Sulphur hexafluoride (SF_6) is used as the tracer gas because of its low toxicity and because it is easy to detect at low concentrations. The steady release method is as follows: a constant flow rate of tracer (q) is injected into an airflow (Q), and at a given distance from the emission point (L_m) and at a given time (t_m) the measurement[4] of the concentration (c) is made, as shown in Fig. 3.14 (Higgins and Shuttleworth 1958; Thimons and Kissell 1974).

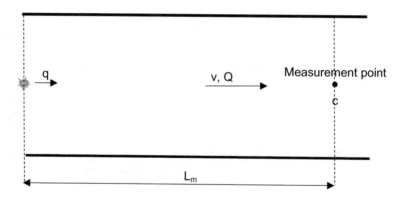

Fig. 3.14 System of measuring flow rate in a mine gallery by the dilution method

---

[4]This conditions aim for thorough mixing and equilibrium.

The distance (L_m) from the emission to the measurement point is determined by means of Eq. 3.10:

$$L_m = \frac{44A}{O\sqrt{\lambda}}$$ (3.10)

where

- A: Area of the gallery (m^2),
- O: Perimeter of the gallery (m), and
- λ: Coefficient of friction.

The start time for measurements (t_m) corresponds to Eq. 3.11:

$$t_m = \frac{2.6L_m}{v}$$ (3.11)

where

v is the air speed in the gallery (m s^{-1}).

The concentration of the tracer gas can be calculated as:

$$c = \frac{q}{Q + q}$$

Since:

$$q \ll Q$$

Therefore, the rate of air flowing through the gallery can be approximated as (Eq. 3.12):

$$Q = \frac{q}{c}$$ (3.12)

A variant of the previous method is applied to measuring leaks in airlocks. Figure 3.15 illustrates the methodology (d'Albrand 1976).

Using the preceding expression, c_1, c_2 and c_3 are assigned as the concentrations at points 1, 2 and 3, respectively. The flow rate at point 1 in Fig. 3.15 will thus be:

$$Q_1 = \frac{q}{c_1}$$

At point 2:

$$Q_2 = \frac{q}{q\left(\frac{1}{c_2} - \frac{1}{c_1}\right)}$$

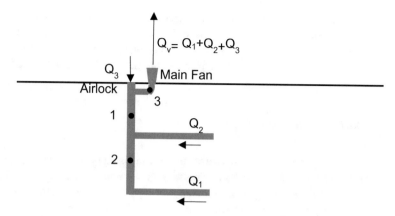

Fig. 3.15 Detecting leaks in airlocks

Hence, in the airlock itself Eq. 3.13 applies:

$$Q_3 = \frac{q}{q\left(\frac{1}{c_3} - \frac{1}{c_2}\right)} \qquad (3.13)$$

3.7 Measurements in Ducts

Although some of the equipment used for measurements in galleries can also be used in ducts, the two commonest devices for measurements in ducts are static pressure probes and Pitot tubes (Fig. 3.16). Further information about their fundamentals can be found in Sect. 1.3.5.

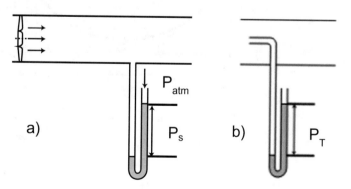

Fig. 3.16 **a** Static pressure probe and **b** Pitot tube

As in mine galleries, an approximation that should not yield an error greater than 10% involves measuring the speed at the midpoint (v_D) of the duct and multiplying it by a coefficient that in this case is usually 0.85 (Oller 2004). However, to obtain greater precision, several different procedures, as described below, can be used.

3.7.1   Ducts of Circular Section

Classic methods employing concentric circles delimiting sectors of equal surface area and the log-linear method can be used. The sampling points indicated in Fig. 3.17 correspond to the *Tchebycheff logarithmic method* (log–Tchebycheff method), which can be used in any measurements with a Pitot tube. The number of points to be considered varies in accordance with the diameter of the duct. Thus, for ducts with a diameter less than or equal to 25.4 cm, six measuring points are used, while for larger ducts, from 8 to 10 are employed (Table 3.1). The speed of fluid going through the duct is obtained from the arithmetic mean of all measurements (Macferran 1999).

Fig. 3.17 Location of the sampling points following the log-Tchebycheff six-point rule for ducts of circular section. ANSI/ASHRAE Standard Modification 111-2008

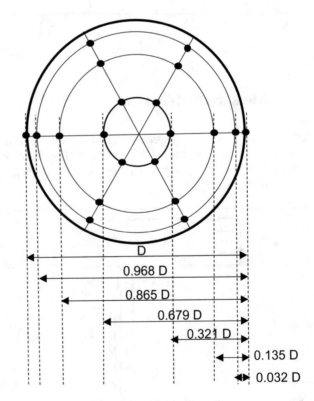

Table 3.1 Log-Tchebycheff rule for ducts of circular section

Number of points	Position relative to inner wall
6	0.032, 0.135, 0.321, 0.679, 0.865, 0.968
8	0.021, 0.117, 0.184, 0.345, 0.655, 0.816, 0.883, 0.981
10	0.019, 0.153, 0.217, 0.361, 0.639, 0.783, 0.847, 0.923, 0.981

3.7.2 Ducts of Rectangular Section

A variant of the log–Tchebycheff method can be used for rectangular ducts (Table 3.2 and Fig. 3.18).

Thus:

- Less than 76.2 cm: 5 points,
- 76.2–91.44 cm: 6 points, and
- Greater than 91.44 cm: 7 points.

In some texts on fluid mechanics, ducts are broken down into between 16 and 64 rectangles with the same area (Fig. 3.19) and the average is calculated. With this configuration, a minimum of 25 points should ideally be selected (Macferran 1999).

Table 3.2 Log-Tchebycheff rule for ducts of rectangular section

Number of points	Position relative to inner wall
5	0.074, 0.288, 0.500, 0.712, 0.926
6	0.061, 0.235, 0.437, 0.563, 0.765, 0.939
7	0.053, 0.0203, 0.366, 0.500, 0.634, 0.797, 0.947

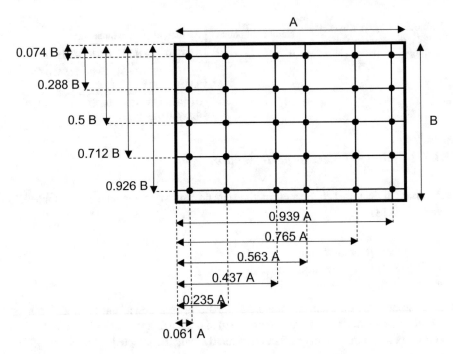

Fig. 3.18 Location of sampling points following the log-Tchebycheff six-point rule for ducts of rectangular section. Modified from ANSI/ASHRAE Standard 111-2008. The number of points that must be established along each side of the duct depends on the length of that side

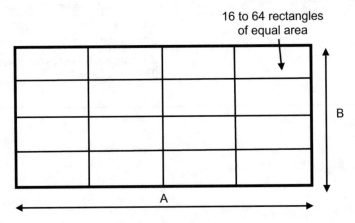

Fig. 3.19 Traversing a duct of rectangular section. Break-down into rectangles of equal area

References

ANSI/ASHRAE Standard 111. (2008). *Measurement, testing, adjusting, and balancing of building HVAC systems.*

Commission Regulation (EU) No. 847/2012 of 19 September 2012 amending Annex XVII to Regulation (EC) No 1907/2006 of the European Parliament and of the Council on the Registration, Evaluation, Authorisation and Restriction of Chemicals (REACH) as regards mercury.

d'Albrand, N. (1976). Mesures d'aérage. In *Aérage.* Document SIM N3. Revue de l'Industrie Minérale. Mine 2–76.

Hall, C. J. (1981). *Mine ventilation engineering* (Chap. 8). New York, NY: Society of Mining Engineers of The American Institute of Mining, Metallurgical and Petroleum Engineers, Inc.

Higgins, J., & Shuttleworth, S. E. H. (1958). A tracer gas technique for the measurement of airflow in headings. *Colliery Engineering, 35*(417), 483–487.

Luque Cabal, V. (1988). *Manual de ventilación de minas.* Madrid: Pedeca, Sociedad Cooperativa. Limitada.

Macferran, E. L. (1999). Equal area vs. Log Tchebycheff. *HPAC Heating, Piping and Air Conditioning Engineering, 71*(12), 26–31.

Oller, N. C. (2004). *NTP 668: Medición del caudal en sistemas de extracción localizada.* Madrid: Ministerio de Trabajo y Asuntos Sociales.

Prosser, P. E. B. S., & Loomis, P. E. I. M. (2004). Measurement of frictional pressure differentials during a ventilation survey. In *Mine Ventilation: Proceedings of the 10th US/North American Mine Ventilation Symposium,* Anchorage, Alaska (pp. 59–66).

Simode, E. (1962). Les mesures d'aérage. In *Aérage.* Document SIM N1. *Revue de l' Industrie Minérale.*

Thimons, E. D., & Kissell, F. N. (1974). *Tracer gas as an aid in mine ventilation analysis* (No. BM-RI-7917). Washington, D.C.: Bureau of Mines.

Bibliography

Elger, D. F., LeBret, B. A., Crowe, C. T., & Robertson, J. A. (2016). *Engineering fluid mechanics* (pp. 170–185). Hoboken, New Jersey: Wiley.

Nakayama, Y. (2018). *Introduction to fluid mechanics.* Kidlington, Oxfordshire: Butterworth-Heinemann.

Shaughnessy, E. J., Katz, I. M., & Schaffer, J. P. (2005). *Introduction to fluid mechanics* (Vol. 8). New York: Oxford University Press.

Smits, A. J. (2000). *A physical introduction to fluid mechanics.* Hoboken, NJ: Wiley.

Young, D. F., Munson, B. R., Okiishi, T. H., & Huebsch, W. W. (2010). *A brief introduction to fluid mechanics.* Hoboken, New Jersey: Wiley.

Chapter 4
Mine Ventilation Networks

4.1 Introduction

Mining ventilation networks are made up of the galleries and ducts through which the air is distributed inside the exploitations. The ultimate aim in these calculations is the dimensioning of the main fan which will supply fresh air to the mine. Here two aspects should be taken into account: the quantity of air to be supplied and the overall resistance of the mine. The first should be sufficient to guarantee people can breathe, the operation of diesel machines, the dilution of fumes and dust, as well as the correct environmental conditions of temperature and humidity for the miners' work. The overall air flow rate requirements for a modern block caving mine are usually about 600 m^3 s^{-1}, while the highest are usually about 3000 m^3 s^{-1}.

Once the quantity has been estimated, the overall resistance of the mine must be estimated. For this purpose, laws similar to those used in the resolution of electrical circuits are applied. Their application requires the use of a large number of equations which are solved using numerical methods, thus computer programs to avoid tedious manual calculation are employed. The bases for the informed management of these computer tools are set out in this chapter.

4.2 Friction Losses: Atkinson Expression

If we start from the Darcy–Weisbach equation (Eq. 1.21):

$$\Delta H_f = \frac{fL}{D}\frac{v^2}{2g} \qquad (1.21)$$

where

© The Editor(s) (if applicable) and The Author(s), under exclusive license
to Springer Nature Switzerland AG 2020
C. Sierra, *Mine Ventilation*, https://doi.org/10.1007/978-3-030-49803-0_4

- ΔH_f: Head loss (meters of air column),
- f: Darcy–Weisbach friction coefficient (dimensionless),[1]
- v: Air speed in the airway ($m^3 \, s^{-1}$),
- L: Airway length (m), and
- D: Airway diameter (m).

Bearing in mind that the expression is given in column meters,[2] it can be transformed into SI units by applying Eq. 1.7:

$$P = H_f \rho g \tag{1.7}$$

Thus, Eq. 4.1 is obtained:

$$\Delta P = \frac{fL}{D}\frac{v^2}{2}\rho \tag{4.1}$$

If the flow is considered to be turbulent, and the geometry of the cross sections could vary, it is advisable to work with hydraulic diameters (D_h) (Eq. 1.19) instead of diameters (D):

$$D_h = \frac{4A}{O} \tag{1.19}$$

where

A: Cross-sectional area of the airway (m^2), and
O: Perimeter of the cross section (m).

Substituting the expression of D_h in Eq. 4.1, one obtains Eq. 4.2:

$$\Delta P = \frac{fL}{8}\frac{v^2 O}{A}\rho \tag{4.2}$$

Bearing in mind that the air density is $1.2 \, kg \, m^{-3}$ and that $\frac{1}{8} \cdot 1.2 = 0.15$ we obtain Eq. 4.3:

$$\Delta P = 0.15\frac{f\,O\,L}{A}v^2 \tag{4.3}$$

If we consider the relation (Eq. 4.4):

$$0.15\,f = K_{1.2} \tag{4.4}$$

---

[1]Since the diameters of our airways are going to be high and the wind speeds significant, the Re will be also high. If we add to this, the roughness will also be high then we will be on the right side of Moody's diagram, where the coefficient of friction is independent of Re.

[2]In the case of ventilation, usually air column.

A new parameter is obtained, which is called the *Atkinson friction factor* ($K_{1.2}$) and is a function of Darcy's friction factor[3] (f) (Eq. 4.4), which, in turn, can be obtained by Eqs. 1.25 to 1.29.[4]

According to the paragraphs above, the equation for pressure losses can be written as (Eq. 4.5):

$$\Delta P = K_{1.2} \frac{O\,L}{A} v^2 \tag{4.5}$$

Nevertheless, expressing the flow as a function of the cross section and air speed, the *Atkinson equation* (1862) (Eq. 4.6) is obtained:

$$\Delta P = K_{1.2} \frac{O\left(L + L_{eq}\right)Q^2}{A^3} \tag{4.6}$$

This expression states that the friction pressure losses in the airway are a function of the fluid velocity, the characteristics of its inner surface ($K_{1.2}$) and its dimensions (O, L, L_{eq} and A).

where

- ΔP: Pressure difference (Pa), (inch water);
- $K_{1.2}$: Atkinson friction factor for 1.2 g cm^{-3} air density (N s^2 m^{-4} ó kg m^{-3}), (lb min^2 ft^{-4});
- O: Perimeter of the airway (m), (ft);
- L and L_{eq}: Length and equivalent length (m), (ft);
- Q: Volumetric air flow rate (m^3 s^{-1}), (ft^3 s^{-1});
- A: Cross-sectional area of the airway (m^2), (ft^2); and
- ρ: Air density (kg m^{-3}), (lb ft^{-3}).

The airflow is affected by the average air density, which for the purposes of the above calculations has been considered to be 1.2 kg m^{-3}. So, if you wish to correct possible density variations due to altitude or temperature, you can use a corrected coefficient (K_c), which is expressed as (Eq. 4.7):

$$K_c = K_{1.2} \frac{\rho}{1.2\left[\frac{kg}{m^3}\right]} \tag{4.7}$$

---

[3]The "f" used in this text corresponds to the friction factor of Darcy's law, sometimes denoted as f_D. Although the former is the most widespread, it is possible to find bibliographic sources using Fanning's friction factor (f_F), for example, Table 5.1 in McPherson (1993), p. 138. The relationship between the two is: $f_D = 4 f_F$. So in this case: $K_{1.2} = 0.6 f_F$.

[4]Note that f also depends on roughness. The same values can be found for different types of mining air conduits, for example, in Montecinos and Wallace (2010).

We finally obtain (Eq. 4.8):

$$\Delta P = K_{1.2} \left(\frac{\rho}{1.2 \left[\frac{kg}{m^3} \right]} \right) \frac{O \left(L + L_{eq} \right) Q^2}{A^3} \tag{4.8}$$

Exercise 4.1 Consider there to be a 6.5 m × 4.5 m rectangular airway of 800 mm of absolute roughness traversed by an airflow of 30 m^3 s^{-1}. You are asked to:

(a)   Determine the Atkinson friction factor of the airway.
(b)   Comment on the extent to which the results obtained are close to those expected in a mining gallery.

Data:
Nikuradse expression:

$$\frac{1}{\sqrt{f}} = -2.0 \log \left(\frac{\frac{\varepsilon}{D_h}}{3.71} \right)$$

where

- ε: Absolute roughness,
- D_h: Hydraulic diameter (m),
- Air density: $1.16 \frac{kg}{m^3}$, and
- Air dynamic viscosity: $1.86 \cdot 10^{-5} N \cdot \frac{s}{m^2}$.

Solution

(a)   The hydraulic diameter is calculated as:

$$D_h = \frac{4 \cdot 6.5 \cdot 4.5}{2 \cdot 6.5 + 2 \cdot 4.5} m = 5.318 \, m$$

The average air speed can be calculated as:

$$v = \frac{Q}{A} = \frac{30 \frac{m^3}{s}}{6.5 \cdot 4.5 \, m^2} = 1.025 \frac{m}{s}$$

With this value and the dynamic viscosity of the air, we obtain:

$$Re = \frac{\rho D_h v}{\mu} = \frac{1.16 \frac{kg}{m^3} \cdot 5.318 \, m \cdot 1.025 \frac{m}{s}}{1.86 \times 10^{-5} N \frac{s}{m^2}} = 33,9951.72$$

Thus indicating fully turbulent flow.
The relative roughness is:

$$\frac{\varepsilon}{D_h} = \frac{800 \times 10^{-3}\,\text{m}}{5.318\,\text{m}} = 0.1504$$

The Re indicates that turbulent flow and since the conduit is rough, the Nikuradse formula applies. Thus, substituting in Nikuradse expression we have:

$$\frac{1}{\sqrt{f}} = -2.0\log\left(\frac{0.1504}{3.71}\right)$$

$$f = 0.0967$$

Then the Atkinson friction factor is:

$$K_{1.2}\left[\frac{\text{N}\,\text{s}^2}{\text{m}^4}\right] = 0.15f = 0.0145$$

(b) In mining galleries, conditions $f > 0.005$ and Re $> 10^6$ usually occur. Under these conditions, if we look for example at the Moody diagram, the coefficient of friction becomes independent of Re.

Exercise 4.2 Calculate the pressure loss taking place due to friction in an 85-m long raise of circular cross section and 1.5 m diameter when an anemometer installed in it registers an airflow rate of 50 m³ s⁻¹. Assume that surface of the raise is:

(a) Shotcrete $\left(K_{1.2} = 0.004\,\frac{\text{N}\,\text{s}^2}{\text{m}^4}\right)$, and

(b) Bare rock $\left(K_{1.2} = 0.015\,\frac{\text{N}\,\text{s}^2}{\text{m}^4}\right)$.

The air density is 1.1 kg m⁻³.

Solution

(a) Applying the Atkinson equation with $L_{eq} = 0$ (there are no accessory elements or relevant cross-sectional changes):

$$\Delta P = K_{1.2}\frac{O(L + L_{eq})Q^2}{A^3}\left(\frac{\rho}{1.2\left[\frac{\text{kg}}{\text{m}^3}\right]}\right)$$

For shotcrete:

$$\Delta P = 0.004\frac{\text{N}\,\text{s}^2}{\text{m}^4}\frac{4.71\,\text{m}\cdot(85)\cdot\left(50\frac{\text{m}^3}{\text{s}}\right)^2}{\left(1.76\,\text{m}^2\right)^3}\left(\frac{1.1}{1.2}\right) = 665.18\,\text{Pa}$$

(b)   For the bare rock:

$$\Delta P = 0.015 \frac{\mathrm{N\,s^2}}{\mathrm{m^4}} \frac{4.71\,\mathrm{m} \cdot (85) \cdot \left(50\frac{\mathrm{m^3}}{\mathrm{s}}\right)^2}{(1.76\,\mathrm{m^2})^3} \left(\frac{1.1}{1.2}\right) = 2494.43\,\mathrm{Pa}$$

Note that the relationship between the two pressure losses is the same as the one between their Atkinson friction factors.

Exercise 4.3 A circular duct is traversed by a flow of 180 m^3 s^{-1} of air under a pressure difference between its extremes of 2000 Pa. Determine the flow through a circular duct of the same material as the previous one, half radius and double length, when the pressure difference between the ends of the duct becomes 7000 Pa. Solve this using the Atkinson equation.

Solution

Particularizing the Atkinson expression for a circular duct, one has:

$$\Delta P = K \frac{O L Q^2}{A^3} = \frac{2\pi r L Q^2}{\left(\pi r^2\right)^3} = \frac{2\pi r L Q^2}{\pi^3 r^6} = \frac{2 L Q^2}{\pi^2 r^5}$$

Applying it to the two proposed cases 1 and 2:

$$\Delta P_1 = \frac{2 L_1 Q_1^2}{\pi^2 r_1^5}$$

$$\Delta P_2 = \frac{2 L_2 Q_2^2}{\pi^2 r_2^5}$$

Dividing both expressions and simplifying:

$$\frac{\Delta P_1}{\Delta P_2} = \frac{\frac{2 L_1 Q_1^2}{\pi^2 r_1^5}}{\frac{2 L_2 Q_2^2}{\pi^2 r_2^5}} = \frac{\frac{L_1}{r_1^5} Q_1^2}{\frac{L_2}{r_2^5} Q_2^2}$$

You have:

$$\Delta P_2 = \frac{\frac{L_2}{r_2^5} Q_2^2}{\frac{L_1}{r_1^5} Q_1^2} \Delta P_1$$

Particularizing:

$$\Delta P_2 = \frac{2^6 Q_2^2}{Q_1^2} \Delta P_1$$

So finally:

$$Q_2 = 42.09 \frac{m^3}{s}$$

4.3 Concept of Resistance of an Airway

The Atkinson equation can be expressed as a function of the square of the airflow rate, and a coefficient (R) that groups together all the geometric parameters (O, L, A), the properties of the conduit (K) and of the density of the fluid (ρ) (Eq. 4.9):

$$R = K_{1.2} \frac{O(L + L_{eq})}{A^3} \left(\frac{\rho}{1.2 \left[\frac{kg}{m^3} \right]} \right) \tag{4.9}$$

So the Atkinson equation would appear as (Eq. 4.10). This expression is called the mine characteristic curve as is discussed in detail in Sect. 4.4:

$$\Delta P = RQ^2 \tag{4.10}$$

Thus, R is usually called the *aerodynamic resistance* (*Atkinson resistance*) of the airway, which expressed in SI units is:

$$R = \frac{\Delta P}{Q^2} = \frac{1 \frac{N}{m^2}}{\left(1 \frac{m^3}{s} \right)^2} = \frac{N s^2}{m^8}, \frac{kg}{m^7}$$

This unit is called *gaul* and is defined as: "Resistance of the airway which causes a pressure loss of 1 Pa when crossed by a quantity of 1 m^3 s^{-1} at the standard air density of 1.2 kg m^{-3}". For the sake of simplicity, the gaul can often be seen as Pa Q^{-2}.

In Germany, the *weisbach* (Wb) is frequently used. Its definition is analogous to the previous one with the particularity that instead of using the N m^{-2}, the kgf m^{-2} is employed, so 9.8 gauls = 1 Wb.

In spite of the diffusion that the SI currently has, the use of the English unit or *atkinson* is still frequent.[5] Analogous to the previous one, it is defined as: "Resistance of the airway which causes a pressure loss of 1 lbf ft^{-2} when crossed by an air quantity of 1000 ft^3 s^{-1} at the air density of 0.075 lbm ft^{-3}", therefore:

---

[5] Not to be confused with Atkinson friction factor.

$$R = \frac{\Delta P}{Q^2} = \frac{1 \frac{\text{lbf}}{\text{ft}^2}}{\left(1000 \frac{\text{ft}^3}{\text{s}}\right)^2} = \frac{1 \frac{\text{lbf}}{\text{ft}^2}}{1000^2 \frac{\text{ft}^6}{\text{s}^2}} = \frac{1 \text{ lbf s}^2}{1000^2 \text{ft}^8}$$

In this case, the conversion to the SI would be:

$$1 \, Atkinson = \frac{1 lbf \cdot s^2}{1000^2 ft^8} = \frac{1 \cancel{lbf} \cdot s^2}{1000^2 \cancel{ft^8}} \frac{4.448 N}{1 \cancel{lbf}} \frac{\cancel{(1 ft)^8}}{(0.3048 \, m)^8} = 0.0597 \frac{N \cdot s^2}{m^8}$$

One historical unit, still of some use, is the *murgue* (μ). This is defined as: "Resistance of a airway which causes a pressure loss [6] of 1 gf m^{-2} when traversed by an airflow rate of 1 m^3 s^{-1}". Thus, 1 kμ = 1 Wb.

All in all, the equivalence between all the units is:

1 atkinson = 0.0597 gaul = 0.0061 kμ.

Exercise 4.4   Based on the above definitions, find the conversion factor from murgues to gauls.

Solution

Let us start by applying the definition of murgue:

$$R = \frac{\Delta P}{Q^2} = \frac{1 \frac{\text{gf}}{\text{m}^2}}{\left(1 \frac{\text{m}^3}{\text{s}}\right)^2} = \frac{1 \text{ gf s}^2}{\text{m}^8}$$

Its conversion to the SI will, therefore, be:

$$1 \, murgue = \frac{1 gf \cdot s^2}{m^8} = \frac{1 \cancel{gf} \cdot s^2}{m^8} \frac{1 \cancel{Kgf}}{1000 \, \cancel{gf}} \frac{9.8 \, N}{1 \cancel{Kgf}} = 9.8 \cdot 10^{-3} \frac{N \cdot s^2}{m^8}.$$

Typical Values of Atkinson Friction Factors (Total Coefficients)

Although specific measurements of friction factors at each mine are required, preliminary studies can be based on existing literature. Thus, a good reference may be Prosser and Wallace (1999), where the authors make their own measurements and compare them with those found in classical literature (Table 4.1).

---

[6] gf = gram force.

Table 4.1 Comparison between Atkinson friction factor in literature. Taken from Prosser and Wallace (1999)

Airway type	Atkinson friction factor K (kg m^{-3})		
	Prosser and Wallace (1999)	McPherson (1993, pp. 134–140)	Hartman et al. (1997, pp. 155–157)
Rectangular airway free of obstacles (soft rock with bolts and limited mesh)	0.0075	0.009	0.0080
Rectangular airway with some irregularities (soft rock with bolts and limited mesh)	0.0087	0.009	0.0091
Metal mine drift (metal frames and bolts, limited mesh)	0.010	0.0120	0.0269
Metal mine ramp (metal frames support and bolts, limited mesh)	0.013	N/A	0.0297
Metal mine beltway (large section, bolting and mesh)	0.015	N/A	N/A
Circular raise (raiseboring, including inlet and outlet losses)	0.0050	0.004	0.0028
Rectangular raise (Alimak method, without wood support with rock bolt and mesh)	0.0129	0.014	N/A
Gallery excavated with TBM (bolting and mesh)	0.0050	0.0055	0.0037

Values for Walls and Floors

When estimating a coefficient of friction for a mine gallery, it should be borne in mind that there is more wall surface than floor surface. Some tables already consider this fact by providing a general value (e.g. McElroy 1935), while others provide differentiated values for walls and floor (e.g. Simode 1976). Normally the weight to be given to each of them is 70% and 30%, respectively, so (Eq. 4.11):

$$f = 0.7\, f_{\text{wall}} + 0.3\, f_{\text{floor}} \tag{4.11}$$

There is also a scale effect, i.e. a larger gallery has a smaller friction factor. This fact is of particular relevance to small cross sections, as the coefficient grows potentially as the cross section decreases. Therefore, when using old tables, it must be taken into account that they were derived when the mining galleries were much smaller than the present ones. Thus, values of two-thirds of those in the tables are closer to present needs (de la Vergne 2008). For standardization purposes, many tables (e.g. Simode 1976; Carrasco et al. 2011) refer to sections of 10 m^2 (Table 4.2). If the gallery is different from 10 m^2, the values can be corrected using Eq. 4.12:

Table 4.2 Main friction coefficients. Modified from Simode (1976) and Carrasco et al. (2011)

Element	Coefficient of friction
Floor	f_{floor}
Smooth (concrete)	0.025
Well trimmed	0.058
Regular	0.084
Irregular	0.108
Gallery wall	f_{wall}
Bare rock	
Well trimmed	0.058
Medium	0.084
Irregular	0.108
Bolted	
Well trimmed	0.058
Medium	0.084
Irregular	0.108
Wire mesh	0.130
Coated	
Smooth concrete	0.022
Brick lined	0.025–0.040

$$f_x = \frac{f_{10}}{(0.75 + 0.25 \log A)^2} \qquad (4.12)$$

where

- f_x: Coefficient of friction for an airway of A cross section,
- f_{10}: Coefficient of friction of an airway of 10 m^2 of cross section, and
- A: Cross-sectional area of the airway (m^2).

Note that while there is a simple connection between $K_{1.2}$ and f ($0.15\,f = K_{1.2}$), this is not the case for $R_{1.2}$. This is so because, by definition $R = K_{1.2}\frac{O(L+L_{eq})}{A^3}$, and therefore depends on geometric parameters of the gallery, such as the cross-sectional area and the perimeter.

Question 4.1 Given the von Karman equation for rough conduit:

$$\frac{1}{\sqrt{f_x}} = 1.76 + 2.0\log\left(\frac{r}{\varepsilon}\right)$$

(a) Demonstrate that the relation between the friction factor for any section (f_x) and the friction factor a $10\ m^2$ gallery (f_{10}) is given by the expression:

$$f_x = \frac{f_{10}}{(0.75 + 0.25\log A)^2}$$

Assume circular cross section for the gallery and absolute roughness (ε) of $0.15\ m$.

(b) Determine how many times larger is the coefficient of friction of a gallery of $4\ m^2$ than that of a gallery of $12\ m^2$.

Answer

(a) At the turbulence level of a mine gallery, the friction factor is independent of Re, and therefore we can employ:

$$\frac{1}{\sqrt{f_x}} = 1.76 + 2.0\log\left(\frac{r}{\varepsilon}\right)$$

Particularizing for (a) f_{10} and A_{10} and (b) f_x and A, and calculating the quotient, for a circular cross section we have that:

$$\frac{\frac{1}{\sqrt{f_{10}}}}{\frac{1}{\sqrt{f_x}}} = \frac{1.76 + 2.0\log\left(\frac{\sqrt{A_{10}}}{\varepsilon\sqrt{\pi}}\right)}{1.76 + 2.0\log\left(\frac{\sqrt{A}}{\varepsilon\sqrt{\pi}}\right)}$$

Thus, for $A_{10} = 10\ m^2$ and $\varepsilon = 0.15\ m$:

$$\frac{f_x}{f_{10}} = \left(\frac{1.76 + 2.0\log\left(\frac{\sqrt{10}}{0.15\sqrt{\pi}}\right)}{1.76 + 2.0\log\left(\frac{\sqrt{A}}{0.15\sqrt{\pi}}\right)}\right)^2$$

Operating we get:

$$f_x = \frac{f_{10}}{(0.75 + 0.25\log A)^2}$$

(b) Using the above expression, we have:

$$f_{12} = \frac{f_{10}}{(0.75 + 0.25\log 12)^2}$$

$$\frac{f_4}{f_{12}} = \frac{(0.75 + 0.25 \log 12)^2}{(0.75 + 0.25 \log 4)^2} = \frac{1.04}{0.81}$$

$$\frac{f_4}{f_{12}} = 1.28$$

Exercise 4.5 A test was carried out to determine the resistance to the passage of air in a group of homothetic galleries, of equal coefficient of friction, supported by yieldable steel arches with steel wire mesh and equipped with rails and pipes. Data from these trials are shown in table:

Cross-sectional area, A (m^2)	Atkinson's hectometric resistance, R_{100} (μ)
20	0.28
15	0.59
13	0.94
8	2.9
6	6

You are asked to:

(a) Estimate the value of the aerodynamic resistance for a hectometer of gallery of 10 m^2 of cross-sectional area.

(b) Estimate the value of the aerodynamic resistance for a hectometer of gallery of 1 m^2 of cross-sectional area.

(c) Comment on the extent to which these data conform to the theory exposed in the chapter.

Solution

(a) The previous data can be represented in a Cartesian coordinate system offering the exponential equation of the figure.

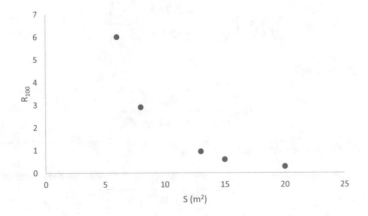

The data can be linearized by taking logarithms of both R_{100} and S, then:

$\log_{10} A$	$\log_{10} R_{100}$
1.30103	−0.55284
1.176091	−0.22915
1.113943	−0.02687
0.90309	0.462398
0.778151	0.778151

Ploting again:

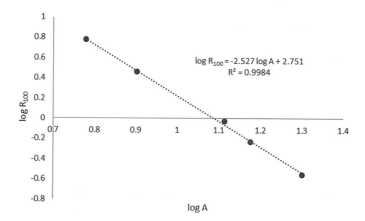

Then, for $A = 10 \text{ m}^2$:

$$\log R_{100}(A = 10) = -2.527 \log(10) + 2.751 = 0.224$$

So, $R_{100}(A = 10) = 1.67$.

(b) Proceeding in the same way for $A = 1 \text{ m}^2$:

$$\log R_{100}(A = 1) = -2.527 \log(1) + 2.751 = 2.751$$

With what:

$$R_{100}(A = 1) = 563.64$$

The quotient between the two is:

$$\frac{R_{100}(A = 1)}{R_{100}(A = 100)} = \frac{563.64}{1.67} = 337.51$$

Note that we have performed inference for an area of 1 m^2 which lies at a region of the function where it grows exponentially. It is, therefore, to be expected that the results obtained will be accompanied by a large error.

(c)   If we start from the definition of aerodynamic resistance, we have:

$$R = K \frac{OL}{A^3} = 0.15 f \frac{OL}{A^3}$$

Since we have a hectometric (100 m) resistance:

$$R_{100} = 0.15 f \frac{O\,100}{A^3} = 15 f \frac{O}{A^3}$$

If a circular airway is assumed, then:

$$O = 2\pi R$$

$$A = \pi R^2$$

Therefore:

$$O = 2\sqrt{\pi A}$$

Substituting in the expression of the aerodynamic resistance, we have:

$$R_{100} = 15 f \frac{2\sqrt{\pi A}}{A^3} = 53.17 f A^{-2.5}$$

Since in our case the airway is not circular, it must be corrected with a form coefficient (φ), thus:

$$R_{100} = 53.17 \varphi f A^{-2.5}$$

If we also take into account the presence of obstacles by means of an E coefficient, then:

$$R_{100} = 53.17 \varphi E f A^{-2.5}$$

Taking logarithms, we obtain:

$$\log R_{100} = \log 53.17 \varphi E f - 2.5 \log A$$

which comes very close to the experimental regression line obtained in a):

$$\log R_{100} = 2.751 - 2.527 \log A$$

4.4 Characteristic Curve

As already seen in the Atkinson equation, the pressure loss in a conduit depends on its aerodynamic resistance (R), certain geometric parameters of the conduit, and air density (Eq. 4.10). In addition, it can also be deduced that the friction loss in an airway varies in direct proportion to the square of the airflow rate. This is why the graphical representation of Eq. 4.10, called the *characteristic curve*,[7] is shaped like a parabola. This parabola will be more close to the ordinate axis, the greater the value of this resistance (Fig. 4.1).

Although in general for mine ventilation problems, it will take the form of a parabola, a clarification is required. This is due to the fact that the expression does not present the same form for the different fluid circulation regimes. In fact, it can be generalized as (Eq. 4.13):

$$P = R\, Q^{\alpha} \quad 1 < \alpha < 2 \tag{4.13}$$

Hence, for the turbulent flow of ventilation systems α approaches 2, so the expression becomes a parable. While for the laminar flow α is close to 1, so the relation between P and Q tends to be a straight line. For its part, in the case of the transitional flow, α takes values of between 1 and 2 (Fig. 4.2). Ostermann (1960) indicates values of α of 2.3 for stopes, of between 1.75 and 1.9 for ventilation doors, 2.2 for underground hoisting shafts, and about 1.85 for mine galleries with wind speeds of 0–3 m s^{-1} and about 2 for greater speeds.

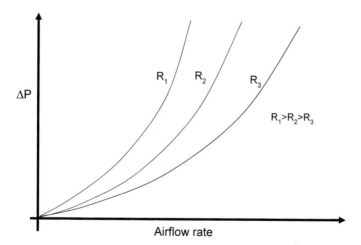

Fig. 4.1 Characteristic curves of various mines with different values of resistance (R)

---

[7]The characteristic curve can correspond to the mine as a whole, to an airway or to any gallery the mine.

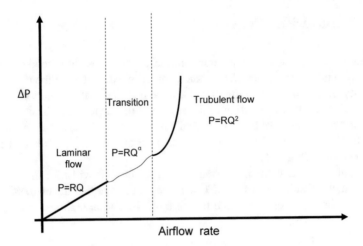

Fig. 4.2 Characteristic curve for different air circulation regimes

Exercise 4.6 The air quantity flowing through a duct is 2.8 m³ s⁻¹ when the static pressure difference between its two ends is of 15 Pa. What air quantity would flow through that same duct if the static pressure difference between its ends was of 20 Pa? Assume that the flow is turbulent.

Solution

Given that the flow is turbulent $\alpha = 2$. Therefore:

$$\Delta P_1 = R\, Q_1^2$$

$$\Delta P_2 = R\, Q_2^2$$

$$\frac{\Delta P_1}{\Delta P_2} = \frac{Q_1^2}{Q_2^2}$$

$$Q_2 = \sqrt{\frac{Q_1^2\, \Delta P_2}{\Delta P_1}} = \sqrt{\frac{\left(2.8\frac{m^3}{s}\right)^2 \cdot 20\,\text{Pa}}{15\,\text{Pa}}} = 3.23\frac{m^3}{s}$$

4.5 Shock Pressure Losses

As has already been seen the total pressure loss that takes place through shock can be obtained from Eq. 1.30:

$$\Delta P_{\text{shock}} = X\, P_v \qquad\qquad (1.30)$$

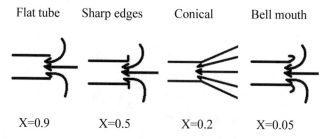

Fig. 4.3 Shock loss coefficients for various shapes for the air inlet of a conduit

where

- X: Shock loss factor (dimensionless), and
- P_v: Velocity pressure (Pa).

The coefficient is influenced by the angle of change of direction, the configuration, the abruptness of the change, the radius of curvature, the ratio radius/width, the roughness and the speed. Some coefficients for the case of air inlets are given in Fig. 4.3.

As in the previous case, losses due to abrupt section changes can be obtained as the product of a shock loss coefficient (X) times the dynamic pressure (P_v). The loss coefficients, as a function of the cross-sectional area of the two ducts, are shown for sudden cross-sectional changes in Fig. 4.4.

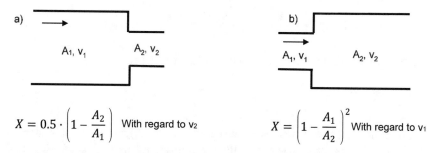

Fig. 4.4 Loss coefficients for sudden: **a** contraction and **b** expansion. Note that they may refer to different velocity pressures (inlet (v_1) or outlet (v_2))

The most common shock loss coefficients for the case of ducts are presented in Table 1.1 of Chap. 1. As already indicated in the same chapter, it is commonly considered that the above losses are static pressure losses, whereas they are in fact energy losses and thus total pressure losses. In fact, the loss is only of the static kind if there is no cross-sectional change and thus speed variation between the inlet and outlet of the component (friction in ducts, joints and so on).

Exercise 4.7 Calculate shock losses in a circular ventilation shaft, 3 m in diameter, abruptly widening to 5 m. The airflow rate through it is 100 m³ s⁻¹.

Solution

$$X = \left[1 - \left(\frac{3}{5} \right)^2 \right]^2 = 0.4096$$

$$v_1 = \frac{Q}{A_1} = \frac{100 \, \frac{m^3}{s}}{\frac{\pi 3^2}{4} \, m^2} = 14.147 \frac{m}{s}$$

$$P_v = \frac{1}{2} \cdot 1.2 \frac{Kg}{m^3} \cdot \left(14.147 \frac{m}{s} \right)^2 = 120.08 \, Pa$$

$$P_{shock} = 0.4096 \cdot 120.08 = 49.18 \, Pa$$

4.6   Shock Resistance and Equivalent Length

The equivalent length is the length of a straight airway of the same cross section as that of the airway in which the fitting is installed which produces the same pressure losses as the fitting.

In order to express this mathematically, we can start from Eq. 1.30 and work it out as follows:

$$\Delta P_{shock} = X\rho \frac{v^2}{2} = X\rho \frac{\left(\frac{Q}{A} \right)^2}{2} = \frac{X\rho}{2A^2} Q^2 = R_{shock} Q^2$$

Therefore, the value of the shock resistance is (Eq. 4.14):

$$R_{shock} = \frac{X\rho}{2A^2} \tag{4.14}$$

If, in addition, the Atkinson resistance equation (Eq. 4.9) is used and it is adapted only for the case of shock losses through the equivalent length of shock (L_{eq}) one has: (Eq. 4.15):

$$R_{shock} = \frac{KOL_{eq}}{A^3} \left(\frac{\rho}{1.2 \left[\frac{kg}{m^3} \right]} \right) \tag{4.15}$$

Then equalizing the two expressions of resistances and clearing the equivalent length, we have Eq. 4.16:

$$L_{eq} = \frac{1.2\left[\frac{kg}{m^3}\right] X}{2K} \frac{A}{O} \qquad (4.16)$$

By introducing the concept of hydraulic diameter (Eq. 1.19), the mathematical definition of equivalent length (Eq. 4.17) is finally obtained:

$$L_{eq} = \frac{1.2\left[\frac{kg}{m^3}\right] X}{8K} D_h \qquad (4.17)$$

A summary of the most common equivalent lengths is given in Table 4.3.

Table 4.3 Equivalent lengths for various elements. Modified from Hartman et al. (1997, p. 162)

Element	Equivalent length (m)
Turns	1–45
Door	20
Inlet	6
Outlet	20
Gradual contraction	1
Gradual expansion	1
Abrupt contraction	3
Abrupt expansion	6
Shaft with skip (20% of the section)	30
Shaft with skip (50% of the section)	150

Exercise 4.8 A rectangular tunnel of cross section $5 \times 3 \text{ m}^2$ and 500 m in length changes of direction by means of a $90°$ arch of 3 m radius as shown in the figure. The tunnel is in good condition but shows significant wall irregularities, so its Atkinson friction coefficient is estimated at $0.012 \text{ N s}^2 \text{ m}^{-4}$. In addition, the shock loss coefficient in the curve is assumed 0.86. If 80 $\text{m}^3 \text{ s}^{-1}$ of air whose density is 1.15 kg m^{-3}, they are expected to flow through it. Calculate (omitting inlet and outlet shock losses):

(a) The aerodynamic resistance due to friction,
(b) The aerodynamic resistance of the curve union,
(c) The equivalent length of the elbow,
(d) The total aerodynamic resistance in the tunnel, and
(e) The total pressure drop in the tunnel.

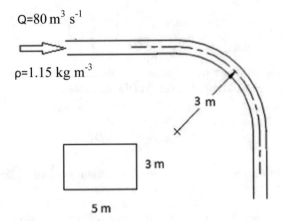

Solution

(a) Resistance due to friction:

$$R = \frac{KOL}{A^3}\left(\frac{\rho}{1.2\left[\frac{kg}{m^3}\right]}\right) = \frac{0.012\frac{Ns^2}{m^4}\cdot 16\,m\cdot 500\,m}{15^3 m^6}\cdot\frac{1.15}{1.2} = 0.0273\frac{N\,s^2}{m^8}$$

(b) Resistance due to shock in the curve union:

$$X = 0.86$$

$$R_{shock} = \frac{X\rho}{2A^2} = \frac{0.86\cdot 1.15\frac{kg}{m^3}}{2\cdot(15\,m^2)^2} = 0.00219\frac{N\,s^2}{m^8}$$

(c) The hydraulic diameter (D_h) is determined as:

$$D_h = \frac{4A}{O} = \frac{4\cdot(5\cdot 3)\,m^2}{(2\cdot 5 + 2\cdot 3)m} = 3.75\,m$$

The equivalent length is, therefore:

$$L_{eq} = \frac{1.2\left[\frac{kg}{m^3}\right]X}{8K}D_h = \frac{1.2\frac{kg}{m^3}\cdot 0.86}{8\cdot 0.012\frac{kg}{m^3}}\cdot 3.75\,m = 40.31\,m$$

(d) The total aerodynamic resistance (R_T) is obtained as:

$$R_T = \frac{KO(L + L_{eq})}{A^3}\left(\frac{\rho}{1.2\left[\frac{kg}{m^3}\right]}\right)$$

$$= \frac{0.012 \frac{\text{kg}}{\text{m}^3} \cdot 16\,\text{m} \cdot (500 + 40.31)\,\text{m}}{(15\,\text{m}^2)^3} \cdot \frac{1.15}{1.2}$$

$$= 0.02945 \frac{\text{N s}^2}{\text{m}^8}$$

And also as:

$$R_T = (0.0273 + 0.00219) \frac{\text{N s}^2}{\text{m}^8} = 0.02945 \frac{\text{N s}^2}{\text{m}^8}$$

(e) Finally, the total pressure loss (omitting inlet and outlet losses) can be obtained as:

$$\Delta P = R_T\, Q^2 = 0.02945 \frac{\text{N s}^2}{\text{m}^8} \cdot \left(80 \frac{\text{m}^3}{\text{s}}\right)^2 = 188.48\,\text{Pa}$$

4.7 Equivalent Orifice

The *equivalent orifice* is a concept used to characterize ducts, galleries or mines according to their ease of ventilation. Thus, the larger the section of the equivalent orifice compared to that of the airway, the easier it will be to ventilate it. The concept could be enunciated for any airway as the surface of the opening offering to the flow passing through it,[8] the same resistance as the path traveled by air in the airway. A more technical alternative is its definition as the section of the thin-walled orifice (S) through which flows the same quantity (Q) as through the airway, if the pressure difference on both sides of the thin wall (ΔP) coincides with that provided by the fan. In the actual case of an entire mine, the equivalent orifice is the area of the hypothetical circular thin-walled orifice located in the main fan drift through which the main fan is capable of extracting the same quantity when the airlock at the top of the shaft is open (Fig. 4.5b), as from the mine when the airlock at the top of the shaft is closed (Fig. 4.5a).

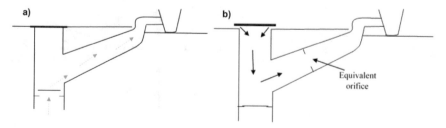

Fig. 4.5 Measurement of the equivalent orifice of a mine in the main fan drift

---

[8] Assumed, generally, this flow at standard conditions.

Therefore, the equivalent orifice replaces the entire airway, as it concerns the pressures and quantities supplied by the fan. The equivalent orifice of an entire mine allows for their classification. Thus, mines with an equivalent orifice of more than 5 m² are generally considered to be very easy to ventilate, orifices of 2 m² are within the norm in European mines and orifices smaller than 1.5 m² indicate mines difficult to ventilate.

Demonstration of the Equation of the Equivalent Orifice

The definition of the equivalent orifice assumes stationary, laminar, incompressible and non-viscous fluid. Under these premises, consider that a fluid exits a thin-walled reservoir through an orifice of area A and in which P_1, v_1 and P_2, v_2 are the pressures and velocities on its both sides (Fig. 4.6).

Applying the Bernoulli equation to both sides of the wall (1 and 2), where z_i is their height and P_i are the static pressures, we have:

$$\frac{P_1}{\rho g} + \frac{v_1^2}{2g} + z_1 = \frac{P_2}{\rho g} + \frac{v_2^2}{2g} + z_2$$

If contour conditions are considered:

- $z_1 = z_2$
- $P_1 - P_2 = \Delta P$
- $v_1 = 0$ (if there is an assumed significant distance from the wall).

When air flows through an orifice, the pathlines contract due to inertia to pass through the orifice. This causes the contracted fluid vein (*vena contracta*) to have a smaller section than the orifice. In the *vena contracta* the fluid velocity is the maximum, and therefore the static pressure is the lowest. After this, a slow expansion of the fluid begins. Normally, the area of the *vena contracta* (A_c) is about 65% of the orifice area (A), therefore:

Fig. 4.6 Pathlines of the flow through a thin-walled orifice

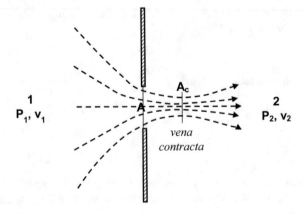

$$A_C = 0.65A$$

As the flow rate exiting the orifice is:

$$Q = A_C v_2$$

Therefore:

$$v_2 = \frac{Q}{0.65A}$$

According to the Bernoulli equation, we have:

$$\frac{P_1}{\rho} - \frac{P_2}{\rho} = \frac{v_2^2}{2}$$

Which can be rearranged as:

$$P_1 - P_2 = \frac{1}{2}\rho v_2^2$$

Therefore:

$$P_1 - P_2 = \frac{1}{2}\rho\left(\frac{Q}{0.65A}\right)^2$$

Then:

$$\Delta P = 1.42\left(\frac{Q}{A}\right)^2$$

Which corresponds to the family of parabolas in Fig. 4.7:

$$\Delta P = \frac{1.42}{A^2}Q^2$$

Solving for A (equivalent orifice), we obtain Eq. 4.18:

$$A = \frac{1.2}{\sqrt{\Delta P}}Q \tag{4.18}$$

Where the units are those of the SI.

In the expression deduced the energy losses in the orifice have not been taken into account. These can be of two types: *permanent* and *recoverable*. The actual pressure diagram is as shown in Fig. 4.8. Thus, the air as it approaches the wall undergoes a slight increase in static pressure (due to the decrease in speed), despite the general

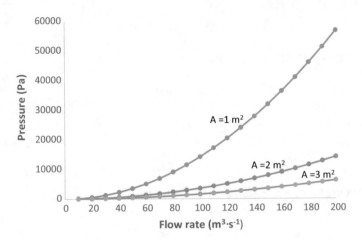

Fig. 4.7 Flow rates generated by different pressure differences for three equivalent orifice areas. Note the conceptual similarity with Fig. 4.1

tendency for the system to lose pressure due to friction. After the orifice, we find the *vena contracta*. In it, the fluid accelerates due to the smaller cross section, which results in an increase in speed at the expense of a reduction of the static pressure.

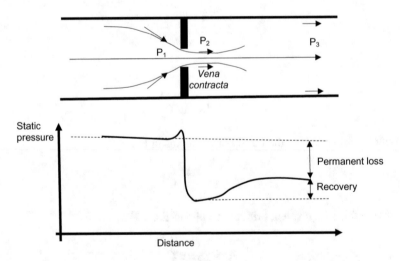

Fig. 4.8 Recovery and permanent pressure losses for a stream crossing an orifice

After the *vena contracta*, the air stream expands so that the speed decreases and the static pressure increases again. If the static pressure measurement is carried out at a sufficient distance from the wall, the speed will have recovered again to the pre-orifice values. This measurement represents the permanent losses of the system.

For a more detailed review of the empirical literature on these types of barriers, see Crane Company (2013).

Question 4.2 It is necessary to double the air quantity in a mine of 3 m² of equivalent orifice. Explain the effect that this has on the following:

(a) Pressure to be communicated by the fan, and
(b) Fan power.

Answer

(a)

For the first conditions, we have:

$$A_1 = \frac{1.2}{\sqrt{\Delta P_1}} Q_1$$

Similarly, for the second conditions:

$$A_2 = \frac{1.2}{\sqrt{\Delta P_2}} Q_2$$

According to the statement $A_1 = A_2 = A = 3$ m², $Q_1 = Q$ and $Q_2 = 2Q_1 = 2Q$. Then, substituting the values and equalizing in the previous expresions:

$$A = \frac{1.2}{\sqrt{\Delta P_1}} Q = \frac{1.2}{\sqrt{\Delta P_2}} 2Q$$

So:

$$\sqrt{\frac{\Delta P_2}{\Delta P_1}} = 2$$

Therefore:

$$\frac{\Delta P_2}{\Delta P_1} = 4$$

Thus, according to the equivalent orifice expression, if you want to double the airflow rate, you have to quadruple the pressure difference.

(b) The power is proportional to the product $\Delta P\, Q$, then you have:

$$Pw_1 = \Delta P_1 Q_1$$
$$Pw_2 = \Delta P_2 Q_2 = 8Pw_1$$

In order to double the flow, the fan power needs to be increased by 8 times.

Exercise 4.9 The main fan of a mine supplies an air flow rate of 50 m^3 s^{-1} to overcome a total pressure loss of 3 kPa in the circuit (excluding outlet losses). Determine the equivalent orifice of the mine and make some indications on its ease of ventilation.

Solution

By direct application of the expression of the equivalent orifice, we have:

$$A = \frac{1.2}{\sqrt{\Delta P}} Q$$

$$A = \frac{1.2}{\sqrt{3000 \, \text{Pa}}} \cdot 50 \frac{m^3}{s} = 1.07 \, m^2$$

As a result, it is difficult to ventilate this mine.

4.8   Resistances in Series

In a *series arrangement*, the circulating airflow (Q) is the same in all resistances and the pressure loss (ΔP) is the sum of the losses in each one of them (Fig. 4.9).
So, mathematically:

$$\Delta P = \Delta P_1 + \Delta P_2 + \Delta P_3$$

Hence, applying Eq. 4.13 for each pressure differential, we have:

$$R_{eq} Q^2 = R_1 Q_1^2 + R_2 Q_2^2 + R_3 Q_3^2$$

Given that:

$$Q = Q_1 = Q_2 = Q_3$$

Then:

$$R_{eq} Q^2 = R_1 Q^2 + R_1 Q^2 + R_1 Q^2$$

Fig. 4.9 Association of aerodynamic resistances in series

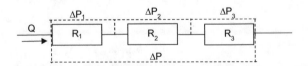

Therefore:

$$R_{eq} = R_1 + R_2 + R_3$$

Thus, generalizing for n resistances (Eq. 4.19):

$$R_{eq} = \sum_{i=1}^{n} R_i \qquad\qquad (4.19)$$

Therefore, in a series circuit of n resistances, the equivalent resistance (R_{eq}) is the sum of the individual resistances (R_i).

4.9   Resistances in Parallel

In a *parallel circuit*, the total circulating airflow rate (Q) is the sum of the flows of each of the branches, and the pressure loss (ΔP) in each branch is the same (Fig. 4.10).
So, mathematically:

$$Q = Q_1 + Q_2 + Q_3$$

Therefore, applying Eq. 4.13 to each branch, we have:

$$\sqrt{\frac{\Delta P}{R_{eq}}} = \sqrt{\frac{\Delta P_1}{R_1}} + \sqrt{\frac{\Delta P_2}{R_2}} + \sqrt{\frac{\Delta P_3}{R_3}}$$

Given that:

$$\Delta P = \Delta P_1 = \Delta P_2 = \Delta P_3$$

Fig. 4.10 Association of aerodynamic resistances in parallel

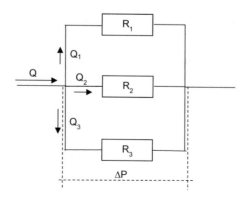

Therefore:

$$\sqrt{\frac{\Delta P}{R_{eq}}} = \sqrt{\frac{\Delta P}{R_1}} + \sqrt{\frac{\Delta P}{R_2}} + \sqrt{\frac{\Delta P}{R_3}}$$

Finally:

$$\frac{1}{\sqrt{R_{eq}}} = \frac{1}{\sqrt{R_1}} + \frac{1}{\sqrt{R_2}} + \frac{1}{\sqrt{R_3}}$$

Thus, generalizing for n resistances (Eq. 4.20):

$$\frac{1}{\sqrt{R_{eq}}} = \sum_{i=1}^{n} \frac{1}{\sqrt{R_i}} \qquad (4.20)$$

Therefore, in a parallel arrangement, the reciprocal of the square root of the equivalent resistance (R_{eq}) is the sum of the reciprocals of the square roots of the n resistances of each branch (R_i).

Question 4.3 Comment on the expression: "Air always tends to flow through the shortest path".

Answer

For example, one can think of a system formed by a duct that splits into two branches (A and B) of different lengths that converge again forming the parallel system depicted in the figure.

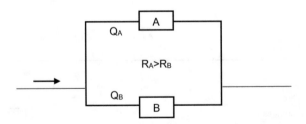

Without varying more parameters, the longest path (A) will have more resistance as resistance is proportional to the length, thus resulting in R_A being larger than R_B.

Given that:

$$Q_A = \sqrt{\frac{\Delta P}{R_A}}$$

$$Q_B = \sqrt{\frac{\Delta P}{R_B}}$$

Then, $Q_A < Q_B$.

Exercise 4.10 The figure corresponds to the vertical cross section of an underground mine.

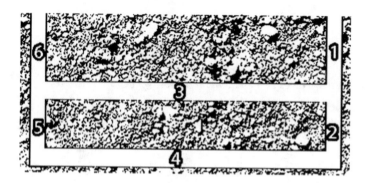

The resistances of airways 1–6 are:

Element	Resistance $\left(\frac{N\,s^2}{m^8}\right)$
1	20
2	15
3	35
4	30
5	18
6	27

Calculate:

(a) The equivalent resistance of the mine.
(b) The loss of pressure between the mouths of the downcast and upcast shafts if an airflow of 150 m³ s⁻¹ is registered in airways 1 and 6.
(c) Comment if the results obtained are possible in a real mine.

Solution

(a) First, the network can be represented schematically in terms of a circuit with resistances 1–6, thus:

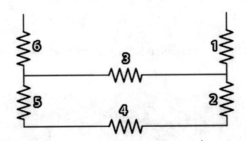

Since the resistances 2, 4 and 5 are in series, we have:

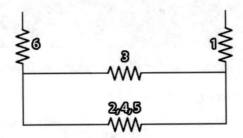

Hence:

$$R_{245} = R_2 + R_4 + R_5 = (15 + 30 + 18)\frac{N\,s^2}{m^8} = 63\frac{N\,s^2}{m^8}$$

R_{245} are in parallel with R_3:

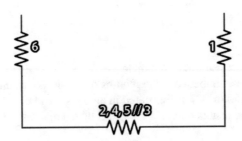

Thus:

$$\frac{1}{\sqrt{R_{2453}}} = \frac{1}{\sqrt{63\frac{N\,s^2}{m^8}}} + \frac{1}{\sqrt{35\frac{N\,s^2}{m^8}}}$$

Therefore:

$$R_{2453} = 11.49\frac{N\,s^2}{m^8}$$

Once this is done, we see that R_{2453} is in series with both R_1 and R_6 such that, to calculate the equivalent resistance for the network of airways (R_{eq}):

$$R_{eq} = R_1 + R_6 + R_{2453} = (20 + 27 + 11.49)\frac{N\,s^2}{m^8} = 58.49\frac{N\,s^2}{m^8}$$

(b) The pressure loss between the inlet and the outlet is defined by Atkinson's equation:

$$\Delta P = R_{eq}Q^2 = 58.49\frac{N\,s^2}{m^8}\left(150\frac{m^3}{s}\right)^2 = 1.31\,\text{MPa}$$

(c) This exercise is designed to establish the importance of verifying whether results make physical sense. Here, although calculations are correct, the results make no technical sense. The calculated pressure of 13 atm required to ventilate the mine, is absolutely unfeasible. To understand this, consider the following:

1. No fan would supply such pressure; the air inside the network would simply become compressed.
2. The proposed pressure would put at risk the lives of workers.
3. Atkinson's equation would, in fact, no longer apply under these operating conditions.

The inconsistency observed here between calculation and operating reality may be due to two considerations: the airflow recorded is excessive and/or the resistances of the mine airways have been incorrectly estimated The first is the one which most influences the working pressure of the fan. To illustrate this, consider that in order to double the airflow rate into any mine, the pressure provided by the main fan must be quadrupled. Moreover, the airflow rate stated in the exercise is not excessive for a main fan as normal flow rates range between 150 and 350 $m^3\,s^{-1}$. However, as far as the resistances are concerned, this is a very high total resistance for a very simple scheme. In fact, it is unlikely that it could be overcome even by the use of booster fans to assist the main fan. It seems, therefore that this last consideration is the likely source of the inconsistencies highlighted.

Exercise 4.11 The figure represents a mine ventilation network. It is known that resistances of the inlet (R_1) and outlet airways (R_2) are 0.25 N s^2 m^{-8}, respectively, and that the pressure difference between the inlet (i) and the outlet (o) is 600 Pa.

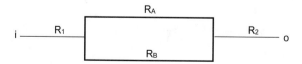

It is known also that the resistance of parallel branches A and B are $R_A = 0.7$ N s^2 m^{-8} and $R_B = 3.1$ N s^2 m^{-8}, respectively. Determine the volumetric airflow rates traversing each airway.

Solution

Let us start by calculating the equivalent resistance (R_3) for parallel branches A and B:

$$\sqrt{\frac{1}{R_3}} = \sqrt{\frac{1}{R_A}} + \sqrt{\frac{1}{R_B}}$$

Substituting values we have that:

$$\sqrt{\frac{1}{R_3}} = \sqrt{\frac{1}{0.7\frac{N\,s^2}{m^8}}} + \sqrt{\frac{1}{3.1\frac{N\,s^2}{m^8}}}$$

Therefore:

$$R_3 = 0.3216\,\frac{N\,s^2}{m^8}$$

Then, the total resistance (R_T) is:

$$R_T = R_1 + R_3 + R_2$$

$$R_T = (0.25 + 0.3216 + 0.25)\frac{N\,s^2}{m^8} = 0.8217\frac{N\,s^2}{m^8}$$

Thus, the total airflow is (Q_T):

$$Q_T = \sqrt{\frac{\Delta P_T}{R_T}} = \sqrt{\frac{600\,Pa}{0.8217\frac{N\,s^2}{m^8}}} = 27.022\,\frac{m^3}{s}$$

According to the mass conservation principle:

$$Q_T = Q_A + Q_B = 27.022\frac{m^3}{s}$$

The pressure loss in the parallel branch (with equivalent resistance R_3) can be calculated as:

$$\Delta P_3 = R_3 Q_T^2 = 0.3216\frac{N\,s^2}{m^8}\left(27.022\frac{m^3}{s}\right)^2 = 234.83\,Pa$$

Given that pressure loss is the same for all parallel branches:

$$\Delta P_3 = \Delta P_A = \Delta P_B$$

Therefore, we obtain, for the system:

$$Q_A + Q_B = 27.022 \frac{m^3}{s}$$

$$0.7 \frac{N s^2}{m^8} Q_A^2 = 234.83 \, Pa$$

Solving this we get:

$$Q_A = 18.322 \frac{m^3}{s}$$

$$Q_B = 8.70 \frac{m^3}{s}$$

Exercise 4.12 The figure shows the resistances of airways in a ventilation network.

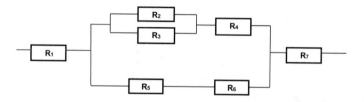

It is known that the resistances (R_i) have the values listed in the table and that the total pressure loss in the system is 1.2 kPa.

Resistance (R_i)	Value ($\frac{N s^2}{m^8}$)
R_1	4
R_2	12
R_3	13
R_4	3
R_5	6.5
R_6	6
R_7	2

Determine:

(a) The equivalent resistance of the network.
(b) The total airflow rate entering the system.
(c) Pressure loss and corresponding airflow rate for each airway.

Solution

(a) Calculation of the equivalent resistance.

$R_2 - R_3$ are in parallel:

$$\frac{1}{\sqrt{R_{23}}} = \frac{1}{\sqrt{R_2}} + \frac{1}{\sqrt{R_3}}$$

$$\frac{1}{\sqrt{R_{23}}} = \frac{1}{\sqrt{12\frac{N\,s^2}{m^8}}} + \frac{1}{\sqrt{13\frac{N\,s^2}{m^8}}}$$

$$R_{23} = 3.121$$

$R_{23} - R_4$ are in series:

$$R_{234} = R_{23} + R_4 = (3.121 + 3)\frac{N\,s^2}{m^8} = 6.121\frac{N\,s^2}{m^8}$$

$R_5 - R_6$ are in series:

$$R_{56} = R_5 + R_6 = (6.5 + 6)\frac{N\,s^2}{m^8} = 12.5\frac{N\,s^2}{m^8}$$

$R_{234} - R_{56}$ are in parallel:

$$\frac{1}{\sqrt{R_{23456}}} = \frac{1}{\sqrt{R_{234}}} + \frac{1}{\sqrt{R_{56}}} = \frac{1}{\sqrt{6.121\frac{N\,s^2}{m^8}}} + \frac{1}{\sqrt{12.50\frac{N\,s^2}{m^8}}}$$

$$R_{23456} = 2.121\frac{N\,s^2}{m^8}$$

$R_1 - R_{23456} - R_7$ are in series:

$$R_{eq} = R_{1234567} = R_1 + R_{23456} + R_7 = (4 + 2.121 + 2)\frac{N\,s^2}{m^8} = 8.121\frac{N\,s^2}{m^8}$$

(b) Once the equivalent resistance of the circuit (R_{eq}) has been calculated and since the total pressure loss (ΔP_T) is known, the total airflow rate (Q_T) can be calculated as:

$$Q_T = \sqrt{\frac{\Delta P_T}{R_{eq}}} = \sqrt{\frac{1200\,Pa}{8.12\frac{N\,s^2}{m^8}}} = 12.16\frac{m^3}{s}$$

(c) Calculation of pressure losses for the airway resistances in series.

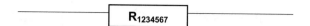

Reversing the procedure used to find R_{eq}, we can break this resistance down:

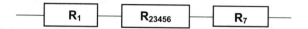

And knowing that the total airflow rate through the network is Q_T and that this will be the same through all resistances in series, then:

$$\Delta P_1 = R_1\, Q_T^2 = 4\frac{N\,s^2}{m^8}\left(12.16\frac{m^3}{s}\right)^2 = 591.46\,Pa$$

$$\Delta P_{23456} = R_{23456}\, Q_T^2 = 2.12\frac{N\,s^2}{m^8}\left(12.16\frac{m^3}{s}\right)^2 = 313.48\,Pa$$

$$\Delta P_7 = R_7\, Q_T^2 = 2\frac{N\,s^2}{m^8}\left(12.16\frac{m^3}{s}\right)^2 = 295.73\,Pa$$

R_{23456} can be further broken down, thus:

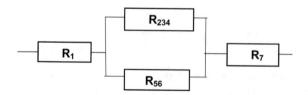

Given that $\Delta P_{234} = \Delta P_{56}$, the respective airflow rates can be calculated:

$$Q_{234} = \sqrt{\frac{\Delta P_{234}}{R_{234}}} = \sqrt{\frac{313.48\,Pa}{6.12\frac{N\,s^2}{m^8}}} = 7.16\frac{m^3}{s}$$

$$Q_{56} = \sqrt{\frac{\Delta P_{56}}{R_{56}}} = \sqrt{\frac{313.48\,Pa}{12.5\frac{N\,s^2}{m^8}}} = 5.0\frac{m^3}{s}$$

Breaking down R_{234} reveals a series arrangement with a common flow rate Q_{234}. Thus, the pressure losses at R_{23} and R_4 can be calculated:

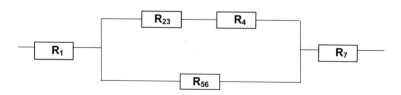

$$\Delta P_{23} = R_{23}\, Q_{234}^2 = 3.12 \frac{\mathrm{N\,s^2}}{\mathrm{m^8}} \left(7.16\frac{\mathrm{m^3}}{\mathrm{s}}\right)^2 = 159.83\,\mathrm{Pa}$$

$$\Delta P_4 = R_4\, Q_{234}^2 = 3\frac{\mathrm{N\,s^2}}{\mathrm{m^8}} \left(7.16\frac{\mathrm{m^3}}{\mathrm{s}}\right)^2 = 153.64\,\mathrm{Pa}$$

If R_{23} is broken down we see that $\Delta P_{23} = \Delta P_2 = \Delta P_3$ because we have a parallel arrangement of resistances. Thus, it is possible to calculate the airflow rates traversing R_2 and R_3:

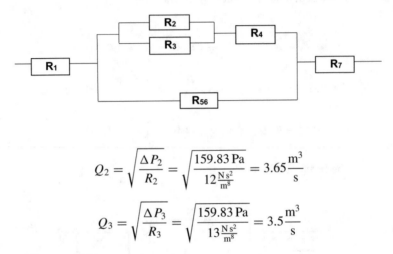

$$Q_2 = \sqrt{\frac{\Delta P_2}{R_2}} = \sqrt{\frac{159.83\,\mathrm{Pa}}{12\frac{\mathrm{N\,s^2}}{\mathrm{m^8}}}} = 3.65\frac{\mathrm{m^3}}{\mathrm{s}}$$

$$Q_3 = \sqrt{\frac{\Delta P_3}{R_3}} = \sqrt{\frac{159.83\,\mathrm{Pa}}{13\frac{\mathrm{N\,s^2}}{\mathrm{m^8}}}} = 3.5\frac{\mathrm{m^3}}{\mathrm{s}}$$

Finally, the resistance R_{56} can be broken down. As R_5 and R_6 are in a series arrangement, the airflow Q_{56} is the same through both and pressure losses ΔP_5 and ΔP_6 can be calculated, thus:

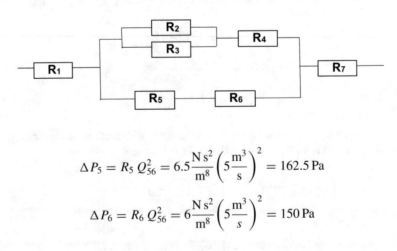

$$\Delta P_5 = R_5\, Q_{56}^2 = 6.5\frac{\mathrm{N\,s^2}}{\mathrm{m^8}} \left(5\frac{\mathrm{m^3}}{\mathrm{s}}\right)^2 = 162.5\,\mathrm{Pa}$$

$$\Delta P_6 = R_6\, Q_{56}^2 = 6\frac{\mathrm{N\,s^2}}{\mathrm{m^8}} \left(5\frac{\mathrm{m^3}}{\mathrm{s}}\right)^2 = 150\,\mathrm{Pa}$$

Summary table:

Resistance	Airflow rate ($m^3\ s^{-1}$)	Pressure loss (Pa)
R_1	12.16	591.46
R_2	3.65	159.83
R_3	3.50	159.83
R_4	7.16	153.64
R_5	5.00	162.5
R_6	5.00	150.0
R_7	12.16	295.73

Question 4.4 Demonstrate that for a parabolic relationship between the airflow rate and the pressure difference, the total equivalent orifice of a parallel circuit can be obtained as the sum of the equivalent orifices of each of its branches. Use the following expression for the equivalent orifice (A) of a ventilation network.

$$A = \frac{1.2}{\sqrt{\Delta P}} Q$$

Answer

Let us consider a system with three resistances in parallel. In this case:

$$Q_T = Q_1 + Q_2 + Q_3$$

The airflow rate through an equivalent orifice is:

$$Q = \frac{A\sqrt{\Delta P_T}}{1.2}$$

Thus:

$$\frac{A_T\sqrt{\Delta P_T}}{1.2} = \frac{A_1\sqrt{\Delta P_1}}{1.2} + \frac{A_2\sqrt{\Delta P_2}}{1.2} + \frac{A_3\sqrt{\Delta P_3}}{1.2}$$

Since we are dealing with a parallel circuit:

$$\Delta P_T = \Delta P_1 = \Delta P_2 = \Delta P_3$$

Simplifying terms, we get that:

$$A_T = A_1 + A_2 + A_3$$

4.9.1   Splitting Air Currents

Diving the total airflow through several branches of a network is a requirement in most ventilation regulations as it permits the creation of separated ventilation districts thus increasing the overall safety of the network. Moreover, it is a common strategy by which to reduce the pressure that the main fan must supply. This is a natural consequence of the fact that the pressure drop is equal across multiple airways in parallel. Parallel branches can be created in multiple locations of the mine but are mostly located close to the intake and the return shafts. Figure 4.11 illustrates a system with one, two and three parallel airways.

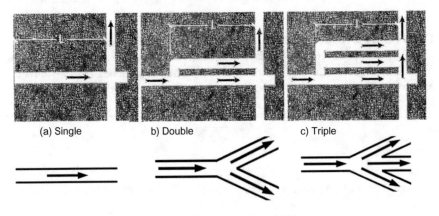

Fig. 4.11   Scheme for **a** one, **b** two and **c** three airways in parallel

The equivalent resistance of the single system (Fig. 4.11a) is:

$$R_{eq} = \frac{K\,O\,(L + L_{eq})}{A^3}$$

Assuming each branch has the same length and cross section then their resistances will be equal, R_1.

In this way, the equivalent resistance of the double arrangement (Fig. 4.11b) is:

$$\frac{1}{\sqrt{R_{eq}}} = \frac{1}{\sqrt{R_1}} + \frac{1}{\sqrt{R_1}} = \frac{2}{\sqrt{R_1}}$$

$$\frac{1}{R_{eq}} = \frac{4}{R_1}$$

For the triple system (Fig. 4.11c) we have:

$$\frac{1}{\sqrt{R_{eq}}} = \frac{1}{\sqrt{R_1}} + \frac{1}{\sqrt{R_1}} + \frac{1}{\sqrt{R_1}}$$

$$\frac{1}{\sqrt{R_{eq}}} = \frac{3}{\sqrt{R_1}} \rightarrow \frac{1}{R_{eq}} = \frac{9}{R_1}$$

Generalizing this expression for a system with n parallel airways each of resistance R_i, the equivalent resistance for the network (R_{eq}) can be expressed as:

$$\frac{1}{R_{eq}} = \frac{n^2}{R_i}$$

Rearranging this we have (Eq. 4.21):

$$R_{eq} = \frac{R_i}{n^2} \tag{4.21}$$

It follows that the greater the number of parallel airways, the lower the equivalent resistance of the complete circuit, and therefore the more economic it is to ventilate.[9]

Question 4.5 A ventilation system is changed from one comprising a single inlet to another comprising five parallel airways each identical to the original inlet. How does the equivalent resistance of the system change?

Answer

Using the expression deduced above:

$$R_{eq} = \frac{R_i}{n^2}$$

Substituting:

$$R_{eq} = \frac{R_i}{25}$$

So the resistance turns out to be 25 times smaller.

Question 4.6 The resistance of a ventilation system with three identical inlets is 15 $\frac{N s^2}{m^8}$. What would be the resistance of the system if it had 8 parallel inlets of analogous characteristics to the three original inlets?

Answer

R_i is the resistance of an individual airway in both systems so, calculating for the initial 3-inlet system:

---

[9]Note that shock losses are ignored in these calculations.

$$R_{eq} = 15 = \frac{R_i}{n^2}; \ R_i = 15 \cdot 3^2 = 135$$

Then for the 8-inlet system with the same R_i:

$$R_{eq} = \frac{135}{8^2}$$

$$R_{eq} = 2.1 \frac{\mathrm{N\,s^2}}{\mathrm{m^8}}$$

4.10   Complex Networks

The analysis of ventilation networks requires a formal definition of the following concepts:

- *Node or junction*: Point where three or more branches converge.
- *Branch*: Airway that joins two nodes.
- *Mesh or loop*: Closed contour (formed by three or more branches).
- *Circuit*: A set of nodes, branches and meshes that form a structure or ventilation system.

For every polyhedron (both irregular and regular), there is a mathematical relationship between the number of faces, vertices and edges, given by Euler's theorem. This theorem as applied to ventilation networks can be expressed as Eq. 4.22:

$$M = R - N + 1 \tag{4.22}$$

where

- N: Number of nodes,
- R: Number of branches, and
- M: Number of meshes.

4.10.1   Kirchhoff's Laws

The key principles used in the analysis of ventilation networks are the conservation of mass and the conservation of energy. These are applied through Kirchhoff's laws.

Kirchhoff's First Law (Law of Nodes or Law of Continuity)

Kirchhoff's first law is an application of the law of conservation of mass.

According to this law, the airflow rate that leaves a node must be equal to the airflow rate that enters it, mathematically (Eq. 4.23):

Fig. 4.12 Representation of a node

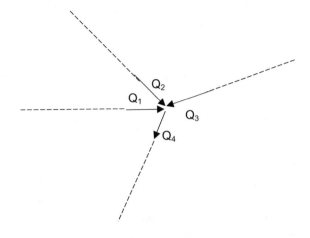

$$\sum_{i=1}^{k} Q_i = 0 \tag{4.23}$$

where

- k: Number of branches meeting at a node, and
- Q_i: Airflow rate in each branch.

Which applied to the particular case of Fig. 4.12:

$$Q_1 + Q_2 + Q_3 - Q_4 = 0$$

Kirchhoff's Second Law (Loop Rule)

Kirchhoff's second law is an application of energy conservation to a loop. Accordingly, the sum of the pressure drops within a closed circuit must be zero.

To express this mathematically, we need to go through the following steps:

- An arbitrary direction of travel around the mesh (clockwise or counterclockwise) has to be selected. In the case of complex networks, it is highly recommended that the direction of travel selected is the same in all meshes.
- Similarly, the direction of airflow should be established for each branch.
- There is a positive friction loss if the direction of airflow in a branch coincides with the direction of travel around the mesh, and negative friction loss in the case where it opposes the direction of travel around the mesh. Therefore, in each branch without a fan whose arbitrary airflow direction coincides with the direction of travel around the mesh we have (Eq. 4.24):

$$\Delta P_i = R_i Q_i^2 = R_i Q_i |Q_i| \tag{4.24}$$

- A pressure source (fan or natural ventilation) is said to be negative if the direction of the airflow that it creates in the branch coincides with the direction of travel around the mesh. This must be so since pressure losses and pressure gains in the system must have opposite signs. Therefore, in each branch with a fan whose arbitrary airflow direction coincides with the direction of travel (Eq. 4.25):

$$\sum_{i=1}^{N_r}(R_i Q_i |Q_i| - P_v) = 0 \qquad (4.25)$$

where

- N_r: Number of branches in each mesh,
- P_v: Pressure created by the fan (or natural ventilation in each branch),
- R_i: Resistance of each branch, and
- Q_i: Airflow rate in each branch.

Note that in the expression, Q_i^2 appears as $Q_i \cdot |Q_i|$ in order to preserve the sign associated with the airflow Q_i An alternative approach would be to give the sign directly to the whole term $R_i Q_i^2$. This option, however, generates errors during iterative calculations as signs can change.

Exercise 4.13   In the ventilation network shown in the figure, we know the resistances of the branches (R_{ij}, N s^2 m^{-8}), the inlet airflow rate ($Q_e = 150$ m^3 s^{-1}) and the airflow rate traversing the branch containing the fan (50 m^3 s^{-1}).

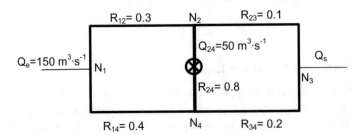

Determine[10]:

(a) Airflow rate at the outlet of the circuit;
(b) Airflow rate in each branch;
(c) The pressure being supplied by the fan in branch 2–4 if the system is in equilibrium and the airflow rate in this branch is 50 m^3 s^{-1};
(d) Fan pressure and power required to meet the conditions of 50 m^3 s^{-1} on branch $N_4 - N_2$; and
(e) Total useful power used for ventilation.

---

[10]This is one of the simplest ventilation systems that can be solved directly using Kirchhoff's laws. Variants of this network can be found resolved in McPherson (1993), p. 219 and Tuck (2011).

Solve by assuming the directions of circulation indicated by the arrows firstly in Fig. 1 and secondly in Fig. 2.

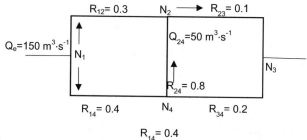

Fig. 1

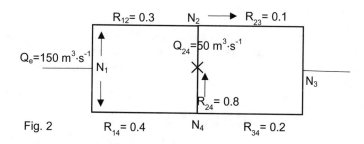

Fig. 2

Note: The numbers in the subscripts of the resistances do not indicate the direction of the airflow, only the connection between nodes.

Solution

- Resolving for the direction of travel shown in Fig. 1:

  (a)   Using the principle of mass conservation: The airflow rate (Q_s) leaving the circuit is the same as the airflow rate entering the circuit, and therefore equal to 150 m³ s⁻¹.

(b) and (c)   We must apply both of Kirchhoff's laws here. The steps are as follows:

1. Identify branches (R) and nodes (N) and determine the number of meshes (M) by means of Euler's equation.

   $M = R - N + 1$
   Branches(R_{ij}): $R_{12}, R_{13}, R_{23}, R_{24}, R_{34}$, therefore 5.
   Nodes (N_i): N_1, N_2, N_3, N_4, thus 4.
   In which case:
   $M = 5 - 4 + 1 = 2 \rightarrow (M_1$ and $M_2)$
   Thus we will need to solve for two meshes by means of Kirchhoff's second law.

2. Establish an arbitrary direction for the airflows in the network respecting Kirchhoff's first law and using the values stated in the initial problem:

3. Apply Kirchhoff's first law:

Node N_1:

$$Q_e = Q_{12} + Q_{14}$$

- $Q_{14} = 150 - Q_{12}$

Node N_2:

- $Q_{23} = Q_{12} + 50$

Node N_4:

$$50 = Q_{14} + Q_{34}; 50 = (150 - Q_{12}) + Q_{34}$$

- $Q_{34} = Q_{12} - 100$

Node N_3[11]:

$$Q_{23} = 150 + Q_{34}$$

4. Establish the arbitrary direction of travel around each mesh. Initially, we shall assume the circulation directions in the two meshes are opposed. This is done with the intention of observing the implications of the final solution to this problem.

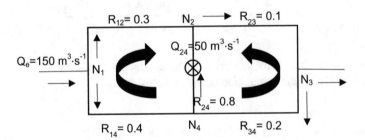

5. Apply Kirchhoff's second law:

Following the usual convention, a positive sign is used, indicating a pressure drop, when the direction of the airflow through the branch coincides with the circulation direction assigned to the mesh.

---

[11] Note that in a mesh, there are $N - 1$ independent nodes and in the example, the equation for node N_4 corresponds to the combination of the equations of nodes N_2 and N_3.

Mesh M_1

Branch	Pressure drop	Fan
$N_1 - N_2$	$0.3\, Q_{12}^2$	No
$N_2 - N_4$	$-0.8 \cdot 50^2$	$+P_v$
$N_1 - N_4$	$-0.4\,(150 - Q_{12})^2$	No
Mesh M_1 $\sum H_{ij} = 0$	$-0.1\, Q_{12}^2 + 120\, Q_{12} - 11000 + P_v = 0$	

Mesh M_2

Branch	Pressure drop	Fan
$N_2 - N_3$	$-0.1\,(Q_{12} + 50)^2$	No
$N_4 - N_3$	$-0.2\,(Q_{12} - 100)^2$	No
$N_4 - N_2$	$-0.8 \cdot 50^2$	$+P_v$
Mesh M_2 $\sum H_{ij} = 0$	$-0.3\, Q_{12}^2 + 30\, Q_{12} - 4.250 + P_v = 0$	

Mesh M_1: $-0.1\, Q_{12}^2 + 120\, Q_{12} - 11000 + P_v = 0$
Mesh $M_2 \cdot (-1)$: $+0.3\, Q_{12}^2 - 30\, Q_{12} + 4250 - P_v = 0$

Adding the expressions for M_1 and M_2 allows us to solve for P_v:

$$M_1 + M_2 :\quad +0.2\, Q_{12}^2 + 90\, Q_{12} - 6750 = 0$$

Note that the solution requires a sign change in the equation derived for M_2. This operation is equivalent to changing the direction of circulation around mesh M_2 (from counterclockwise to clockwise), as shown in the figure below:

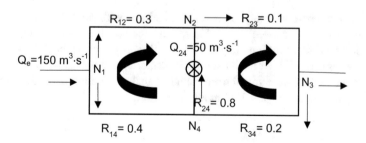

According to this approach, the fan is positive in one mesh and negative in the other since it is located in the branch common to both. In more complex systems, this approach (clockwise of travel in all meshes) facilitates calculations.

Solving the system, we have:

$Q_{12} = 65.47 \text{ m}^3 \text{ s}^{-1}$ ($515.5 \text{ m}^3 \text{ s}^{-1}$ is discarded because it has no physical meaning)

$Q_{14} = 150 - Q_{12} = 84.53 \text{ m}^3 \text{ s}^{-1}$

$Q_{23} = Q_{12} + 50 = 115.47 \text{ m}^3 \text{ s}^{-1}$

$Q_{34} = Q_{12} - 100 = \mathbf{-34.53 \text{ m}^3 \text{ s}^{-1}}$

Discussion

The airflow rate in branch R_{34} ($Q_{34} = -34.53 \text{ m}^3 \text{ s}^{-1}$) is in the opposite direction to that initially assumed. Thus, we can conclude that sign given to the monomial R_{ij} Q_{ij}^2 was incorrect. Since the initial directions of the airflows within the network are, in principle, unknown, this is a very important problem to take into account.

Alternative resolution: Methodology employing absolute values

The approach is the same as described above but the monomial $R_{ij} Q_{ij}^2$ derived using Kirchhoff's second law is written in the form $R_{ij} |Q_{ij}| Q_{ij}$. This formulation makes the equation robust to any initial choice of circulation direction.

In this way, alternative expressions for the pressure changes in M_1 and M_2 are obtained which are summarized in the tables below. Once again the equation for M_2 is multiplied by (-1) and summed with that for M_1 to eliminate the variable P_v.

Mesh 1 (M_1)	Corrected approach
$+R_{12} Q_{12}^2$	$+R_{12} Q_{12} |Q_{12}|$
$-R_{24} Q_{24}^2$	$-R_{24} Q_{24} |Q_{24}|$
$-R_{14} (Q_e - Q_{12})^2$	$-R_{14} (Q_e - Q_{12}) |Q_e - Q_{12}|$

$(-1) \cdot$ Mesh 2 (M_2)	Corrected approach
$+R_{23} (Q_{12} + Q_{24})^2$	$+R_{23} (Q_{12} + Q_{24}) |Q_{12} + Q_{24}|$
$+R_{34} ((Q_{24} - Q_e) + Q_{12})^2$	$+R_{34} (Q_{24} - Q_e + Q_{12}) |Q_{24} - Q_e + Q_{12}|$
$+R_{24} Q_{24}^2$	$+R_{24} Q_{24} |Q_{24}|$

Unlike the previous method, here, we end up with an equation effectively containing two unknowns, the variable Q_i and the absolute value of Q_i. This, of course, has no direct solution. To get over this difficulty, we use the Solver Solution Search System from Excel for its simplicity. Entering the values from the problem into the equation, we have found from the mesh analysis and imposing the condition that $Q_{12} \leq Q_e$, (that is $Q_{12} \leq 150$) at all times, that:

	Mesh 1	Mesh 2	$M_1 + M_2$
f (solver):	−3216.83	3216.83	0.00
Q_{12}	**68.80**		

It can be observed that, by making the problem independent of the selection of a correct initial direction of circulation, the solution obtained, $Q_{12} = 68.80$, differs only slightly from the value of $Q_{12} = 65.47$ found using the previous method.

- Resolution for airflow directions shown in Fig. 2.

1. A clockwise direction of travel is selected for both meshes and the direction of flow through branch 34 is corrected. Thus, the scheme is as follows:

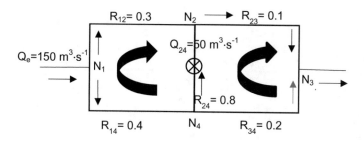

2. Kirchhoff's first law is then applied:

Node N_1:

$$Q_e = Q_{12} + Q_{14} \rightarrow Q_{14} = 150 - Q_{12}$$

Node N_2:

$$Q_{23} = Q_{12} + 50$$

Node N_3:

$$Q_{23} + Q_{34} = 150 \rightarrow Q_{34} = 150 - Q_{23} = 100 - Q_{12}$$

Node N_4[12]:

$$Q_{14} = Q_{24} + Q_{34} \rightarrow Q_{34} = Q_{14} - Q_{24} = 150 - Q_{12} - 50 = 100 - Q_{12}$$

3. Kirchhoff's second law is applied. As before, a positive sign (+) indicating a pressure drop, is used when the direction of the airflow in the branch coincides with the direction of travel around the mesh, and a negative sign (−) is used when they are opposed. Thus:

---

[12]Note that this last equation is redundant as it is obtained by the combination of the equations of nodes N_2 and N_3. This is because in a mesh of N nodes there are $N - 1$ independent nodes.

Mesh M_1

Branch	Pressure drop	Fan
Q_{12}	$0.3\,Q_{12}^2$	No
Q_{24}	$-0.8 \cdot 50^2$	$+P_v$
Q_{14}	$-0.4\,(150 - Q_{12})^2$	No
Mesh M_1 $\sum H_{ij} = 0$	$-0.1\,Q_{12}^2 + 120\,Q_{12} - 11000 + P_v = 0$	

Mesh M_2

Branch	Pressure drop	Fan
$N_2 - N_3$	$0.1\,(Q_{12} + 50)^2$	No
$N_3 - N_4$	$-0.2\,(100 - Q_{12})^2$	No
$N_4 - N_2$	$0.8 \cdot 50^2$	$-P_v$
Mesh M_2 $\sum H_{ij} = 0$	$-0.1\,Q_{12}^2 + 50\,Q_{12} + 250 - P_v = 0$	

Mesh M_1: $-0.1\,Q_{12}^2 + 120\,Q_{12} - 11000 + P_v = 0$
Mesh M_2: $-0.1\,Q_{12}^2 + 50\,Q_{12} + 250 - P_v = 0$

$M_1 + M_2$: $+0.2\,Q_{12}^2 + 170\,Q_{12} - 10750 = 0$

Solving the system we have that:
$Q_{12} = 68.80$ m^3 s^{-1} (781.2 m^3 s^{-1} is discarded as it has no physical meaning)
$Q_{14} = 150 - Q_{12} = 81.20$ m^3 s^{-1}
$Q_{24} = 50$ m^3 s^{-1}
$Q_{23} = Q_{12} + 50 = 118.80$ m^3 s^{-1}
$Q_{34} = Q_{12} - 100 = 31.20$ m^3 s^{-1}

All the calculated values are positive thus there are no contradictions between the circulation directions assigned to the different branches. Thus, they are interpreted as the correct solution according to Kirchhoff's equations.

A methodology based on absolute values

Proceeding in an analogous manner to the previous section in which we considered absolute values, we have:

M_1	M_2
$+R_{12}\,Q_{12}\,|Q_{12}|$	$+R_{23}\,(Q_{12} + Q_{24})\,|(Q_{12} + Q_{24})|$
$-R_{24}\,Q_{24}\,|Q_{24}|$	$-R_{34}\,(Q_e - Q_{24} - Q_{12})\,|Q_e - Q_{24} - Q_{12}|$
$-R_{14}\,(Q_e - Q_{12})\,|Q_e - Q_{12}|$	$+R_{24}\,Q_{24}\,|Q_{24}|$

Entering the appropriate values into the resulting equation and imposing the condition $Q_{12} \leq Q_e = 150$, the solution is obtained:

	Mesh 1	Mesh 2	$M_1 + M_2$
f (solver):	−3216.83	3216.83	0.00
Q_{12}	**68.80**		

Observe that the result we find here is equal to that obtained using the same method but considering airflow directions as shown in Fig. 1. Once Q_{12} is obtained, the rest of the airflow rates can be calculated directly through the relations obtained using Kirchhoff's first law for the nodes.

Conclusion: The appearance of negative airflow rates implies that the direction initially assigned to the flow is not correct. Indeed, even the absolute value of the solution obtained cannot be trusted. To be sure that the correct values have been found, all results obtained must be positive. For this reason, when a negative value is obtained, the direction of airflow circulation should be reversed and the calculations must be repeated. Only when the validity of a solution for the airflow rate has been checked can other parameters be calculated.

(c) Calculation of the pressure supplied by the fan in branch 2–4.

$\Delta P_v = 3\,216.8$ Pa (calculated with mesh M_1 equation).

(d) Characteristics of the fan required in branch R_{24}.

Bearing in mind that $Q_{24} = 50$ m^3 s^{-1} and $\Delta P_v = 3\,216.8$ Pa, the useful power is $Pw = 3\,216.8$ Pa $\cdot$ 50 m^3 s$^{-1} = 160.8$ kW

(e) Total power of the ventilation system.

Calculation of the total power that will be necessary in this network is summarized in the table:

Mesh	Branch ij	R_{ij}	Q_{ij} (m^3 s^{-1})	$\Delta P_{ij} = R_{ij}\,Q_{ij}^2$ (Pa)	Useful power (kW)	Notes
	14	0.4	81.2	2637.1	214.1	
M_1	12	0.3	68.8	1420.2	97.7	
	24	0.8	50.0	2000.0	100.0	
	34	0.2	31.2	194.6	6.1	
M_2	23	0.1	118.8	1411.5	167.7	
	24	0.8	50.0	–	–	(*)
M_1 y M_2	–	–	–	**Total:**	585.6	

(*) This is common to both meshes and has already been considered in calculations for M_1

4.10.2 Newton's Method

The *Newton–Raphson method*, widely used for the solution of fluid network problems, is an iterative method of calculation used to find the roots of a real non-linear function or a system of non-linear functions. It is often found under the method names of Newton, Newton–Raphson, and Newton–Fourier, after the scientists and mathematicians associated with its development.

The method can be applied to any continuous, differentiable function, $f(x)$ that intersects the x-axis at least once. The method allows us to approach the value of x for which $f(x) = 0$ through a series of calculations of the tangent to $f(x)$.

Given that the curve $y = f(x)$ crosses the x-axis at some point, $(x_0, y_0 = 0)$, we know that the line tangent to the curve has the same gradient as the curve at that point and is, therefore, equal to its derivative (Fig. 4.13). Thus:

$$f'(x_1) = \frac{y_1 - y_0}{x_1 - x_0}$$

Since $y_1 = f(x_1)$, and $y_0 = f(x_0) = 0$ (the intersection with the x-axis), substituting and solving for x_2 gives the ratio:

$$x_2 = x_1 - \frac{f(x_1)}{f'(x_1)}$$

We can generalize this to produce the following iteration formula (Eq. 4.26)[13]:

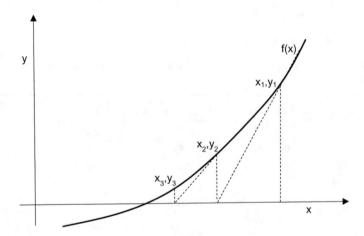

Fig. 4.13 Showing how Newton's method converges through a series of calculations of the tangent to the curve $y = f(x)$

---

[13] In this formula, x_n is the desired solution (where the function $f(x)$ intersects the x-axis). Ideally, each iteration will produce a result closer to this solution.

Fig. 4.14 Second-degree polynomial with two roots $(a, 0)$ and $(b, 0)$

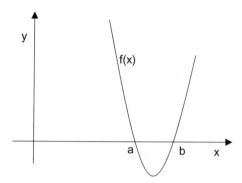

$$x_n = x_{n-1} - \frac{f(x_{n-1})}{f'(x_{n-1})} \tag{4.26}$$

That is, the next estimate for $f(x) = 0$ is obtained taking the difference between the value of a seed point $(x_n - 1)$ and the ratio of the value of $f(x)$ at that seed point and its derivative at that same point. For the series to converge, the starting point, $x_{n=1}$, must belong to the existing field of the function.

It should be noted that the fact that the method finds a solution does not necessarily indicate that it is a physically meaningful solution to any problem under investigation since the value found depends on the type of function under analysis and the seed values used. For example, taking the function shown in Fig. 4.14, if Newton's method is used with a starting value $(0 < x_{n=1} > a)$ it will locate the root $(a, 0)$; while if it is used with a starting value belonging to the interval $(a < x_{n=1} > b)$, depending on whether this value occurs before the minimum value or after, it could converge on either root $(a, 0)$ or $(b, 0)$.

This highlights that, when solving real-life problems, there is a technical necessity to identifying points close to the solution or having a system for checking solutions either graphically or by some other independent numerical method.

Finally, it must be remembered that Newton's method cannot be applied if $f'(x) = 0$, since the tangent will be horizontal and therefore the function will not cut the x-axis.

Note: The choice of the function to be optimized must fulfill the following conditions:

1. Be related to the process to be studied.
2. Include all known variables together with those to be determined and reflect the relationship between them.
3. Include as few as possible unknown variables.

In addition:

4. The function should be as simple as possible to facilitate calculations, for example, $Z = X^2 + Y^2$. If functions of the type $Z = X Y$ are considered, the solution may prove difficult to find as sign changes in one of the terms will cause

such functions to oscillate about the x-axis. Functions of the type $Z = X^2 Y^2$, although more stable, are, however, very sensitive to the initial value chosen.

5. The function must be balanced, or compensated, with regards to what each of the terms represents. Thus, it would not be correct to look for a solution to our problem using a function of the type $Z = X \ln(Y)$, since this would give one of the meshes more weight than the other.

6. The chosen function should avoid errors as far as possible. Thus, when considering functions containing sums it should be remembered that a positive value for X may occur together with a negative value for Y giving a solution of zero (or close to zero). Such as solution would have no physical meaning. This situation is avoided by squaring X and Y which always gives positive values.

Exercise 4.14 Taking the network shown in Exercise 4.13, and removing the fan from branch 2–4, calculate the airflow rates in each branch using Newton's method.[14]

Using Kirchhoff's first law:		
Node	Equation	Substituted variable
N_1	$Q_a + Q_{14} = Q_{12}$	$Q_{14} = Q_{12} - Q_a$
N_3	$Q_{23} = Q_s$ (or Q_a) $+ Q_{34}$	$Q_{34} = Q_{23} - Q_a$
N_2	$Q_{12} = Q_{24} + Q_{23}$	$Q_{24} = Q_{12} - Q_{23}$
Equations in the nodes with the initial condition: $\lvert Q_a \rvert = \lvert Q_s \rvert$ Using Kirchhoff's second law:		
Mesh 1: $R_{12} \lvert Q_{12} \rvert Q_{12} + R_{24} \lvert Q_{24} \rvert Q_{24} + R_{14} \lvert Q_{14} \rvert Q_{14} = 0$		
Mesh 2: $R_{23} \lvert Q_{23} \rvert Q_{23} + R_{34} \lvert Q_{34} \rvert Q_{34} - R_{24} \lvert Q_{24} \rvert Q_{24} = 0$		
Substituting variables:		
Mesh 1:	$R_{12} \lvert Q_{12} \rvert Q_{12} + R_{24} \lvert Q_{12} - Q_{23} \rvert Q_{12} - Q_{23} + R_{14} \lvert Q_{12} - Q_a \rvert (Q_{12} - Q_a) = 0$	
Mesh 2:	$R_{23} \lvert Q_{23} \rvert Q_{23} + R_{34} \lvert Q_{23} - Q_a \rvert (Q_{23} - Q_a) - R_{24} \lvert Q_{12} - Q_{23} \rvert (Q_{12} - Q_{23}) = 0$	
Note that to avoid losing negative signs during the squaring process we have used absolute values		
Generating the function: $Z = X^2 + Y^2$, wherein:		
$X = R_{12} \lvert Q_{12} \rvert Q_{12} + R_{24} \lvert Q_{12} - Q_{23} \rvert (Q_{12} - Q_{23}) + R_{14} \lvert Q_{12} - Q_a \rvert (Q_{12} - Q_a) = 0$		
$Y = R_{23} \lvert Q_{23} \rvert Q_{23} + R_{34} \lvert Q_{23} - Q_a \rvert (Q_{23} - Q_a) - R_{24} \lvert Q_{12} - Q_{23} \rvert (Q_{12} - Q_{23}) = 0$		
The solution of this function can then be found through a numerical optimization method, where Q_{12} and Q_{23} are the variables to be optimized		

Variable	Value	Solution	Branch	Quantity ($m^3 \ s^{-1}$)
$Q_{12} =$	80.70		$Q_a = Q_s$	150.00

(continued)

[14]By modifying the flow rate Q_{24}, the system becomes unsolvable by traditional methods. Usually, a system is generated in $x, y, x \cdot y, x^2, y^2$, with roots from the above variables, which requires the use of numerical calculation methods

(continued)

Using Kirchhoff's first law:				
$Q_{23} =$	87.10		Q_{14}	−69.30
			Q_{12}	80.70
$X^2 =$	0.00		Q_{24}	−6.40
$Y^2 =$	0.00		Q_{34}	−62.90
$Z = X^2 + Y^2 =$	0.00		Q_{23}	87.10
			$-Q_{24}$	6.40

The above calculations were made using the Excel Solver to find the value 0. The total power of the net is 406.9 kW.

4.10.3 Hardy–Cross Method

The *Hardy–Cross method* (Cross 1936) is a numerical calculation procedure that has its origin in the field of structural engineering. It is an iterative method developed to solve calculations for fluid flow networks. The method has the advantage of being self-correcting and will converge even where initial conditions are poorly specified.

Demonstration

Taking a mesh for which the flows of each of the branches have been estimated, let Q_i be the initial airflow rate assigned to the ith branch of the network and where ΔQ is the correction made to the flow rate after initial calculations Then, the corrected flow rate will be:

$$Q = Q_i + \Delta Q$$

This new value of flow rate, Q, becomes the value Q_i for the following iteration and ΔQ is the positive or negative correction that must be made in all branches of the mesh. Thus, for the next iteration, the pressure drop for one branch (ΔP_i) will be:

$$\Delta P_i = R_i (Q_i + \Delta Q)^2$$

That can be written as:

$$\Delta P_i = R_i Q_i^2 \left(1 + \frac{\Delta Q}{Q_i}\right)^2$$

By expanding the binomial, we get:

$$\Delta P_i = R_i Q_i^2 \left[1 + 2\frac{\Delta Q}{Q_i} + \left(\frac{\Delta Q}{Q_i}\right)^2 \right]$$

The value ΔQ is normally small and tends to decrease as the correct solution is approached so, after several iterations, the term $\left(\frac{\Delta Q}{Q_i}\right)^2$ tends to 0. Thus:

$$\Delta P_i = R_i Q_i^2 \left(1 + 2\frac{\Delta Q}{Q_i} \right)$$

If the principle of energy conservation is applied to the mesh, we have that:

$$\Delta P_i = \sum_{i=1}^{N_r} R_i Q_i^2 + 2\Delta Q \sum_{i=1}^{N_r} R_i Q_i$$

Given that:

$$\Delta P_i = 0$$

Therefore, solving for ΔQ:

$$\Delta Q = -\frac{\sum_{i=1}^{N_r} R_i Q_i^2}{2 \sum_{i=1}^{N_r} R_i Q_i}$$

As in previous examples, the direction of airflow in the network must be taken into account and since the airflow term is squared the expression is better formulated as (Eq. 4.27):

$$\Delta Q = -\frac{\sum_{i=1}^{N_r} R_i Q_i |Q_i|}{2 \sum_{i=1}^{N_r} R_i Q_i} \tag{4.27}$$

This then, is the correction to airflow rate made at each iteration of the Hardy–Cross method.

Procedure

1. The network is divided into a number of closed meshes. The branches are usually named with the subscripts corresponding to the nodes they connect. For instance, the branch that connects the node N_2 with the node N_3 will be named Q_{23} or Q_{32}.
2. For each mesh, a direction of travel around the mesh is chosen. This may be either clockwise or counterclockwise, but preferably the first. Each duct is then assigned an initial flow rate, respecting the law of conservation of the mass at each node. Conventionally, the clockwise direction is positive, so that the flow

rates are positive if they circulate in this direction and negative if they do so in the opposite direction.

3. The equivalent resistance of each duct is determined if it is not already known. These resistances will always have positive values.
4. The pressure drop for each duct is calculated according to $\Delta P_i = R_i Q_i^2$.
5. For the above calculations, the sign of ΔP_i is lost due to squaring. This can be remedied by assigning ΔP_i the same sign as that of the initial value of Q_i.
6. The sum of the losses for each mesh is determined with its corresponding sign.
7. The quotient of the loss (ΔP_i) and its corresponding flow rate (Q_i) is calculated. Note that the sign of this value will always be positive.
8. The sum of the above ratios is determined for each mesh since the system corrects the calculation on a mesh by mesh basis.
9. Apply the correction formula for each mesh. That is, for each mesh calculate the quotient ΔQ according to Eq. 4.27.
10. Apply the correction to the initial flow rate values. This produces new flow rate values that are used for the following iteration.
11. Repeat until ΔQ is very small or until the results of successive iterations show little change.

Exercise 4.15 For the network described in Exercise 4.14 (i.e. that used in Exercise 4.13 with the fan eliminated, from branch 2–4), calculate the airflow rate in each branch using the Hardy–Cross method.

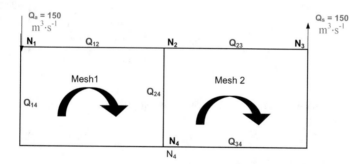

Solution

In accordance with the principle of conservation of mass, the output flow rate, when the circuit is stabilized, is equal to the input flow rate, then:

$$Q_a = Q_s = 150 \, \mathrm{m^3 \, s^{-1}}$$

According to this, the first iteration is done with initial values of estimated airflow rates, avoiding the use of any values, since the convergence towards the final solution could take too long. The elected values $(Q_{ij}: -60, 90, 30, \text{etc.})$ must comply with Kirchhoff's first law.

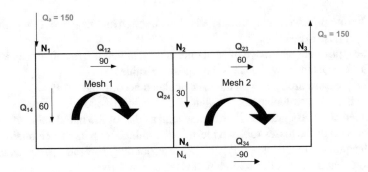

Airflow rates in each branch:
   Iteration 1.

Iteration $(n°)$	Mesh (M)	Branch (ij)	Q_{ij} (m³ s⁻¹)	Resistance (R_{ij})	$\Delta P_{ij} = R_{ij} Q_{ij}^2$ (with the sign of the airflow rate)	$\Delta P_{ij}/Q_{ij}$	Correction (ΔQ)
1	M_1	14	−60	0.4	−1440	24	−11.4
		12	90	0.3	2430	27	
		24	30	0.8	720	24	
	Σ				1710	75	
	M_2	34	−90	0.2	−1620	18	20.625
		23	60	0.1	360	6	
		24	−30	0.8	−720	24	
	Σ				−1980	48	

The values obtained for ΔQ_{ij} in each mesh are applied to correct the flows in the following calculation:
   Example for Q_{14}

Initial airflow rate: -60 m³ s⁻¹
Correction for M_1: $\Delta Q_1 = -11.4$ m³ s⁻¹
Corrected flow rate for the following calculation: -60.0 m³ s⁻¹ $+ (-11.4)$ m³ s⁻¹ $= -71.4$ m³ s⁻¹

Note that since Q_{14} belongs only to M_1, only the correction calculated for this mesh need be used.

For the branches that are common to two meshes, airflow rates must be corrected using the values of ΔQ found for both meshes as follows:

Example for Q_{24}

Initial flow rate: $30.0 \text{ m}^3 \text{ s}^{-1}$
Correction for M_1: $\Delta Q_1 = -11.4 \text{ m}^3 \text{ s}^{-1}$
Correction for M_2: $\Delta Q_2 = 20.625 \text{ m}^3 \text{ s}^{-1}$
Corrected flow rate: $30.0 \text{ m}^3 \text{ s}^{-1} + (-11.4) \text{ m}^3 \text{ s}^{-1} - (20.625) \text{ m}^3 \text{ s}^{-1} = -2.0 \text{ m}^3 \text{ s}^{-1}$

The sign (+ or −) is related to the direction of travel around the mesh. Thus, a branch shared by two meshes will have normally positive flow rates in one and negative in the adjacent one. This is so because the same direction of travel is given to all the meshes (normally clockwise).

Iteration 2.

Iteration ($n°$)	Mesh	Branch (ij)	Q_{ij} (m^3 s^{-1})	Resistance (R_{ij})	$\Delta P_{ij} = R_{ij} Q_{ij}^2$ (with the sign of the airflow rate)	$\Delta P_{ij}/Q_{ij}$	Correction (ΔQ_M)
2	M_1	14	−71.4	0.4	−2039.184	28.56	1.758429475
		12	78.6	0.3	1853.388	23.58	
		24	−2	0.8	−3.2	1.6	
	Σ				−188.996	53.74	
	M_2	34	−69.4	0.2	−963.272	13.88	6.593797791
		23	80.6	0.1	649.636	8.06	
		24	2	0.8	3.2	1.6	
	Σ				−310.436	23.54	

If we look at iterations 19 and 20, it can be seen that the correction is less than 10^{-15}.

Table of corrections for iterations 19 and 20:

Iteration ($n°$)	Mesh	Correction (ΔQ_M)
19	M_1	-6.22757×10^{-17}
	M_2	-2.01795×10^{-15}
20	M_1	-1.86827×10^{-16}
	M_2	-1.74889×10^{-15}

This shows the rapid convergence of the method. The following table compares the results for 19, 20 and 700 iterations.

Mesh	Branch	R_{ij}	Q_{ij}		
			19 iter.	20 iter.	700 iter.
M_1	14	0.400	−69.30	−69.30	−69.30
	12	0.300	80.70	80.70	80.70
	24	0.800	−6.40	−6.40	−6.40
M_2	34	0.200	−62.90	−62.90	−62.90
	23	0.100	87.10	87.10	87.10
	24	0.800	6.40	6.40	6.40

It can be seen that there is no need to go beyond iterations 19 and 20 since there is no difference between the correction values found at each of these steps: the system has found a solution.

4.10.4   Wood–Charles Method

This model also called the *Linear Theory Method (LTM)*. Like the previous models described, it is based on the equations of conservation of mass and energy. The key feature of this algorithm is that it linearizes the quadratic expression obtained by applying Kirchoff's second law (energy conservation), see Eq. 4.24, before attempting an iterative solution (Wood and Charles 1972). The rationale for this begins with an expression for the pressure in the ith branch as follows (Eq. 4.28):

$$P_{i(n)} = R_i \left| Q_{i(n-1)} \right| Q_{i(n)} = K_{(n)} Q_{i(n)} \tag{4.28}$$

where

- $P_{i(n)}$: Pressure loss of the ith branch for the nth iteration.
- R_i: Resistance of the ith branch.
- $Q_{i(n-1)}$: Estimated airflow rate or seed value for the ith branch.
- $Q_{i(n)}$: Airflow rate to be calculated for the nth iteration in the ith branch.
- $K_{(n)}$: Proportionality constant for the nth iteration. It acts as a constant for a specific calculation, but it is recalculated in each iteration.
- n: Iteration number ($n > 1$).

In the first iteration ($n = 1$), the value $Q_{i(o)}$ can be estimated, although it is common to assume it is unity,[15] i.e. $Q_{i(o)} = 1$. This means that pressure in the ith–branch for iteration $n = 1$ is:

$$P_{i(1)} = R_i \cdot 1 \cdot Q_{i(1)}$$

---

[15]In fact, in the original article by Wood and Charles (1972) it was 1 cfs (28.03 ls^{-1}). The model does require the use of particular initial values, but in order not to introduce arbitrary divergences, these must be the same for all branches.

In other words, it is a linear equation in Q_i, which allows, together with the mass conservation equations, forms a system of linear equations where the number of equations is equal to the number of unknowns, and is, therefore, solvable.

The procedure is then repeated substituting the value calculated for $Q_{i(1)}$ into the above equation. Consequently, for the second iteration we have:

$$P_{i(2)} = \left(R_i Q_{i(1)} \right) Q_{i(2)}$$

Which gives us the system of linear equations for the second iteration.

For those branches whose calculated $R_i Q_{i(n)}$ is lower than the real one (initially unknown) the method will correct them to a higher flow rate (lower resistance implies greater flow rate). The reverse is also true: for those branches whose calculated $R_i Q_{i(n)}$ is higher than the actual one, the system will correct to a lower flow rate. Successive iterations compensate for these deviations. It is, therefore, advisable that, from the second iteration onwards, the starting flow rate for the nth iteration be calculated as the average of the previous two values for Q_i.[16] This procedure has the joint advantages of converging quickly to the final solution, and of not needing to assume an initial distribution of flows. In addition, as opposed to the Hardy–Cross method, corrections in this method are made to individual branches rather than meshes.

Exercise 4.16 Again, for the scheme in Exercise 4.14, calculate the airflow rate in each branch using the Wood and Charles linear method.

Solution

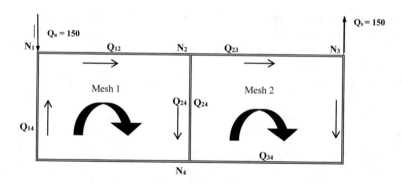

With the values of the exercise and the directions of travel indicated, the following equations can be written:

---

[16]The practice of using the average of the two previous solutions is normal in this type of oscillating solution-seeking system because the average value, while not the solution, will be closer to the solution than the two extreme values from which it is calculated.

Node N_1 : $Q_{14(n)} - Q_{12(n)} = -150$
Node N_2 : $Q_{12(n)} - Q_{23(n)} - Q_{24(n)} = 0$
Node N_3 : $-Q_{34(n)} + Q_{23(n)} = 150$
Mesh M_1 : $0.4|Q_{14(n-1)}|Q_{14(n)} + 0.3|Q_{12(n-1)}|Q_{12(n)} + 0.8|Q_{24(n-1)}|Q_{24(n)} = 0$
Mesh M_2 : $-0.8 \cdot |Q_{24(n-1)}|Q_{24(n)} + 0.2|Q_{34(n-1)}|Q_{34(n)} + 0.1|Q_{23(n-1)}|Q_{23(n)} = 0$

Then, for iteration 1, represented in matrix form:

Branch (ij)	14	12	24	34	23		
R_{ij}	0.4	0.3	0.8	0.2	0.1		
Q_{io}	1	1	1	1	1		
Node (N)/Matrix (M)	Q_{14}	Q_{12}	Q_{24}	Q_{34}	Q_{23}		Independent term
N_1	1	-1	0	0	0	=	-150
N_2	0	1	-1	0	-1	=	0
N_3	0	0	0	-1	1	=	150
M_1	0.4	0.3	0.8	0	0	=	0
M_2	0	0	-0.8	0.2	0.1	=	0

The system can be solved as:

$$\left[A_{ij}\right]\left[Q_i\right] = \left[B_j\right] \rightarrow \left[Q_i\right] = \left[A_{ij}\right]^{-1}\left[B_j\right]$$

where

- $[A_{ij}]$: Square matrix of the coefficients. It has $M + N - 1$ rows and columns.
- $[A_{ij}]^{-1}$: Inverse matrix of A_{ij}.
- $[Q_j]$: Column matrix of the airflow rates in each branch (values to be calculated). It has $M + N - 1$ rows.
- $[B_j]$: Colum matrix of the independent terms. It has $M + N - 1$ rows.

Therefore:

Inverse matrix						Solution
0.5643564	0.2376238	0.1584158	1.0891089	0.7920792	$Q_{14} =$	-60.89
-0.435644	0.2376238	0.1584158	1.0891089	0.7920792	$Q_{12} =$	89.11
-0.118812	-0.207921	-0.138614	0.2970297	-0.693069	$Q_{24} =$	-2.97
-0.316832	-0.554455	-0.70297	0.7920792	1.4851485	$Q_{34} =$	-57.92
-0.316832	-0.554455	0.2970297	0.7920792	1.4851485	$Q_{23} =$	92.08

Iteration 2
    The seed value for the airflow rate in the second iteration is $Q_{14(1)} = -60.89$. In this way, for example, Q_{14}, es $Q_{14(o)} = -60.89$. In order to avoid errors arising from

sign changes, once again we use absolute values, thus: $K_{(n)} = R_i |Q_{i(n-1)}|$, then: R_{14} $|Q_{14(1)}| = 0.4 |-60.89| = 24.3564$.

The system of equations for iteration 2 will be:

Branch	14	12	24	34	23	
R_{ij}	0.4	0.3	0.8	0.2	0.1	
$Q_{i(1)}$	−60.89	89.11	−2.97	−57.92	92.08	

Variable	$Q_{14(2)}$	$Q_{12(2)}$	$Q_{24(2)}$	$Q_{34(2)}$	$Q_{23(2)}$		Independent term
N_1	1	−1	0	0	0	=	−150
N_2	0	1	−1	0	−1	=	0
N_3	0	0	0	−1	1	=	150
M_1	24.3564	26.7327	2.3762	0	0	=	0
M_2	0	0	−2.3762	11.5842	9.2079	=	0

Inverse matrix						Solution
0.5423583	0.0400687	0.022324	0.0187894	0.0019271	$Q_{14(2)} =$	−78.01
−0.457642	0.0400687	0.022324	0.0187894	0.0019271	$Q_{12(2)} =$	71.99
−0.410704	−0.861477	−0.479966	0.0168622	−0.041433	$Q_{24(2)} =$	−10.39
−0.046938	−0.098454	−0.49771	0.0019271	0.04336	$Q_{34(2)} =$	−67.62
−0.046938	−0.098454	0.5022896	0.0019271	0.04336	$Q_{23(2)} =$	82.38

In order to ensure rapid convergence, from the third iteration onwards, the seed value $Q_{i(n)}$ is found from an average of the previous two values. For example, the value for $Q_{14(3)}$ will be:

$$Q_{14(3)} = [-60.89 + (-78.01)]/2 = -69.45$$

Then, since $R_{14} = 0.4$, we have:

$$R_{14} \cdot Q_{13(3)} = 0.4| - 69 \cdot 45| = 27.779$$

Iteration 3

Branch	14	12	24	34	23		
R_{ij}	0.4	0.3	0.8	0.2	0.1		
$Q_{i(2)}$	−69.45	80.55	−6.68	−62.77	87.23		
							Independent term

(continued)

(continued)

Variable	$Q_{14(3)}$	$Q_{12(3)}$	$Q_{24(3)}$	$Q_{34(3)}$	$Q_{23(3)}$		
N_1	1	−1	0	0	0	=	−150
N_2	0	1	−1	0	−1	=	0
N_3	0	0	0	−1	1	=	150
M_1	27.7792	24.1656	5.3438	0	0	=	0
M_2	0	0	−5.3438	12.5537	8.7232	=	0
	Inverse matrix						Solution
	0.50585	0.07598	0.04483	0.01779	0.00357	$Q_{14(3)} =$	−69.15
	−0.49415	0.07598	0.04483	0.01779	0.00357	$Q_{12(3)} =$	80.85
	−0.39496	−0.73854	−0.43575	0.01422	−0.03471	$Q_{24(3)} =$	−6.12
	−0.09920	−0.18549	−0.51942	0.00357	0.03828	$Q_{34(3)} =$	−63.03
	−0.09920	−0.18549	0.48058	0.00357	0.03828	$Q_{23(3)} =$	86.97

...

Iteration 19

Branch	14	12	24	34	23		
R_{ij}	0.4	0.3	0.8	0.2	0.1		
$Q_{i(18)}$	−69.29	80.71	−6.39	−62.89	87.11		
	$Q_{11(19)}$	$Q_{12(19)}$	$Q_{13(19)}$	$Q_{21(19)}$	$Q_{22(19)}$		Independent term
N_1	1	−1	0	0	0	=	−150
N_2	0	1	−1	0	−1	=	0
N_3	0	0	0	−1	1	=	150
M_1	27.715424	24.213432	5.1149014	0	0	=	0
M_2	0	0	−5.114901	12.578987	8.7105066	=	0
	Inverse matrix						Solution
	0.5055491	0.0735746	0.0434719	0.0178403	0.0034559	$Q_{11(19)} =$	−69.312
	−0.494451	0.0735746	0.0434719	0.0178403	0.0034559	$Q_{12(19)} =$	80.688
	−0.398669	−0.746964	−0.441347	0.0143844	−0.035086	$Q_{13(19)} =$	−6.402
	−0.095782	−0.179462	−0.515181	0.0034559	0.0385419	$Q_{21(19)} =$	−62.910
	−0.095782	−0.179462	0.4848186	0.0034559	0.0385419	$Q_{22(19)} =$	87.090

Iteration 20:

Branch	14	12	24	34	23		
R_{ij}	0.4	0.3	0.8	0.2	0.1		
$Q_{i(19)}$	−69.32	80.68	−6.41	−62.91	87.09		
	$Q_{14(20)}$	$Q_{12(20)}$	$Q_{24(20)}$	$Q_{34(20)}$	$Q_{23(20)}$		Independent term
N_1	1	−1	0	0	0	=	−150

(continued)

(continued)

N_2	0	1	−1	0	−1	=	0	
N_3	0	0	0	−1	1	=	150	
M_1	27.7271	24.2047	5.1241	0	0	=	0	
M_2	0	0	−5.1241	12.5825	8.7087	=	0	
	Inverse matrix						Solution	
	0.50542	0.07367	0.04354	0.01784	0.00346	$Q_{14(20)} =$	−69.282	
	−0.49458	0.07367	0.04354	0.01784	0.00346	$Q_{12(20)} =$	80.718	
	−0.39864	−0.74664	−0.44124	0.01438	−0.03507	$Q_{24(20)} =$	−6.390	
	−0.09594	−0.17969	−0.51522	0.00346	0.03853	$Q_{34(20)} =$	−62.892	
	−0.09594	−0.17969	0.48478	0.00346	0.03853	$Q_{23(20)} =$	87.108	

Since no significant variations in the solution are observed in the last two iterations, it is decided that 20 iterations are sufficient. The final value is found by taking the average of the last two iterations:

Mesh	Airflow rate	Iteration 19	Iteration 20	Mean
M_1	Q_{14}	−69.31	−69.28	−69.30
	Q_{12}	80.69	80.72	80.70
	Q_{24}	−6.40	−6.39	−6.40
M_2	Q_{34}	−62.91	−62.89	−62.90
	Q_{23}	87.09	87.11	87.10
	Q_{24}	6.40	6.39	6.40

4.10.5 The Newton–Raphson Method Applied to Systems of Equations

A technique based on a recursive solution of the Taylor expansion of a function around a point can be generalized to a system of n functions with n unknowns. Thus, given a system of functions (say, the equilibrium equations for an air distribution system in a mine):

$$F_1(Q_1, Q_2 \ldots, Q_n) = 0$$
$$F_2(Q_1, Q_2, \ldots, Q_n) = 0$$
$$\ldots$$
$$F_n(Q_1, Q_2, \ldots, Q_n) = 0$$

The solution can be found by using Taylor's expansion from a starting point $Q_{i,0}$ ($Q_{1,0}$, $Q_{2,0}$, ..., $Q_{n,0}$). Then, replacing the differentials $dQ_{i,0}$ by finite differences $\Delta Q_{i,0}$ and disregarding terms degree greater than one, we have:

$$F_1(Q_{1,0}, Q_{2,0}, \ldots Q_{n,0},) + \frac{\partial F_1}{\partial Q_1}\Delta Q_{1,0} + \frac{\partial F_1}{\partial Q_2}\Delta Q_{2,0} + \cdots + \frac{\partial F_1}{\partial Q_n}\Delta Q_{n,0} = 0$$

$$F_2(Q_{1,0}, Q_{2,0}, \ldots Q_{n,0}) + \frac{\partial F_2}{\partial Q_1}\Delta Q_{1,0} + \frac{\partial F_2}{\partial Q_2}\Delta Q_{2,0} + \cdots + \frac{\partial F_2}{\partial Q_n}\Delta Q_{n,0} = 0$$

$$\ldots \qquad\qquad \ldots \qquad\qquad \ldots$$

$$F_n(Q_{1,0}, Q_{2,0}, \ldots Q_{n,0},) + \frac{\partial F_n}{\partial Q_1}\Delta Q_{1,0} + \frac{\partial F_n}{\partial Q_2}\Delta Q_{2,0} + \cdots + \frac{\partial F_n}{\partial Q_n}\Delta Q_{n,0} = 0$$

Which can be expressed in matrix form:

$$\begin{bmatrix} \frac{\partial F_1}{\partial Q_1} & \frac{\partial F_1}{\partial Q_2} & \cdots & \frac{\partial F_1}{\partial Q_n} \\ \vdots & & \ddots & \vdots \\ \frac{\partial F_n}{\partial Q_1} & \frac{\partial F_n}{\partial Q_2} & \cdots & \frac{\partial F_n}{\partial Q_n} \end{bmatrix} \begin{bmatrix} \Delta Q_{1,0} \\ \cdots \\ \Delta Q_{n,0} \end{bmatrix} = - \begin{bmatrix} F_1 \\ \cdots \\ F_n \end{bmatrix} \tag{4.29}$$

Or more compactly (Eq. 4.30)[17]:

$$[Ji_{ij}][\Delta Q_{i,0}] = -[F_i] \tag{4.30}$$

Then, to find ΔQ_i:

$$[J_{ij}]^{-1}[J_{ij}][\Delta Q_{i,0}] = -[J_{ij}]^{-1}[F_i]$$

Hence:

$$[\Delta Q_i, 0] = -[J_{ij}]^{-1}[F_i] \tag{4.31}$$

where

$$[J_{ij}]^{-1}[J_{ij}] = [I]$$

In this way, if the functions F_1, F_2, ..., F_n and the Jacobian $[J_{ij}]$ are calculated with the initial values of the variable $Q_{i(0)}$ then the values of the increments $\Delta Q_{i(0)}$ can be obtained by solving the system of matrix equations. This allows us to obtain the values of $\Delta Q_{i(0)}$ ($i = 1, 2, \ldots n$) and thence a new values for the flow rates $Q_{i(1)} = Q_{i(0)} + \Delta Q_{i(0)}$. This process is repeated with each new set of values for the flow rates being closer to the real solution than the last set until the difference between two successive sets of values is less than a preselected differential (Δ).

---

[17] $[J_{ij}]$ is the Jacobian matrix, that is, the matrix of first-order partial derivatives of the system of functions, $F_{(1,2\ldots n)}$.

Application of the Model to Ventilation Networks

The application of the system to the solution of ventilation networks has similarities with the previous cases explored, starting as it does from the principles of conservation of mass (Eq. 4.23) and energy (Eq. 4.24). In this case, as before, the expression for energy conservation must be written using absolute values (see Eq. 4.25), leading to Eq. 4.32:

$$F(Q_i) = \Sigma R_i |Q_i| Q_i \qquad (4.32)$$

With regards to application of the law of conservation of mass to the nodes, it must be the case that the derivative (Eq. 4.33) must be a constant, k:

$$\frac{\partial F}{\partial Q_i} = \pm k \qquad (4.33)$$

This derivative is necessary to calculate the Jacobian, and therefore, the term $|Q|$ Q in Eq. 4.32 requires further analysis.

Thus, the partial derivative of $F(Q_i)$ with respect to a generic flow rate (Q_i) will be:

$$\frac{\partial F}{\partial Q_i} = R_i |Q_i| + R_i Q_i |Q_i|^{1-1} \frac{\partial |Q_i|}{\partial Q_i}$$

Since we are working with absolute values there are two options: (a) that Q_i is positive and (b) that Q_i is negative.

(a) If $Q_i > 0$, then $|Q_i| = Q_i$, and the partial derivative $\frac{\partial |Q_i|}{\partial Q_i} = +1$

And the equation remains:

$$\frac{\partial F}{\partial Q_i} = R_i |Q_i| + R_i Q_i |Q_i|^{1-1} \cdot 1$$

From which we obtain Eq. 4.34:

$$\frac{\partial F}{\partial Q_i} = 2 R_i |Q_i| \qquad (4.34)$$

(b) If, on the other hand, $Q_i < 0$, then $|Q_i| = -Q_i$, and the partial derivative $\frac{\partial |Q_i|}{\partial Q_i} = -1$

Then we have:

$$\frac{\partial F}{\partial Q_i} = R_i |Q_i| + R_i Q_i |Q_i|^{1-1}(-1)$$

Which, because Q_i can be simplified and then multiplied by (-1) to give Eq. 4.34.

As a final consideration, it should be borne in mind that, in the system to be solved involves n equations with n unknowns, and it will be necessary to calculate functions, their derivatives and the inverse of a matrix. For this reason, it is of the utmost importance to reduce the number of unknowns as much as possible. This can be done by substituting equilibrium values for the nodes into the mesh equations.

Exercise 4.17 For the scheme explored in Exercise 4.14, calculate the flow rate in each branch using the Newton–Raphson method. Solve (a) For the total number of equations and unknowns; (b) Reducing, as far as possible, the number of equations per substitution.

Solution

(a)   For the total of equations.

As shown in the figure, for the solution of Kirchchoff's equilibrium equations, the initial direction of travel in the two meshes is assumed to be clockwise.

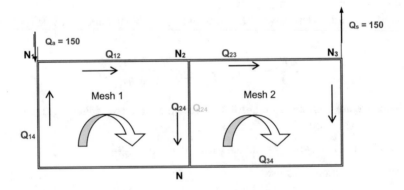

The following system of equations is then derived to describe the network:

Node N_1	$F_1(Q_{ij}) = 0$	$F_1 = Q_{14} - Q_{12} + Q_a = 0$
Node N_2	$F_2(Q_{ij}) = 0$	$F_2 = Q_{12} - Q_{23} - Q_{24} = 0$
Node N_3	$F_3(Q_{ij}) = 0$	$F_3 = Q_{23} - Q_{34} - Q_a = 0$
Node N_4	Not necessary, $(N-1)$ independent nodes	
Mesh 1	$F_4(Q_{ij}) = 0$	$F_4 = R_{14} Q_{14}^2 + R_{12} Q_{12}^2 + R_{24} Q_{24}^2 = 0$
Sign	$F_4(Q_{ij}) = 0$	$F_4 = R_{14} |Q_{14}| Q_{14} + R_{12} |Q_{12}| Q_{12} + R_{24} |Q_{24}| Q_{24} = 0$
Mesh 2	$F_5(Q_{ij}) = 0$	$F_5 = R_{23} Q_{23}^2 + R_{34} Q_{34}^2 - R_{24} Q_{24}^2 = 0$
Sign	$F_5(Q_{ij}) = 0$	$F_5 = R_{23} |Q_{23}| Q_{23} + R_{34} |Q_{34}| Q_{34} - R_{24} |Q_{24}| Q_{24} = 0$

The Jacobian is calculated as:

(a) $\frac{dF_n}{dQ_{ij}} = (\pm)2R_{ij}|Q_{ij}|$ for the mesh equations: second-degree equations in Q.

(b) $\frac{dF_n}{dQ_{ij}} = (\pm)k$, for the node equations: linear equations in Q.

and are shown in the table:

Function	Q_{14}	Q_{12}	Q_{24}	Q_{34}	Q_{23}
F_1	dF_1/dQ_{14}	dF_1/dQ_{12}	dF_1/dQ_{24}	dF_1/dQ_{34}	dF_1/dQ_{23}
F_2	dF_2/dQ_{14}	dF_{12}/dQ_{12}	dF_2/dQ_{24}	dF_2/dQ_{34}	dF_2/dQ_{23}
F_3	dF_3/dQ_{14}	dF_3/dQ_{12}	dF_3/dQ_{24}	dF_3/dQ_{34}	dF_3/dQ_{23}
F_4	dF_4/dQ_{14}	dF_4/dQ_{12}	dF_4/dQ_{24}	dF_4/dQ_{34}	dF_4/dQ_{23}
F_5	dF_5/dQ_{14}	dF_5/dQ_{12}	dF_5/dQ_{24}	dF_5/dQ_{34}	dF_5/dQ_{23}

So, using the values in the example:

Function	Q_{14}	Q_{12}	Q_{24}	Q_{34}	Q_{23}
F_1	1	-1	0	0	0
F_2	0	1	-1	0	-1
F_3	0	0	0	-1	1
F_4	$2 R_{14} \cdot |Q_{14}|$	$2 R_{12} |Q_{12}|$	$2 R_{24} |Q_{24}|$	0	0
F_5	0	0	$2 R_{24} |Q_{24}|$	$2 R_{34} |Q_{34}|$	$-2 R_{23} |Q_{23}|$

The initial numerical values $Q_{ij(o)}$ are chosen arbitrarily but in compliance with the laws of mass and energy conservation at the nodes:

Table of initial values

Branch	14	12	24	34	23
R_{ij}	0.4	0.3	0.8	0.2	0.1
$Q_{ij(o)}$	-60.0	90.0	30.0	-90.0	60.0

These values are then substituted into the matrix function for the system $[F_i]$ and the following results are obtained:

Function matrix for initial values

Function	Q_{14}	Q_{12}	Q_{24}	Q_{34}	Q_{23}	Ind. term. $= Q_a$	F_i
F_1	-60.0	-90.0	0	0	0	$150.0 =$	0.0
F_2	0	90.0	-30.0	0	-60.0	$0 =$	0.0
F_3	0	0	0	90.0	60.0	$-150.0 =$	0.0
F_4	-1440	2430	720	0	0	$0 =$	1710.0
F_5	0	0	-720	-1620	360	$0 =$	-1980.0

The Jacobian matrix is then calculated:

Variable	Q_{14}	Q_{12}	Q_{24}	Q_{34}	Q_{23}
F_1	1	−1	0	0	0
F_2	0	1	−1	0	−1
F_3	0	0	0	−1	1
F_4	48	54	48	0	0
F_5	0	0	−48	36	12

The inverse Jacobian matrix is obtained, and the matrix equation of the system can then be solved: $\left[\Delta Q_{i,0}\right] = -\left[J_{ij}\right]^{-1}\left[F_i\right]$.

The results of these calculations are shown below:

						Solution
0.6190476	0.1904762	0.1428571	0.0079365	0.0039683	$\Delta Q_{14} =$	−5.71
−0.380952	0.1904762	0.1428571	0.0079365	0.0039683	$\Delta Q_{12} =$	−5.71
−0.190476	−0.404762	−0.303571	0.0039683	−0.008433	$\Delta Q_{24} =$	−23.48
−0.190476	−0.404762	−0.553571	0.0039683	0.0124008	$\Delta Q_{34} =$	17.77
−0.190476	−0.404762	0.4464286	0.0039683	0.0124008	$\Delta Q_{23} =$	17.77

The initial, arbitrarily selected, airflow rates are then corrected using the values of ΔQ_{ij} found in this first iteration. The new, modified airflow rates are then:

Branch	14	12	24	34	23
R_{ij}	0.4	0.3	0.8	0.2	0.1
$Q_i + \Delta Q_i$	−65.71	84.29	6.52	−72.23	77.77

Using these values, the entire calculation process is repeated and further new values are obtained for $\Delta Q_{ij}(1, 2, \ldots n)$, allowing the system solution to be approximated.

The numerical results for the second, fourth and fifth iteration are given below. After the fourth iteration results are the same as two decimal places. Thus, we can say that convergence is obtained at the fourth iteration.

Iteration 2:

Matrix of the system function (F_i)							
Function	Q_{14}	Q_{12}	Q_{24}	Q_{34}	Q_{23}	Ind. term = Q_a	F_i
F_1	−65.7	−84.3	0	0	0	150.0 =	0.0
F_2	0	84.3	−6.5	0	−77.8	0 =	0.0
F_3	0	0	0	72.2	77.8	−150.0 =	0.0
F_4	−1727.347	2131.2245	33.985969	0	0	0 =	437.9
F_5	0	0	−33.98597	−1043.496	604.78396	0 =	−472.7

(continued)

(continued)

Matrix of the system function (F_i)

Function	Q_{14}	Q_{12}	Q_{24}	Q_{34}	Q_{23}	Ind. term = Q_a	F_i
Jacobian of the system							

Variable	Q_{14}	Q_{12}	Q_{24}	Q_{34}	Q_{23}		
F_1	1	−1	0	0	0		
F_2	0	1	−1	0	−1		
F_3	0	0	0	−1	1		
F_4	52.571429	50.571429	10.428571	0	0		
F_5	0	0	−10.42857	28.892857	15.553571		

	Inverse matrix calculation						Solution
	0.5288858	0.0756944	0.0492059	0.0089614	0.001703	$\Delta Q_{14} =$	−3.12
	−0.471114	0.0756944	0.0492059	0.0089614	0.001703	$\Delta Q_{12} =$	−3.12
	−0.381583	−0.748648	−0.486667	0.0072584	−0.016844	$\Delta Q_{24} =$	−11.14
	−0.089532	−0.175657	−0.464127	0.001703	0.0185469	$\Delta Q_{34} =$	8.02
	−0.089532	−0.175657	0.5358725	0.001703	0.0185469	$\Delta Q_{23} =$	8.02

Branch	14	12	24	34	23
R_{ij}	0.4	0.3	0.8	0.2	0.1
$Q_i + \Delta Q_i$	−68.83	81.17	−4.62	−64.21	85.79

...

Iteration 4

Branch	14	12	24	34	23
R_{ij}	0.4	0.3	0.8	0.2	0.1
$Q_i + \Delta Q_i$	−69.30	80.70	−6.40	−62.90	87.10

Iteration 5

Branch	14	12	24	34	23
R_{ij}	0.4	0.3	0.8	0.2	0.1
$Q_i + \Delta Q_i$	−69.30	80.70	−6.40	−62.90	87.10

(b) Solution by substitution of variables.

The initial approach is the same as outlined in the previous section. However, since we have two meshes we can simplify matters by select two arbitrary variables (in this case Q_{12}, Q_a) and allow all other variables to depend on these, such that:

Node	Equation	Substituted variable
N_1	$Q_a + Q_{14} = Q_{12}$	$Q_{14} = Q_{12} - Q_a$
N_3	$Q_{23} = Q_a + Q_{34}$	$Q_{34} = Q_{23} - Q_a$
N_2	$Q_{12} = Q_{24} + Q_{23}$	$Q_{24} = Q_{12} - Q_{23}$

Then, substituting for the dependent variables Q_{14}, Q_{34} and Q_{24} in equations F_4 and F_5 above, we obtain the following system of two equations with two unknowns:

$$F_4 = R_{12}\,|Q_{12}|\,Q_{12} + R_{24}\,|Q_{12} - Q_{23}|\,(Q_{12} - Q_{23}) + R_{14}\,|Q_{12} - Q_a|\,(Q_{12} - Q_a)$$
$$F_5 = R_{23}\,|Q_{23}|\,Q_{23} + R_{34}\,|Q_{23} - Q_a|\,(Q_{23} - Q_a) - R_{24}\,|Q_{12} - Q_{23}|\,(Q_{12} - Q_{23})$$

Following the same procedure as for the previous section, we have:
Jacobian of the system

Function	Q_{12}	Q_{23}
F_4	$2\,R_{12}\,|Q_{12}| + 2\,R_{24}\,|Q_{12} - Q_{23}| + 2\,R_{14}$ $|Q_{12} - Q_a|$	$2\,R_{24}\,|Q_{12} - Q_{23}| \cdot (-1)$
F_5	$-2 \cdot R_{24} \cdot |Q_{12} - Q_{23}|$	$2\,R_{23}\,|Q_{23}| + 2\,R_{34}\,|Q_{23} - Q_a| - 2\,R_{24}$ $|Q_{12} - Q_{23}| \cdot (-1)$

Initial resistances and airflow rates

R_{14}	R_{12}	R_{24}	R_{34}	R_{23}	Q_{12}	Q_{23}	Q_a
0.4	0.3	0.8	0.2	0.1	1	1	150

Note that, once again although the seed airflow rates, $Q_{12} = 1$ and $Q_{23} = 1$ have been arbitrarily chosen they do potentially satisfy equilibrium conditions at the nodes. Thus, we have:

Functions matrix $|F_i|$		Jacobian of the system $|J|$				$\Delta Q_i = -J^{-1}\,F$	
Function	Calculation		Q_{12}	Q_{23}	Inverse matrix $|J| =$ $|J|^{-1}$	ΔQ_{ij}	Value
$F_x =$	−8880.1	dF_x/dQ_i	119.8	0	0.0083472	0	ΔQ_{12}	74.124
$F_y =$	−4440.1	dF_y/dQ_i	0	59.8	0	0.0	ΔQ_{23}	74.249
R_{14}	R_{12}	R_{24}	R_{34}	R_{23}	Q_{12}	Q_{23}	Q_a	
0.4	0.3	0.8	0.2	0.1	**75.124**	**75.249**	150	

Solutions for iteration 1

Q_{14}	Q_{12}	Q_{24}	Q_{34}	Q_{23}
−74.876	75.124	−0.125	−74.751	75.249

Functions matrix $|F_i|$		Jacobian of the system $|J|$					$\Delta Q_i = -J^{-1} F$	
Function	Calculation		Q_{12}	Q_{23}	Inverse matrix $|J| = |J|^{-1}$		ΔQ_{ij}	Value
$F_x =$	−549.5	dF_x/dQ_i	105.17479	−0.199664	0.0095081	4.205E−05	ΔQ_{12}	5.247
$F_y =$	−551.28138	dF_y/dQ_i	−0.199664	45.149831	4.205E−05	0.0	ΔQ_{23}	12.233
R_{14}	R_{12}	R_{24}	R_{34}	R_{23}	Q_{12}	Q_{23}	Q_a	
0.4	0.3	0.8	0.2	0.1	**80.372**	**87.482**	**150**	

Solutions for iteration 2

Q_{14}	Q_{12}	Q_{24}	Q_{34}	Q_{23}
−69.628	80.372	−7.111	−62.518	87.482

Matrix of functions $|F_i|$		Jacobian of the system $|J|$					$\Delta Q_i = -J^{-1} F$	
Function	Calculation		Q_{12}	Q_{23}	Inverse matrix $|J| = |J|^{-1}$		ΔQ_{ij}	Value
$F_x =$	−41.8	dF_x/dQ_i	115.30261	−11.37697	0.0088574	0.0018703	ΔQ_{12}	0.325
$F_y =$	24.076097	dF_y/dQ_i	−11.37697	53.880492	0.0018703	0.0	ΔQ_{23}	−0.378
R_{14}	R_{12}	R_{24}	R_{34}	R_{23}	Q_{12}	Q_{23}	Q_a	
0.4	0.3	0.8	0.2	0.1	**80.697**	**87.104**	150	

Solution for iteration 3

Q_{14}	Q_{12}	Q_{24}	Q_{34}	Q_{23}
−69.303	80.697	−6.407	−62.896	87.104

Matrix of functions $|F_i|$		Jacobian of the system $|J|$					$\Delta Q_i = -J^{-1} F$	
Function	Calculation		Q_{12}	Q_{23}	Inverse matrix $|J| = |J|^{-1}$		ΔQ_{ij}	Value
$F_x =$	−0.4	dF_x/dQ_i	114.11222	−10.25162	0.0089188	0.0017307	ΔQ_{12}	0.003
$F_y =$	0.3814563	dF_y/dQ_i	−10.25162	52.830772	0.0017307	0.0	ΔQ_{23}	−0.007
R_{14}	R_{12}	R_{24}	R_{34}	R_{23}	Q_{12}	Q_{23}	Q_a	
0.4	0.3	0.8	0.2	0.1	**80.700**	**87.098**	150	

Solutions for iteration 4

Q_{14}	Q_{12}	Q_{24}	Q_{34}	Q_{23}
−69.300	80.700	−6.398	−62.902	87.098

Matrix of functions $|F_i|$		Jacobian of the system $|J|$					$\Delta Q_i = -J^{-1}F$	
Function	Calculation		Q_{12}	Q_{23}	Inverse matrix $|J| = |J|^{-1}$		ΔQ_{ij}	Value
$F_x =$	0.0	dF_x/dQ_i	114.09626	-10.23624	0.0089196	0.0017287	ΔQ_{12}	0.000
$F_y =$	6.945×10^{-5}	dF_y/dQ_i	-10.23624	52.816727	0.0017287	0.0	ΔQ_{23}	0.000
R_{14}	R_{12}	R_{24}	R_{34}	R_{23}	Q_{12}	Q_{23}	Q_a	
0.4	0.3	0.8	0.2	0.1	**80.700**	**87.098**	150	

Solutions for iteration 5				
Q_{14}	Q_{12}	Q_{24}	Q_{34}	Q_{23}
-69.300	80.700	-6.398	-62.902	87.098

It can be observed that the values of ΔQ_{ij} produced in the 4th iteration are already very close to zero and that by the following iteration they are in fact zero.

4.11   Comparison of the Different Methods

Basha and Kassab (1996) make a general comparison of numerical methods applied to water distribution networks. Maleki and Mozaffari (2016) make a particular study of these methods as applied to mine ventilation networks. The main considerations arising from both studies are set out below:

- Mesh-based methods (Hardy–Cross and Newton–Raphson) require an initial seed value, Q_0, that satisfies equilibrium conditions at the nodes.
- In terms of the initial seed values, the Wood–Charles (linear) method does not require the establishment of a specific initial seed value.
- The success of the Newton–Raphson method is highly dependent on the initial seed value, so the closer the seed value is to an actual solution, the more rapidly convergence is achieved.
- In mining airflow networks, the possibility of selecting an incorrect seed value for the Newton–Raphson method is lower than in gas and water networks. This is because mining networks tend to be less extended.
- In principle, the Wood–Charles (linear) method converges faster than the others; however, it uses larger matrices and sometimes oscillates around the exact solution, thus the advantages of its use are not clear.
- Convergence of the Wood–Charles method is more problematic when the network is very large.

References

Atkinson, J. J. (1862). Gases met with in coal mines, and the general principles of ventilation. *Transactions of the Manchester Geological Society, III*, 218.

Basha, H. A., & Kassab, B. G. (1996). Analysis of water distribution systems using a perturbation method. *Applied Mathematical Modelling, 20*(4), 290–297.

Crane Company. (2013). *Flow of fluids through valves, fittings, and pipe* (No. 410). Engineering Division. Crane Company.

Cross, H. (1936). *Analysis of flow in networks of conduits or conductors*. University of Illinois at Urbana Champaign, College of Engineering. Engineering Experiment Station.

Carrasco, J., Alarcón, D., Albuerne, J., Fernández-Bustillo, E., Fernández, E., Gracía, L., Madera, J. (2011). Manual de ventilación de minas y obras subterráneas. Madrid: Aitemin Centro Tecnológico.

de la Vergne, J. (2008). Ventilation and air conditioning. In *Hard Rock Miner's handbook*, 5th ed. Alberta, Canada: Stantec Consulting Ltd.

Hartman, H. L., Mutmansky, T. M., Ramani, R. V., & Wang, W. J. (1997). *Mine ventilation and air conditioning*, 3rd ed. Wiley Publishing Company.

Maleki, B., & Mozaffari, E. (2016). A comparative study of the iterative numerical methods used in mine ventilation networks. *International Journal of Advanced Computer Science and Applications, 7*(6), 356–362.

McElroy, G. E. (1935). *Engineering factors in the ventilation of metal mines*. U.S. Department of the Interior, Bureau of Mines, Bulletin Number 385.

McPherson, M. J. (1993). *Subsurface ventilation and environmental engineering*. Chapman & Hall.

Montecinos, C., & Wallace, K. (2010). Equivalent roughness for pressure drop calculations in mine ventilation. In *Proceedings of 13th US/North American Mine Ventilation Symposium* (pp. 225–230).

Ostermann, W. (1960). *Mecánica aplicada al laboreo de minas*, 6th ed. (p. 605). Barcelona: Ediciones Omega.

Prosser, B. S., & Wallace, K. G. (1999). Practical values of friction factors. In *Proceedings of 8th US. Mine Ventilation Symposium* (pp. 691–696).

Simode, E. (1976). *Valeurs practiques des résistances*. Document SIM N3. Industrie Minérale. Mine 2–76.

Tuck, M. A. (2011). Mine ventilation. In P. Darling (Ed.), *SME mining engineering handbook* (Vol. 2). Society for Mining, Metallurgy & Exploration (SME).

Wood, D. J., & Charles, C. O. (1972). Hydraulic network analysis using linear theory. *Journal of the Hydraulics Division, 98*(7), 1157–1170.

Bibliography

Bise, C. J. (2003). *Mining engineering analysis*. Society for Mining, Metallurgy & Exploration (SME).

Conde, C., & Winter, G. (1990). Métodos y algoritmos básicos del álgebra numérica. Madrid: Reverté.

Maleki, B., & Mozaffari, E. (2016). A comparative study of the iterative numerical methods used in mine ventilation networks. *International Journal of Advanced Computer Science and Applications, 7*(6), 356–362.

Sereshki, F., Saffari, A., & Elahi, E. (2016). Comparison of mathematical approximation methods for mine ventilation network analysis. *International Journal of Mining Science, 2*(1), 1–14.

Wang, Y. J. (1982). Ventilation Network Theory. In H. L. Hartman (Ed.), *Mine ventilation and air conditioning* (2 ed., pp. 167–195). NY: Wiley-Interscience.

Wang, Y. J., & Hartman, H. L. (1967). Computer solution of three-dimensional mine ventilation networks with multiple fans and natural ventilation. *International Journal of Rock Mechanics and Mining Sciences, 4*(2), 129–154.

Chapter 5
Main Ventilation

5.1 Introduction

Main ventilation, also termed *primary ventilation*, takes air from the outside atmosphere and distributes it to through the whole mine. It differs from secondary ventilation, also known as auxiliary or ancillary ventilation, which will be discussed later in Chap. 8, in that secondary ventilation takes air from the main ventilation circuit conducing it to specific areas of the mine, mainly developing zones such as faces and stopes.

Main ventilation currents can be either natural or created by powerful fans. The energy used in the main ventilation is so high that it can account for 40% of the electrical energy consumed in the mine. For this reason, ventilation calculations are of the utmost importance for the economy of the mine. Since the mine is a continuously evolving system, so is the main ventilation, which must face new challenges every day to adapt to changing environmental and circuit resistance conditions. In this chapter, we deal with the main characteristics and calculations concerning primary ventilation systems.

5.2 Mine Airflow Requirements

Among the main tasks of main ventilation we find:

- Bringing pure air in sufficient quantity to where it is required.
- Diluting, extracting or displacing toxic, asphyxiating and flammable gases.
- Diluting, extracting or displacing dust.
- Reducing or increasing as appropriate, the temperature and humidity of the mining environment.

C. Sierra, *Mine Ventilation*, https://doi.org/10.1007/978-3-030-49803-0_5

- Reducing the transmission of infectious diseases (flu, COVID-19, tuberculosis, etc.) among workers, when necessary.[1]

Table 5.1 Production factor values (α) as a function of the operating method. Taken from Howes (1998)

Exploitation method	$\alpha \left(\frac{m^3}{s \frac{Mt}{year}} \right)$
Block caving	50
Rooms and pillars	75
Sublevel caving	120
Open stoping	
• Large $\left(> 5 \frac{Mt}{año} \right)$	160
• Small $\left(> 5 \frac{Mt}{año} \right)$	240
Cut and fill	
• Mechanized	320
• Manual tools	400

The quantity needed to ventilate the whole mine, which must comply the above-mentioned tasks, can be approximated depending on the average mine production and the mining method (Eq. 5.1) (Howes 1998):

$$Q = \alpha t + \beta \tag{5.1}$$

where

- Q: Airflow rate ($m^3\ s^{-1}$).
- t: Average annual production (Mt año^{-1}).
- α: Production factor $\left(\frac{m^3}{s \frac{Mt}{year}} \right)$. The factor α can be obtained from Table 5.1. From this, the first idea of the precise air quantities for each operating method can be obtained.
- β: Airflow required to ventilate the mining infrastructure ($m^3\ s^{-1}$). This factor depends fundamentally on how the interior haulage and the primary crushing are carried out. Some references for this parameter are:

  - Transport by inclined drift without underground primary fragmentation: 50 $m^3\ s^{-1}$.
  - Skip transport and underground primary fragmentation: 100 $m^3\ s^{-1}$.

The above values increase up to 50% with the presence of conveyor belts, panzers and transfer systems.

---

[1]For instance, silicosis is connected to a higher prevalence of tuberculosis. An example is the transmission of tuberculosis among South African miners (Hermanus 2007).

5.3 Natural Ventilation

Natural ventilation or *natural draught*[2] is the phenomenon by which the air moves without the help of any machinery providing an extrinsic drive. This movement is based mostly on the heating owing to the geothermal gradient that the gases undergo when crossing the mine, which results in a pressure difference between downcast and upcast shafts. Other factors, such as surface winds, air self-compression, heat released by processes of oxidation, hydration and solution/dissociation, and waterfalls –mainly present in the shafts–, can also be added to the above.

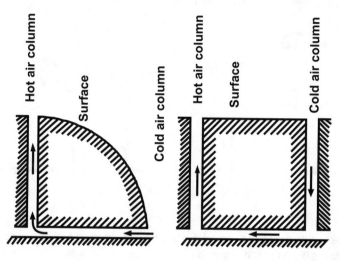

Fig. 5.1 Behavior patterns of the natural ventilation of a mine based on the temperature difference at the mouths of the ventilation shafts

Ventilation shafts and adits which communicate with the surface at different topographic levels favour natural draught (Fig. 5.1). Therefore, if the aim is to benefit air circulation, the mouths of the downcast or intake shaft, and upcast or return shafts, should be constructed at different heights. In the past, if this was not possible, it was common to build high prismatic or cylindrical chimneys well insulated by brick walls from the outside temperature and the wind. These chimneys generated a certain thermal contrast which helped the impulsion.

At certain times of the year, natural ventilation may be interrupted since the temperature difference between shafts is not sufficient to generate airflow. Similarly, the airflow direction is usually not the same throughout the year. This is so because sometimes the air inside the mine may be hotter than the air outside, or vice versa. This is a consequence of seasonal variations between the cold and hot months. Airflow inversions can even occur due to temperature variations between day and night. For

---

[2]*Natural draught flow* and *natural draught pressures*, respectively, refer to the air quantity and the pressure differences caused by natural ventilation.

all this reasons, natural ventilation sometimes assists the main fan and others flows against it.

The most classical approach for the calculation of natural ventilation is to consider the difference in density between the downcast and upcast air currents. However, in reality, natural ventilation is due to a transformation of heat into mechanical energy and is, therefore, a thermodynamic problem. Both approaches are taken into account in this chapter.

Exercise 5.1 Indicate the direction of the air inside the mine for the two conditions in the figure: (a) Outdoor temperature is lower than inside the mine, (b) Outdoor temperature is higher than inside the mine.

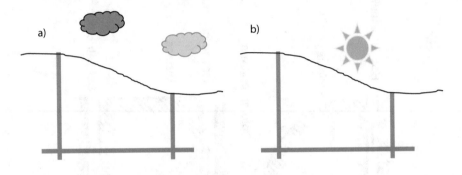

Solution

The qualitative explanation is based on the establishing of a hydrostatic equilibrium of pressures at the height of B and B′.

In the case of Fig. (a), when it is cold outside, the air column BC on the small shaft weighs more than its equivalent in the deep shaft, which is hotter. For this reason, the air circulates from the short to the long shaft.

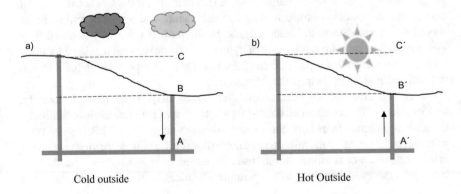

Cold outside   Hot Outside

Conversely, in the case of Fig. (b), the air column in the small shaft (B′C′), is not capable of compensating the analogous column in on the deep shaft. Therefore, air circulates from the long shaft to the short shaft.

5.3.1 Calculation by the Static Method

It is a simplified calculation method. The system in Fig. 5.2, consists of two vertical

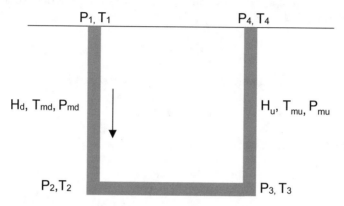

Fig. 5.2 Ventilation scheme for a simplified mine composed of two shafts of equal depth and a gallery

shafts connected by a horizontal gallery.

The following parameters are defined in it, namely:

- T_1: Dry-bulb temperature at the inlet of the downcast (K),
- T_2: Dry-bulb temperature at the bottom of the downcast shaft (K),
- T_3: Dry-bulb temperature at the bottom of the upcast (K),
- T_4: Exhaled dry-bulb temperature (K),
- T_{md}: Average dry-bulb temperature in the downcast shaft (K), and
- T_{mu}: Average dry-bulb temperature in the upcast shaft (K).

The pressure of the air column at the bottom of the downcast and upcast shafts (points 2 and 3) is:

$$P_2 = \rho_{(1-2)}gh_d = \gamma_{(1-2)}h_d; \quad P_3 = \rho_{(3-4)}gh_u = \gamma_{(3-4)}h_u$$

So if $P_2 > P_3$, the air flows from 2 to 3.

The Natural Ventilation Pressure (NVP or P_n) must be (Eq. 5.2):

$$P_n = P_2 - P_3 \qquad\qquad (5.2)$$

So that (Eq. 5.3):

$$P_n = \gamma_{(1-2)}h_d - \gamma_{(3-4)}h_u \qquad\qquad (5.3)$$

Note that specific weights are not constant, that is, they depend on other variables. In order to take this aspect into account, the universal equation of perfect gases can be used: $P V = n R_{mole} T$. This equation can be transformed to operate with mass instead of moles leading to:

$$P V = m R_{mass} T$$

Moreover, in terms of weight, the equation can be written as:

$$P V = W R_w T$$

Rearranging the equation:

$$\frac{W}{V} = \frac{P}{R_w T}$$

Since $\frac{W}{V} = \gamma$ (specific weight), replacing γ in the equation we have:

$$P_2 - P_3 = h_d \frac{P_{md}}{R_{wd} T_{md}} - h_u \frac{P_{mu}}{R_{wu} T_{mu}}$$

where

- P_{md}: Average pressure in the downcast shaft (Pa),
- P_{mu}: Average pressure in the upcast shaft (Pa),
- R_{wd}: Constant per unit mass of gases in the downcast shaft $\left(\frac{J}{K\,kg}\right)$, and
- R_{wu}: Constant per unit mass of the gases in the upcast shaft $\left(\frac{J}{K\,kg}\right)$.

At this point, a certain humidity can be assumed for the intake depending on the climatology of the place. In the case of the return shaft, a humidity close to 100% can be admissible. With both parameters set, it is possible to obtain accurate values for R_{we} and R_{ws} from air data tables. Despite this fact, simplifications can still be made, specifically:

- To use the same value of the constant R for both the intake and return shafts.

- To use the dry air values for R. Therefore, as R_{mass} for dry air is approximately[3] 287 $\frac{J}{K\,kg}$. Referred in weight $R_w = 29.29$ $\frac{J}{N\,K}$.
- If both shafts have the same depth (h), the P_{md} is approximately equal to the P_{mu}, and equal to the average pressure on the shafts (P_m).
- Finally, P_m is about 101 kPa and it is difficult to measure, so in some texts, it is simplified by the value of atmospheric pressure.

So P_n can be calculated as (e.g. Hartman et al. 1997, p. 298):

$$P_n = \frac{h\,P_m}{R}\left(\frac{1}{T_{md}} - \frac{1}{T_{mu}}\right)$$

where substituting, we have (Eq. 5.4):

$$P_n = 0.03415\left[\frac{N\,K}{J}\right] h\,P_m\left(\frac{1}{T_{md}} - \frac{1}{T_{mu}}\right) \tag{5.4}$$

The temperature close to the mouth of the downcast shaft could be quite similar to that of the air outside, thus being instable. Similarly, the temperature of the upcast shaft may resemble largely that of the interior of the mine. So T_{md} and T_{mu} are normally obtained as the mean between the temperatures that exist at the bottom of the shafts and 35 m below their respective mouths (neutral zone).

Equation 5.4 assumes that downcast and upcast shafts have an equal depth. Some considerations must then be made in the event that the mouths of the two shafts are at different heights in the topographic surface. Thus, if air enters through the longest shaft (Fig. 5.3a), the *effective height* is only the depth of the short shaft (h') as the

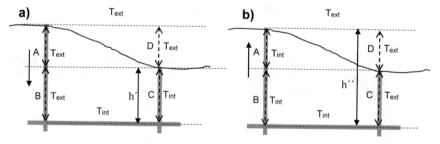

Fig. 5.3 a Air inlet through the long shaft; **b** air inlet through the short shaft

column of air located above it should have approximately the same temperature as the twin air column in the long shaft and both are compensated. Similarly, if air enters through the short shaft, the temperature contrast between columns $A + B$ and $C + D$ is total, and h'' should be used as the effective height (Fig. 5.3b). The exact

---

[3]For water vapour 461 $\frac{J}{kg\,K}$.

solution to this problem is to calculate the air density for the different columns (A, B, C and D), and with them, to establish the equilibrium of pressures.

If the difference of the inverse of temperatures from Eq. 5.4 is operated obtaining the common denominator, we have:

$$P_n = 0.03415 \left[\frac{N\,K}{J}\right] h\ 101,300\ [Pa] \left(\frac{T_{mu} - T_{md}}{T_{md}T_{mu}}\right)$$

Then $T_{md} \cdot T_{mu}$ can be replaced by its possible values, which range[4] within 273–293 K (Eq. 5.5), therefore:

$$P_n = 0.04 \left[\frac{Pa}{C\,m}\right](T_{mu} - T_{md})h \qquad\qquad (5.5)$$

For this reason, McElroy (1935) and Borisov et al. (1976), indicated that 44 Pa of natural ventilation pressure are generated for every 10 °C difference between the average temperature of the shafts for every 100 m of effective height. Some NVP values depending on depth and temperature are given in Table 5.2.

Table 5.2 Natural ventilation pressure (P_n) depending on the average depth of the shafts (h_m) and their average temperature difference (ΔT)

		ΔT (°C)			
		5	10	20	30
h_m (m)	200	50	100	190	280
	500	120	240	500	710
	1000	240	480	950	1400

Question 5.1 Pressures have been measured at each of the points indicated in the mine in the figure. Determine an expression as the sum of the given pressures for the calculation of the NVP of the mine.

---

[4]Note that other temperatures within the range of possible values do not significantly vary the result.

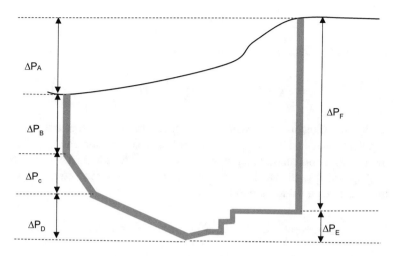

Answer

As can be seen, the NVP is the pressure difference at the bottom of the downcast and upcast shafts. In this case, both come together at the same depth, thus:

$$P_n = \Delta P_A + \Delta P_B + \Delta P_C + \Delta P_D - (\Delta P_E + \Delta P_F)$$

As indicated in the previous question, air can circulate in different directions depending on the difference in temperature between inside and outside the mine, thus changing the sign of NVP.

Exercise 5.2 The ventilation of a mine takes place through two vertical shafts of 600 and 400 m. The conditions in them are monitored automatically, recording average temperatures of 2 °C in the mouth and 9 °C at the bottom for the long shaft, and 17 °C at the mouth and 12 °C at the bottom for the short shaft. It is also known that the pressure at the average depth of both shafts is 108.475 kPa. You are asked to:

(a) Estimate the NVP of the system.
(b) Comment on the direction of the airflow and the temperature measurements provided in the problem statement.

Solution

(a) Firstly, average pressure must be estimated for the midpoint of the air columns in the shafts. This value could be slightly higher than the atmospheric pressure. As this value is supplied, 108.475 kPa is used directly.
    Secondly, the average temperatures of the downcast and upcast shaft are calculated:

$$T_{md} = (5.5 + 273)\,\text{K} = 278.5\,\text{K}$$

$$T_{mu} = (14.5 + 273)\,\text{K} = 287.5\,\text{K}$$

These values can be substituted in Eq. 5.4:

$$P_n = P_2 - P_3 = 0.03415 \left[\frac{\text{N K}}{\text{J}} \right] h\,P_m \left(\frac{1}{T_{md}} - \frac{1}{T_{mu}} \right)$$

An airflow direction can easily be assigned considering that the air outside is colder than inside the shafts. In such a case, the air enters the system through the short shaft, then it is more appropriate to assume a height of 600 m corresponding to the depth of the long shaft. However, it is a common practice to use the mean values, therefore:

$$P_n = 0.03415 \left[\frac{\text{N K}}{\text{J}} \right] \left(\frac{400 + 600}{2} \right) \text{m} \cdot 108{,}475\,\text{Pa} \cdot \left(\frac{1}{278.5\,\text{K}} - \frac{1}{287.5\,\text{K}} \right)$$

$$P_n = 208.2\,\text{Pa}$$

Checking by the approximated formula:

$$P_n = 0.044 \left[\frac{\text{Pa}}{\text{C\,m}} \right] (T_{mu} - T_{md}) \left(\frac{h_d + h_u}{2} \right)$$

We have:

$$P_n = 0.044 \left[\frac{\text{Pa}}{\text{K\,m}} \right] (287.5 - 278.5)\,\text{K} \left(\frac{600 + 400}{2} \right) \text{m} = 198\,\text{Pa}$$

(b)   As a conclusion, it can be deduced that the NVP in this mine is low. This is due to the rapid increase in temperature throughout the long shaft, the low increase in temperature towards the outlet in the short shaft as well as the short effective height of the latter. This first approach points to an unstable equilibrium which makes inversions frequent.[5]

As the direction of the airflow is given by the increase of temperatures in the circuit (from colder to hotter), there would be no discussion about its direction. However, there are grounds for suspecting that some measurements could be incorrect. Thus, temperatures at the bottom of the long shaft and at the mouth of the short shaft could be lower and higher respectively than those supplied by the problem statement.

---

[5]In cold and shallow mines (<400 m), natural ventilation is unreliable. In other words, it does not guarantee that there is an adequate NVP, nor that it takes place in the required direction.

5.3.2 Calculation of the Logarithmic Formula Method

Despite the simplifications in the previous section, pressure does increase with depth, and its increase takes place in a nonlinear manner. In order to consider this fact, the following expression was developed. A height differential (dL), of a dry air column of height h, increases the hydrostatic pressure at its base (A) by a value dP. Therefore, establishing the balance of forces, we have:

$$A\,dP = \gamma\,A\,dL$$

where γ is the specific weight of air. This value can be obtained from the general gas equation, therefore:

$$A\,dP = \frac{P}{R_w T}A\,dL$$

where

R: Constant of gases (per unit mass) $\left(\frac{J}{K\,kg}\right)$.
   Simplified:

$$dP = \frac{P}{R\,T}dL$$

Integrating between the pressure at the top of the downcast shaft (P_1) and the pressure at the bottom (P_2), for a depth of the shaft (h), and considering the temperature as an integration constant (T_{md}), we have:

$$\int_{P_1}^{P_2}\frac{dP}{P} = \int_{0}^{h_d}\frac{dL}{R\,T_{md}}$$

Therefore, the pressure at the base of the downcast shaft is (Eq. 5.6):

$$\log P_2 = \log P_1 + 0.03415\frac{h_d}{T_{md}} \tag{5.6}$$

In a similar way, the pressure at the base of the upcast shaft can be obtained (Eq. 5.7):

$$\log P_3 = \log P_1 + 0.03415\frac{h_u}{T_{mu}} \tag{5.7}$$

where

- h_u: Depth of upcast shaft (m), and
- T_{mu}: Average temperature of the upcast shaft air (K).

Therefore, the Natural Ventilation Pressure (P_n) is (Eq. 5.8):

$$P_n = P_2 - P_3 \qquad\qquad (5.8)$$

Note that the same pressure (P_1) has been assumed outside both shafts. Note also that these calculations have been simplified to a static problem while the problem is actually dynamic due to air movement.

Exercise 5.3 Calculate natural ventilation pressure for the mine in the figure.

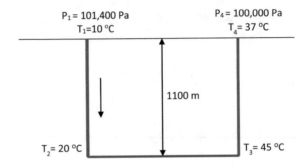

Solution

$$\log P_2 = \log P_1 + 0.03415\frac{h_d}{T_{md}} = \log 101,300 + 0.03415\frac{1100}{\frac{10+20}{2} + 273} = 5.1360$$

$$\log P_3 = \log P_4 + 0.03415\frac{h_u}{T_{mu}} = \log 100,000 + 0.03415\frac{1100}{\frac{37+45}{2} + 273} = 5.1196$$

$$P_n = P_2 - P_3 = (1.3677 - 1.3170) \times 10^5 = 5068.57 \; Pa$$

The value obtained corresponds to a very high NVP.

5.3.3 Calculation by the Thermodynamic Method

The air inside the mine is heated, resulting in an addition of energy which causes potential and kinetic energy changes, as well as friction losses in the air mass. If this energy refers to the unit of air mass, it is called *Natural Ventilation Energy (NVE)*.

The NVE can be simply obtained if the values of the pressure and the specific volume of the airflow are available at the following locations (Fig. 5.4, left), namely: topographic surface (*A*), bottom of the downcast shaft (*B*), exit from the stopes (*C*), exit to the outside of the upcast shaft (*D*).

The above data can be represented on Cartesian axes of absolute pressure versus specific volume ($P - V_e$). In this case, if it is assumed that there is no heat exchange

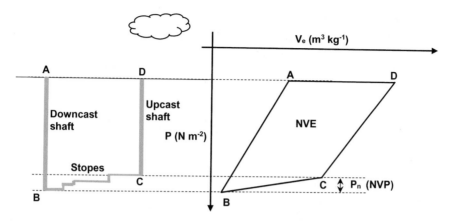

Fig. 5.4 Pressure vs. specific volumen (P–V_e) diagram applied to the simplified schematic of a mine

with the walls of the shafts, the adiabatics AB and CD are formed. In the same way, the isobar corresponding to the external pressure DA is observed. The highest heat incorporation to the air mass (BC curve) takes place in mining stopes. However, in well developed mines it is possible to assume that the heat flow to the air mass is zero, giving rise to an adiabatic process, which is also isothermal ($T_B = T_C$).

The Natural Ventilation Pressure (NVP, P_n), from a static point of view, is the PB-PC difference. The ABCD area corresponds to the NVE (Fig. 5.4, right). If the NVE is multiplied by the density of the air (ρ), the work per unit volume is obtained, or in other words, the NVP (Eq. 5.9):

$$P_n = \text{NVE}\,\rho \qquad (5.9)$$

where

- NVE: Natural Ventilation Energy (J kg^{-1}), and
- ρ: Air Density (kg m^{-3}).

The determination of the pressure at each point can be made by means of an aneroid barometer, while the specific volume can be approximated by the expression (Eq. 5.10):

$$V_e = \frac{287.1\,T}{1000(P - P_v)} \qquad (5.10)$$

where

- T: Dry-bulb temperature (K),
- P: Barometric pressure (kPa), and
- P_v: Vapour pressure of water (kPa).

Bearing in mind the equations for an adiabatic system with self-compression of air (Eq. 2.9), Voropaev (1950) stated that: "The work performed by 1 kg of air in a natural ventilation system can be approximated with the area of a closed contour in a coordinate system depth-temperature (h-T), divided by the absolute temperature of its centroid[6]" (e.g. Novitzky 1962, p. 211) (Eq. 5.11):

$$P_n = \frac{S}{T_c}\rho \qquad\qquad (5.11)$$

where

- P_n: Natural ventilation pressure (NVP) (kp m^{-2}),
- S: Surface of the figure $\left(\frac{\text{kp m}}{\text{kg}}\text{K}\right)$,
- T_c: Centroid temperature (K), and
- ρ: Air density (usually taken as 1.25 kg m^{-3}).

The limitations of the above expression in the presence of a fan have been discussed by Lepikhov (1975).

Surface decomposition can be systematized by the shoelace formula. This procedure consists of defining a polygonal surface from a series of lines determined by pairs of points (x_{i-1}, y_{i-1}), (x_i, y_i) of a Cartesian coordinate system. In this nomenclature, i is the point number. It is important that the last point in the series coincides with the first for the polygonal to be closed. Also, in order to not complicate the method, the coordinates defining the polygonal should not produce line crossings.

The area of each element defined by coordinates is calculated as:

$$A_n = x_i y_{i-1} - x_{i-1} y_i$$

Then the total area of the closed polygonal is:

$$A = \frac{1}{2}\left|\sum A_i\right|$$

The centroid of any triangle is given by the following expression:

$$X_g = \frac{1}{3}(x_1 + x_2 + x_3)$$

$$Y_g = \frac{1}{3}(y_1 + y_2 + y_3)$$

Exercise 5.4 The figure represents the scheme of a simplified mine with sidehill shafts. The data corresponding to the coordinates of each point are included in the table. You are asked to:

---

[6]The centroid or barycenter is a geometric concept which defines the centre of symmetry of a geometric figure. For a body with uniform density, it coincides with the centre of masses.

(a)   Calculate the centre of gravity of the surface h-T by the shoelace formula.
(b)   Determine the NVP by the Voropaev method.

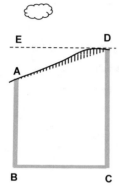

Point	Temp. T (°C)	Depth h (m)
A	6	50
B	10	175
C	16	175
D	15	0
E	5	0

Solution

(a)   First, the h-T diagram is constructed with the data from the statement. Thus, the diagram is broken down into geometric figures of a known centre of gravity:

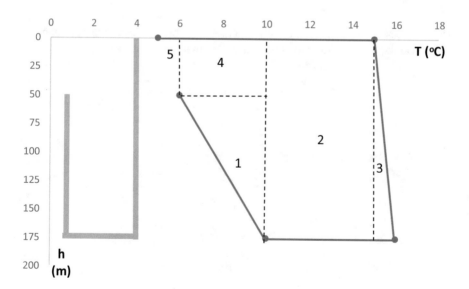

According to the above, the areas would be[7]:

_____

[7]Note that since temperature differences are to be measured, it makes no difference whether the graph is in °C or in K.

Figure					
	1	2	3	4	5
$S_i \left(\frac{Kp\,m}{kg} K \right)$	250	875	87.5	200	25

Then, the total surface $S_T = 1437.5 \frac{kp\,m}{kg} K$.
The coordinates of the centres of gravity of the individual surfaces are then:

Figure					
	1	2	3	4	5
T_i (K)	8.67	12.5	15.33	8	5.67

$$T_c = \frac{\sum_{i=1}^{5} S_i T_i}{S_T} = 11.26\,°C$$

Then the NVP is obtained as:

$$P_n = \frac{S_T}{T_c} \rho = \frac{1437.5 \frac{kp\,m}{kg} K}{(273 + 11.26)\,K} 1.25 \frac{kg}{m^3} = 6.3 \frac{kp}{m^2}$$

(b) The centre of gravity can be also calculated for the above case by the shoelace formula, e.g. for the decomposition of the figure as:

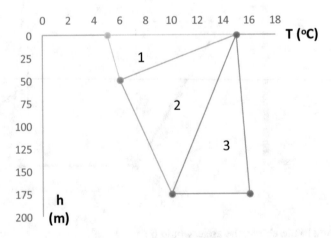

Triangle	T_i (°C)	h_i (m)	Centroid coordinates		$T_{i-1} \cdot h_i - T_i \cdot h_{i-1}$	Area of the triangle
			T_{cg}	h_{cg}		
1	5	0			–	
	15	0			0	
	6	50			750	
	5	0	8.67	16.67	−250	250
2	6	50			–	
	15	0			−750	
	10	175			2625	
	6	50	10.33	75.00	−550	662.5
3	15	0			–	
	10	175			2625	
	16	175			−1050	
	15	0	13.67	116,67	−2625	525
Weighted mean values			**11.26**	80.07	Total area	1437.5

Or, also:

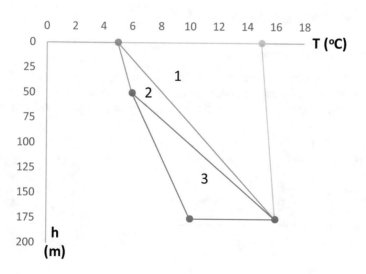

Triangle	T_i (°C)	h_i (m)	Centroid coordinates		$T_{i-1} \cdot h_i - T_i \cdot h_{i-1}$	Area of the triangle
			T_{cg}	h_{cg}		
1	5	0			–	

(continued)

(continued)

Triangle	T_i (°C)	h_i (m)	Centroid coordinates		$T_{i-1} \cdot h_i - T_i \cdot h_{i-1}$	Area of the triangle
			T_{cg}	h_{cg}		
	15	0			0	
	16	175			2625	
	5	0	12.00	58.33	−875	875
2	5	0			−	
	16	175			875	
	6	50			−250	
	5	0	9.00	75.00	−250	187.5
3	6	50			−	
	16	175			250	
	10	175			1050	
	6	50	10.67	133.33	−550	375
Weighted mean values			**11.26**	80.07	Total area	1437.5

5.3.4   Measuring Natural Ventilation

Method of the Ventilation Door

One of the most commonly used methods is to measure when the fan is switched off, the pressure difference on both sides of a closed ventilation door through which the entire mine airflow passes (Fig. 5.5a). This door is usually located at any point

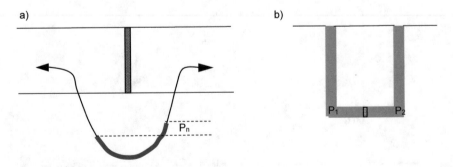

Fig. 5.5   Measure of natural ventilation with the door closed

between the downcast shaft and the upcast shaft. In this case, it is possible to measure the pressure difference between the bottom of both shafts (Fig. 5.5b) (Hatman et al. 1997, p. 302).

Turning the Main Fan Off and On

The natural draught can be estimated from measurements taken in the main fan drift with the main fan off and operating. Thus, when the fan is on and the surface airlock, also termed shaft cover is closed, the resistance is:

$$R = \frac{P_v + P_n}{Q_v^2}$$

where

- P_v: Pressure supplied by the fan, and
- Q_v: Flow rate with the fan running.

Afterwards, the main fan is turned off, the fan drift is closed with a damper, and the shaft cover is open. Then, we have:

$$R = \frac{P_n}{Q_n^2}$$

with:

- Q_n: Flow rate with the main fan off (flow rate supplied by natural ventilation).

  Therefore:

$$\frac{P_v + P_n}{Q_v^2} = \frac{P_n}{Q_n^2}$$

Finally (Eq. 5.12):

$$P_n = P_v \frac{Q_n^2}{Q_v^2 - Q_n^2} \qquad (5.12)$$

Exercise 5.5 The upcast shaft in a mine registers an airflow rate of 800 m³ min⁻¹ when the depression developed by the main fan is 6000 Pa. When the fan stops the flow in the upcast shaft, it reduces to 250 m³ min⁻¹. Determine the value of the NVP in this mine.

Solution

If the fan is on, we have:

$$R = \frac{P_v + P_n}{Q_v^2} = \frac{6000\,\text{Pa} + P_n}{\left(800\frac{\text{m}^3}{\text{min}}\right)^2}$$

When the fan is stopped:

$$R = \frac{P_n}{Q_n^2} = \frac{P_n}{\left(250\frac{m^3}{min}\right)^2}$$

Equalizing:

$$\frac{P_n}{\left(250\frac{m^3}{min}\right)^2} = \frac{6000\,Pa + P_n}{\left(800\frac{m^3}{min}\right)^2}$$

Therefore:

$$P_n = 6000\,Pa\frac{\left(250\frac{m^3}{min}\right)^2}{\left(800\frac{m^3}{min}\right)^2 - \left(250\frac{m^3}{min}\right)^2} = 649.35\,Pa$$

Operating the main fan at two different speeds

In this case, the resistance is the same for both speeds. Therefore, for the main fan drift we have:

$$R = \frac{P_{v_1} + P_n}{Q_1^2} = \frac{P_{v_2} + P_n}{Q_2^2}$$

where the subscripts 1 and 2 represent the two rotation speeds to which the motor fan is subjected.

So, in this case, the P_n will be (Eq. 5.13):

$$P_n = \frac{P_{v_1}Q_2^2 - P_{v_2}Q_1^2}{Q_1^2 - Q_2^2} \tag{5.13}$$

Exercise 5.6 The main fan in a mine moves 1000 m³ min⁻¹ of air operating at a pressure of 8000 Pa. When its rotational speed is reduced, the flow drops to 700 m³ min⁻¹ and the pressure developed is 3500 Pa. Determine the NVP of this mine.

Solution

For the first rotation velocity, we have:

$$R = \frac{P_{v_1} + P_n}{Q_1^2} = \frac{8000\,Pa + P_n}{\left(1000\frac{m^3}{min}\right)^2}$$

For the second velocity:

$$R = \frac{P_{v_2} + P_n}{Q_2^2} = \frac{3500\,\text{Pa} + P_n}{\left(700\,\frac{\text{m}^3}{\text{min}}\right)^2}$$

Equating:

$$\frac{8000\,\text{Pa} + P_n}{\left(1000\,\frac{\text{m}^3}{\text{min}}\right)^2} = \frac{3500\,\text{Pa} + P_n}{\left(700\,\frac{\text{m}^3}{\text{min}}\right)^2}$$

Finally:

$$P_n = 823.53\,\text{Pa}$$

5.4 Forced Ventilation

When air is supplied to a certain space using mechanical devices called fans, a forced ventilation system is being used. Depending on the location of the ventilation shaft with regard to the ore deposit, two types of forced ventilation are distinguished: *central* and *boundary*.

5.4.1 Central Ventilation

In this ventilation system, also referred to as *bidirectional*, downcast and upcast shafts are very close together and located in the centre of the mineral deposit (Fig. 5.6). It is characterized by the movement of air from the centre of the mine to the stopes, and to return through the ventilation galleries to the upcast, which, as has been already indicated, is located in the centre of the mineral deposit. In this case, the upcast shaft can be used for the extraction in parallel with the downcast shaft. It is a common ventilation system in medium-sized mines. Some of the main advantages and disadvantages of this system are set out below (Novitzky 1962, p. 254).

Advantages

- Grouped installations, and
- It is more convenient in the early stages of the mine.

Disadvantages

- This system increases the resistance of the circuit a lot, which is more evident as the mine increases.

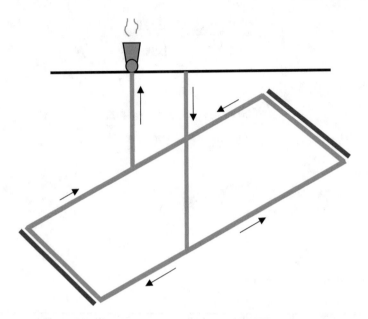

Fig. 5.6 Central ventilation. Modified from Novitzky (1962, p. 300)

- Short-circuits can occur as fresh and foul airways are very close to each other and sometimes they are almost parallel. Short-circuits usually occur near the shafts where the pressure is maximum.[8]
- An accident in a shaft, for example, a collapse can affect the other as both are very close to one another.
- The resistance of the mine can increase when the air has to cover the same distance twice (downcast shaft to stopes and stopes to upcast shaft).

5.4.2 Boundary Ventilation

In *boundary ventilation*, also known as *radial*, *side*, *unidirectional* or *diagonal ventilation*, the air goes from the downcast to the stopes and from the stopes to the upcast, which is located at the limits of the concession (Fig. 5.7). All this takes place without returning to the centre of the mine.

Advantages

- Shafts well separated, which means that an eventual accident in one does not affect the other.

---

[8]*Air short-circuits* can be understood as those roads of low or null resistance where the air flows with more facility unbalancing the systems and the desired circulations of the air in the mine. They can also be produced by the opening of doors which should normally be closed.

Fig. 5.7 Diagonal ventilation. Modified from Fritzsche (1965, p. 712)

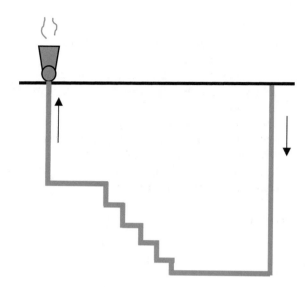

- In this system, several upcast shafts are possible, so the air can enter through one and leave diagonally through several of them. This fact results in higher security of the ventilation system.
- More stable ventilation.
- The same level can be divided and used for the air intake and return.
- Shorter circuit, and therefore, less resistant, thus the mine can be operated with less depression.
- Resistance circuit is most constant as it is independent on whether the work is in the central or border areas of the concession.

Disadvantages

- In mines where firedamp is present and requiring significant ventilation airflows, several upcasts have to be built, with the high costs that this entails. The existence of these shafts is not a complete disadvantage as, in the long run, they also facilitate the movement of personnel, materials and rescue equipment.

5.5 Airflow Generation Systems

5.5.1 Depression System

Generally, the most commonly used system is *depression ventilation*. This configuration has the following advantages and disadvantages (Tien 1978; Hartman et al. 1997, p. 526):

Advantages

- The haulage galleries are kept free of dust and gas, allowing workers to have a cleaner working atmosphere.
- It simplifies the action of the rescue teams if fires or explosions occur. This is because fresh air is in the access route, which makes it easier to instal equipment and materials.
- The system can be more energy efficient. For this purpose, gradual expansion outlets (*evasés*) should be used in the main fans which lower air discharge velocities and improve static efficiency.
- In a mine with firedamp, as this gas has a lower density than air, this system favours its evacuation to the exterior.

Disadvantages

- In winter, the temperatures can be very low in the mouth of the intake shaft making it difficult to work in them, and may even lead to equipment malfunctions due to freezing.
- It may be more difficult to detect a fire on the conveyor belt.
- Dust generated by transport in the downcast, as well as fires in these areas, can reach the stopes.
- Dirty air and dust pass through the fan and may affect its normal functioning.
- Firedamp escapes from the already abandoned areas of the mine can be drawn towards the centre of the mine with the subsequent problems.

5.5.2 Blower System

Blower ventilation causes air movement resulting in an overpressure in the area to be ventilated. It has the following advantages and disadvantages (Tien 1978; Hartman et al. 1997, p. 527):

Advantages

- While the fan is running, coal seams are under high pressure thus reducing the gases emanating from them. Although this fact is debatable, at least in the case of firedamp.
- The systems and areas through which air enters the mine are heated by the main fan, thus preventing their freezing.
- Fire outbreaks are more obvious.
- The fan receives clean air, which extends its useful life.
- These types of fans are cheaper as they have a smaller diffuser.

Disadvantages

- It may hinder mine rescue efforts by tending to push contaminating gases into escape routes.

- Shock pressure losses are higher in blowing than in depression systems.

Taking all this into account, the preferred system for the main ventilation is depression, assisted by several booster fans distributed inside the mine.

5.6 Direction of Transportation

The direction of air circulation is also strongly conditioned by the direction of transport inside the mine. According to this, two systems are distinguished: *anti-tropic* and *homotropic*. In the former, the movement of air and materials transported from the mine to the outside takes place in the opposite direction. Whereas, in the latter, the direction of the airflow and the materials transported outside the mine coincide.

In general, the homotropic is the preferred one, as it raises less dust and heat, and gases generated inside the mine are expelled more easily. Moreover, this system also allows for a better management of eventual fires that take place in the mine slope.

Another factor to take into account is the inclination of the stopes. *Ascensional ventilation* is the most common in inclined stopes (Fig. 5.8). This system is used to:

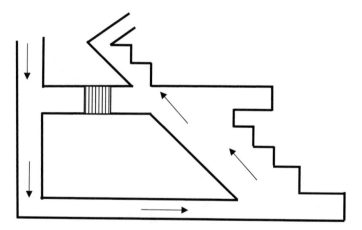

Fig. 5.8 Ascentional air circulation in a mine stope

- Take advantage of the ascensional force due to the heating of the air that takes place in them. This natural draught allows for some air circulation even in the event of the main fan stopping.
- Facilitate the evacuation of firedamp as it is lighter than air.
- Decrease the arrival of dust in the transport galleries.

For all these reasons, this system is compulsory in many regulations mainly in coal mining. In order to make the ventilation ascend, the downcast shaft must be at least as deep as the upcast shaft so as to prevent part of the ventilation from descending.

Descensional ventilation is also possible when a more compact system is needed. In this case, it is normal to also move the materials in the same direction to operate in a homotropic system.

Exercise 5.7 The image represents the ventilation systems of two metal mines. The following manometric pressures have been measured:

Scheme (a): $-1200, 0, -2100, -720, -2400, -1500$, (Pa)
Scheme (b): $-680, 0, 0, 0, -500, 450$ (Pa)

You are asked to:

(a)   Indicate which type of ventilation system each figure corresponds to.
(b)   Place the static and absolute pressures on the graph in their corresponding place.

Scheme (a)

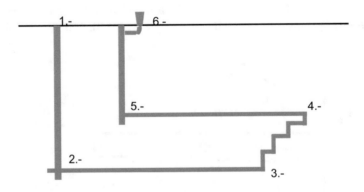

Scheme (b)

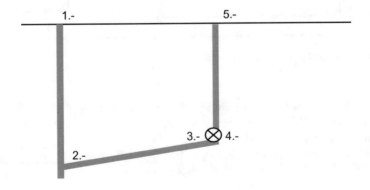

Solution

Scheme (a): Central ventilation. 1: 0, 2: −720, 3: −1200, 4: −1500, 5: −2100, 6: −2400 (Pa)

Scheme (b): Diagonal ventilation. 1: 0, 2: −500, 3: −680, 4: +450, 5: 0 (Pa)

In order to locate the absolute pressures add 101,300 Pa to all the previous values.

5.7 Ventilation on Demand

Ventilation on Demand (VOD) is a system by which ventilation can exactly adapt to real ventilation needs. This means (Dicks and Clausen 2017):

(a) Ventilating only the zones of the mine were ventilation is needed. This is equivalent, in most cases, to the active zones of the mine.

(b) If ventilation in a mine zone is necessary airflow requirements may change over time.

(c) Not all auxiliary systems are required simultaneously.

(d) Probably the original design criteria have been exceeded so changes are necessary. This is the opposite of the "set and forget" policy.

(e) Non-active periods usually require less ventilation.

Ventilation on Demand can be carried out at 5 different levels (Tran-Valade and Allen 2013), namely:

1. *User control*: It is based on the manual control of fans and regulators to be adapted to the ventilation requirements.

2. *Time of day scheduling*: Actions over fans and regulators are not carried out manually but on a pre-set schedule.

3. *Event-based*: Changes in the ventilation system correspond with certain activities and events, e.g. blastings or a mine fire.

4. *Tagging*: The airflow is distributed inside the mine depending on a tag and tracking system which provides real-time location of the personnel and vehicles.

5. *Environmental*: Making use of modern-day computer software to continuously monitor the concentration of gases (*Atmospheric Monitoring System, AMS*). For this system to work, it is essential to have: (a) gas and flow sensors, (b) personnel and machinery tracking devices indicating their location at any given time, (c) fan control systems (motors with variable-frequency drive), and (d) qualified personnel.

The airflow diminution consequence of VOD results in a significant energy consumption reduction (Jahir et al. n.d.). Some studies indicate that VOD can save up to 50% on ventilation costs (Wallace et al. 2015).

5.8   Ventilation and Exploitation Method

The fundamental factors to consider in the selection of an exploitation method are morphological, geotechnical and economic. In addition, as each method must be ventilated differently, ventilation may, in some cases, influence the selection of the method. Any mining engineering textbook in which mining methods are studied will provide a detailed account of the particulars of the ventilation system to be used for each of them. The ventilation of the most common mining methods is explained afterwards for illustrative purposes.

For example, in the case of *pillar-supported methods*, the most characteristic is that of *room and pillars*. This method generates large horizontal extensions, of complex ventilation in which diesel equipment is the most common. The method implies the installation of auxiliary fans in the faces, although it is still frequent to direct the air with line brattices. It shows two variants: the *bidirectional* or *in W* and the *unidirectional* or *in U* (McPherson 1993, p. 106). In the first one, the air enters through the centre of the exploited zone and exits through both sides (Fig. 5.9a). In

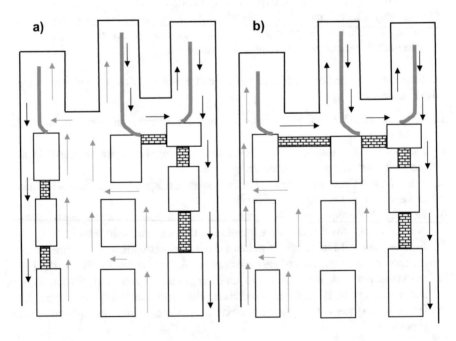

Fig. 5.9 Ventilation of active faces in the operation of rooms and pillars by means of deflectors: **a** bidirectional system (*W*); **b** unidirectional system (*U*)

the second, air enters from one side, is directed one after the other to all the faces, and returns through the other side (Fig. 5.9b). One of the main disadvantages of the *U*-system is that the air is loaded with toxic gases as it sweeps the faces, thereby decreasing its ability to renew the air. The *W* system, on the other hand, requires more line brattices to be executed and the pressure losses are greater.

Another pillar-supported method is the *sublevel stoping*. In this method, the ventilation is achieved by injecting air through the production level[9] where the gases from the *Load Haul Dump (LHD)* diesel chargers accumulate (Fig. 5.10). This air ascends

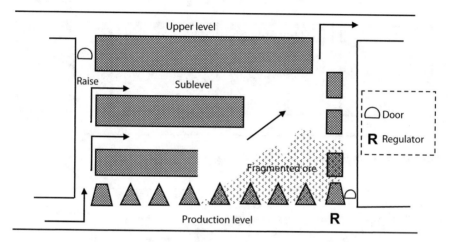

Fig. 5.10 Ventilation air circulation in the sublevel stoping method

mainly through the raises which connect to the upper level and are drifted to the different sublevels where the blasting fumes accumulate. From the sublevels, the air enters the stope and after sweeping, it finally reaches the upper level.

As far as *artificially supported methods* are concerned, cut and fill and shrinkage stoping[10] are some of the most characteristic.

Multiple configurations come under the category of *cut and fill*. In any case, its general layout is very similar to that of the sublevel stoping method. Thus, fresh air is injected from the transport level to the stopes via raises, and from there it reaches the upper level from where it is incorporated into the general mine ventilation (Fig. 5.11).

Regarding *shrinkage stoping*, the ventilation is also very similar to that of the previous methods. The stope is flanked by pillars, which, in turn, are surrounded by two raises: one for access and the other for evacuation. The air ascends from the transport level through the access raise and sweeps the stope (Fig. 5.12). The foul air comes out through the evacuation raise and is conducted towards the upper level where it is incorporated into the general circulation of the mine.

With regard to the *cave mining methods*, as is the case of *panel caving*, we will focus on its current variant: *block caving*, of analogue ventilation system. In order to ensure the significant amounts of precise ventilation flow at the production level,

---

[9]The general principle of ascensional ventilation for the stopes is followed.

[10]This method uses pillars or filled material as a support.

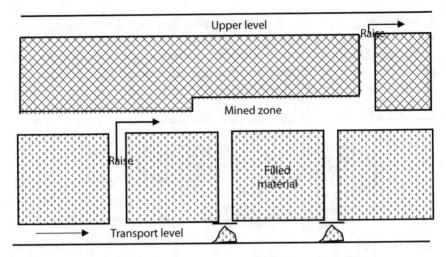

Fig. 5.11 Air circulation in the cut and fill stoping method

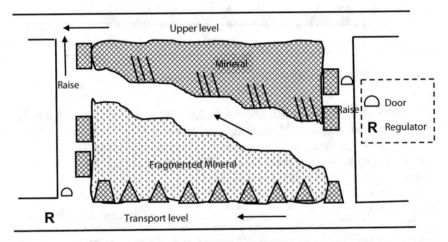

Fig. 5.12 Air circulation in the shrinkage stoping method

these methods have a network of ventilation galleries located about 15–30 m below it (Fig. 5.13). Fresh air is circulated from the injection galleries to the production level through raises. After the sweeping of the production level, the air is conducted again by means of raises to the air extraction level (McPherson 1993, p. 113). An excellent review of block caving ventilation systems can be obtained from Calizaya and Mutama (2004).

Another intermediate method between artificially supported and unsupported is *longwall mining*. The most commonly used ventilation method is the *U-shaped system*. Under this system, the air enters through the entry route to the face, sweeps it, and travels back through the return path to the general circulation (Fig. 5.14).

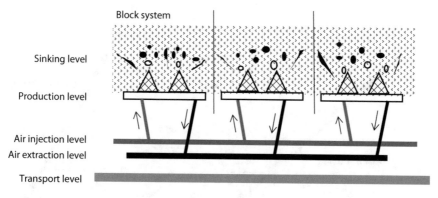

Fig. 5.13 Air circulation in the block caving method

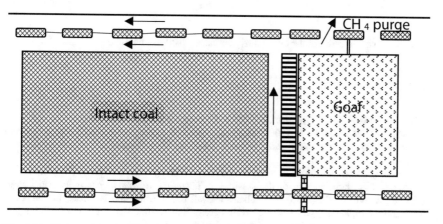

Fig. 5.14 *U*-shaped ventilation for longwall mining with two entry routes

In addition, there is usually a purge system with its own fan which separates CH$_4$ from the previous circuit. A particularly up-to-date review of ventilation techniques in coal mining can be found in Gillies and Wu (2013).

From the point of view of ventilation, coal mines can be divided into *sections* or *ventilating districts*. These are independent ventilation units into which the mine is divided to avoid enrichment in noxious gases and dust, but above all, in firedamp. Their main function is to prevent ventilation air enrichment in the above-mentioned gases as a consequence of it crossing the stopes successively. Districts are considered as being independent when they have only the main inlet and outlet airways in common (e.g. MIE 1985). By main airways, we refer to those galleries or shafts in which circulating air has not already passed through an active stope (Fig. 5.15).

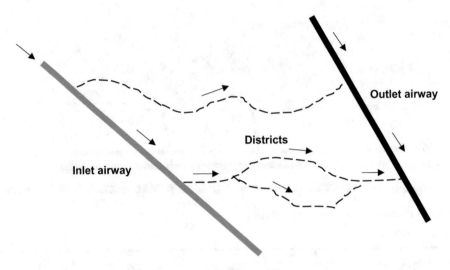

Fig. 5.15 Schematization of the division by ventilating districts of a mine

References

Borisov, S., Klokov, M., & Gornovoi, B. (1976). *Labores mineras*. Moscow: Mir.

Calizaya, F., & Mutama, K. R. (2004, May). Comparative evaluation of block cave ventilation systems. In *Proceedings of the 10th US/North American Mine Ventilation Symposium*, Anchorage, Alaska, USA (pp. 16–19).

Dicks, F., & Clausen, E. (2017). Ventilation on demand. *Mining Report, 153*(4), 334–341.

Fritzsche, C. H. (1965). Ventilación de las minas. In *Tratado de laboreo de minas*. Madrid: Labor.

Gillies, S., & Wu, H. W. (2013). Australian longwall panel ventilation practices. In *13th Coal Operators' Conference*, University of Wollongong, The Australasian Institute of Mining and Metallurgy & Mine Managers Association of Australia (pp. 176–183).

Hartman, H. L., Mutmansky, J. M., Ramani, R. V., & Wang, Y. J. (1997). *Mine ventilation and air conditioning*. Wiley.

Hermanus, M. A. (2007). Occupational health and safety in mining-status, new developments, and concerns. *Journal of the Southern African Institute of Mining and Metallurgy, 107*(8), 531–538.

Howes, M. J. (1998). Ventilation and cooling in underground mines. In J. M. Stellman (Ed.), *Encyclopaedia of occupational health and safety* (Vol. 3). International Labour Organization.

Jahir, T., Zhao, J., Mohammed, M. H., & McCullough, J. D. D. (n.d.). *Using gas monitoring and personnel & vehicle tracking to maximize the benefits of ventilation-on-demand in underground mining operations*. CONSPEC.

Lepikhov, A. G. (1975). A method of calculating the natural ventilating pressure created by a main fan. *Soviet Mining, 11*(6), 691–695.

McElroy, G. E. (1935). *Engineering factors in the ventilation of metal mines* (No. 385). US Government Printing Office.

McPherson, M. J. (1993). *Subsurface ventilation and environmental engineering*. Chapman & Hall.

MIE. (1985). ITC 04.1.01. Labores subterráneas: Clasificación. BOE 224, de 18 de septiembre.

Novitzky, A. (1962). *Ventilación de minas*. Buenos Aires: Yunque.

Tien, J. C. (1978). Pros & cons of underground ventilation system. *Coal Mining & Proceedings*, 110–113.

Tran-Valade, T., & Allen, C. (2013). Ventilation-on-demand key consideration for the business case. In *Convention: Proceedings of the Toronto CIM Conference*, Canada.

Voropaev, A. F. (1950). *Thermal depression in pit ventilation*. Moscow: Izd-vo Akad. Nauk UkrSSR.

Wallace, K., Prosser, B., & Stinnette, J. D. (2015). The practice of mine ventilation engineering. *International Journal of Mining Science and Technology, 25*(2), 165–169.

Chapter 6
Fans and Flow Control Devices

6.1 Introduction

So far, indications have been given of the general movement of air within the mine. However, little has been said about how this movement is induced or how airflow is redirected to specific areas where it is most needed or reduced in others. Fans handle the former and the latter is achieved using airflow control devices. This chapter introduces the reader to the principles of operation and sizing of both systems.

6.2 Fans

A fan is a type of *turbomachine*,[1] which continuously expels air and increases its kinetic energy and pressure. In the past, the volumetric machines (operating by strokes) were common, but at present, their use is null for mine ventilation purposes, being reserved for the production of compressed air for mining equipment actuation. The main purpose of a fan is to generate a high air quantity at a pressure higher than the atmospheric, but not much higher. This makes it possible to consider the air as incompressible which means that their *compression ratio* is <1.15. Fans differ from compressors, as the latter generate much smaller flow rates but at a much higher pressure.

---

[1]Machine capable of exchanging energy with a fluid through the movement of a rotating wheel, usually termed rotor.

C. Sierra, *Mine Ventilation*, https://doi.org/10.1007/978-3-030-49803-0_6

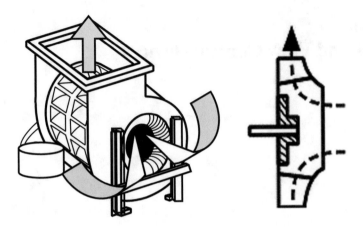

Fig. 6.1 Pathlines in a centrifugal fan

6.2.1 Parts of a Fan

The fans consist of the following parts:

- *Impeller*: Rotating part that communicates pressure and movement to the air. It consists of the blades, which may or may not be adjustable, and their support–normally a shaft.
- *Housing or casing*: The stationary part, protects the impeller and guides the air in its route.
- *Diffusers and/or evasés*: These are the elements of gradually increasing area, which when coupled to the discharge of the fan transform part of the dynamic pressure at the outlet into static pressure.

6.2.2 Classification According to the Direction of the Flow

The most normal classification uses the direction of the air driven out of the fan with regard to the *axis of rotation* of the *blades or vanes*.[2] Accordingly, we have:

Centrifugal Fans

The air leaves the fan in a 90°-direction to the shaft (Fig. 6.1). The very word *centrifugal*, meaning "away from the centre," gives us an idea of it. The most common classification distinguishes centrifugal and axial fans.[3]

---

[2]Other classifications consider the type of drive: direct (if impeller or propeller have a common shaft with the motor) and indirect (via transmission), or the reached pressure (low: <70 Pa, high: >3000 Pa).

[3]In addition to these, there are also cross flow fans (air path normal to the shaft at both the inlet and the outlet) and helico centrifugal fans (air inlet like in axial fans and air outlet like in centrifugal fans).

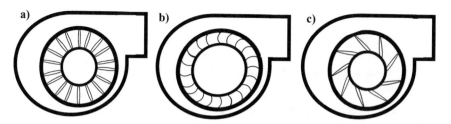

Fig. 6.2 Centrifugal fans: **a** radial, **b** forward-curved and **c** *backward-curved*

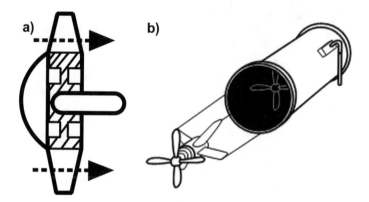

Fig. 6.3 Scheme of a fan: **a** propeller, **b** tubeaxial

Depending on the arrangement of the blades, these type of fans can be divided into the following sub-groups, namely: (a) *straight radial* (Fig. 6.2a), which operate at high speeds, can generate high pressures and are capable of self-cleaning; (b) *forward-curved* (Fig. 6.2b), which tend to operate at lower speed and are smaller; and (c) *backward-curved* (Fig. 6.2c), which usually offer high performances, their power curve is significantly elastic and are the least noisy.

Axial Fans

These consist of a propeller enclosed in a cylindrical casing. They are sometimes referred to as helical fans due to the type of movement they give to the air mass. The following variants exist: (a) *Propellers*, if the housing is short in the flow direction (Fig. 6.3a); (b) *tubeaxials*, longer (Fig. 6.3b); and (c) *vaneaxial*, these are similar to the tubeaxial but incorporate downstream guide vanes to reduce the rotation of the air.

This equipment, when used in tandem,[4] –two consecutive units to give equal flow but higher pressure–counter-rotating arrangement is chosen to limit air swirl.

---

[4]Structure widely used to ventilate civil works tunnels, roads, and in mining when large flow rates and high pressures are needed.

Comparison between Centrifugal and Axial Fans

The main difference is that centrifugal fans deliver low quantities at high pressures, while axial fans provide higher flow rates at lower pressures.

In addition to the above, all things being equal, axial fans as compared with centrifugal:

- Are more compact–smaller, lighter–than centrifugal.
- Tend to be more economical.
- Have fewer breakdowns if they do not use complex motor to fan drive systems.
- Have less inertia, making them easier to start.
- Do not require deep foundations for installation.
- Do not store water–which could worsen their performance–in their interior.
- Are reversible, although if they are reversed they provide less flow and their efficiency is much lower.
- Tend to be more flexible to operating requirements. This is due to the possibility of controlling them by varying both the angle of the blades[5] and the rotation speed of the shaft.

However, all things being equal, axial fans, as compared with centrifugal fans:

- Are usually less energy efficient.
- Are more likely to overload.
- Are less resistant to harsh environments.
- Are noisier.

Mixed Flow Fans

A *mixed flow fan*[6] is similar to an axial fan. However, in the first, the impeller is modified incorporating blades with axial and radial elements that deflect the airflow from its straight path onto a diagonal angle of about 45° relative to the shaft. This effect is obtained by employing a conical impeller that draws the air through expanding channels. Then, the spiral flow created by the impeller is usually transformed into straight flow using straightening vanes located in the casing, thus increasing the static pressure. Mixed flow fans (Fig. 6.4) combine some advantages of axial (high airflow rates) and centrifugal fans (high pressure).

---

[5]This is not usual in centrifugal fans.
[6]Sometimes referred to as diagonal fans.

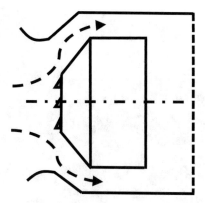

Fig. 6.4 Longitudinal section of a mixed flow fan

6.2.3   Classification According to Function

Main Fans

These large fans are normally located outside the mine and are used to ventilate the whole of it. The most widespread practice is their location outdoors, unless the location area is of difficult access, lacks electrification or is protected from an environmental point of view. In general, the surface location of the main fan facilitates maintenance and emergency access. On the other hand, its underground location makes it difficult to re-establish ventilation after an accident or major breakdown, such as an explosion.[7]

The main fans can be both centrifugal (Fig. 6.5) and axial. Among the possible configurations for the axials are: vertical (Fig. 6.6a), horizontal with horizontal output (Fig. 6.6b), and horizontal with vertical output (Fig. 6.6c). Traditionally, horizontal fans were more common, as they facilitated constructive aspects since the fan was not directly suspended over the shaft. Nowadays, however, vertical configurations are becoming prevalent, as they reduce shock losses due to their lack of elbows, which results in lower energy consumption.

The *roll-out* transfer system is a major step forward in this respect. This new development makes it possible to construct the fan outside the vertical of the shaft as well as to remove it from the shaft in order to carry out repairs (Fig. 6.7). The appearance of these new configurations has led to an increasing number of mines to opt for axial fans over centrifugal fans for the main ventilation.

Main Fan Arrangement

Main fans can be arranged in *series*, in *parallel* or in *combination* (Ramani 1992). In the series layout, one fan is located behind another along the main fan drift (Fig. 6.8a). This system is used when it is necessary to significantly increase the pressure without

---

[7]Fans and their electrical accessories may be a source of fires and explosions in mines.

Fig. 6.5 Panoramic view of a main centrifugal fan. These fans are very appropiated for deep mines in which high pressures are required. Courtesy of Zitron

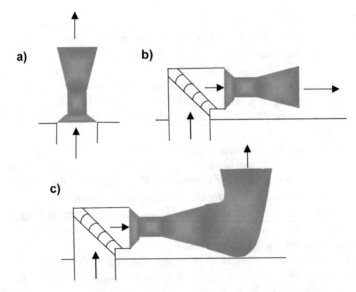

Fig. 6.6 Configuration of the axial fans. Layout: **a** vertical, **b** horizontal with horizontal output, **c** horizontal with vertical output

increasing the airflow rate in the same proportion. These conditions usually occur in the case of mines with high airflow resistance.

In the parallel ventilation system, the fans take air from the fan drift and discharge it into the atmosphere (Figs. 6.8b and 6.9). In this case, the pressure of the system does not increase, however, the flow does. This system can be used in mines with low equivalent resistance.

Fig. 6.7 Roll-out transfer system for a vertical axial fan. Courtesy of Zitron

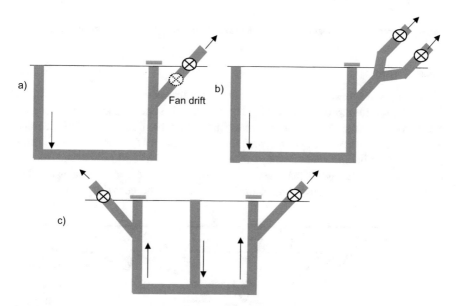

Fig. 6.8 Arrangement of main fans in a mine: **a** series, **b** parallel, **c** combination. Modified from Ramani (1992)

As for the combined arrangement (Fig. 6.8c), it allows each zone of the mine to be ventilated by a specific fan. This ensures that each fan ventilates a shorter circuit lenght. However, in this system, the functioning of each fan influences all the others.

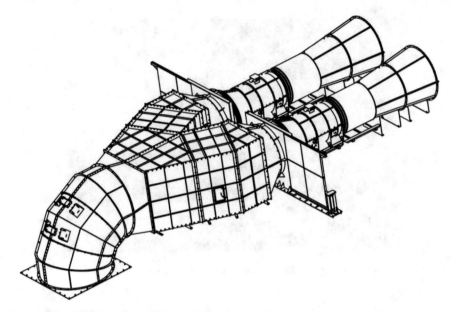

Fig. 6.9 Arrangement of main fans in parallel (Model 3913.00-MON). Courtesy of Zitron

Booster Fans

Booster fans are located inside the mine to assist the main fan in certain locations. They are used to improve the working environment in specific zones when a higher airflow rate or air speed is needed. For example, they are used in areas of the mine where the airflow has been split between several galleries. In this case, if the resistance of one of the galleries is greater, they can help to increase the airflow inside it. Booster fan's main disadvantage is that their installation can decrease airflow rates to other areas[8] where air is also needed. A common installation scheme is shown in Fig. 6.10.

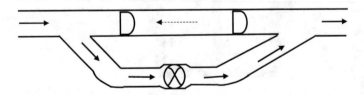

Fig. 6.10 Installation layout for a booster fan

---

[8]In accordance with the principle of conservation of mass, the increase of the airflow rate in one of the galleries reduces the airflow rates in the adjacent ones.

Exercise 6.1 The main mine fan shown in the figure operates at a pressure of 1.5 kPa. Of all this pressure, 0.5 kPa is used in overcoming the combined resistance of the two parallel galleries (A and B). The air quantity circulating through gallery A is 20 m³ s⁻¹ and through gallery B 15 m³ s⁻¹. You are asked to determine the maximum pressure at which the booster fan can operate before the airflow inside gallery A becomes zero (critical pressure). Assume that after installation of the booster fan, the main fan continues to operate at the same pressure.

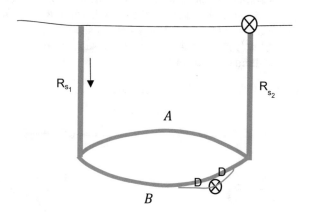

Solution

First, the resistance of both shafts $\left(R_{s_1-s_2}\right)$ can be calculated:

$$R_{s_1-s_2} = \frac{P_{s_1-s_2}}{Q_T^2} = \frac{(1500-500)\,\text{Pa}}{\left((20+15)\,\frac{\text{m}^3}{\text{s}}\right)^2} = 0.816\,\frac{\text{N s}^2}{\text{m}^8}$$

Then, the resistance of gallery B can be obtained in the same way:

$$R_B = \frac{P_B}{Q_B^2} = \frac{500\,\text{Pa}}{\left(15\,\frac{\text{m}^3}{\text{s}}\right)^2} = 2.22\,\frac{\text{N s}^2}{\text{m}^8}$$

When the air quantity through gallery A is zero, the pressure difference between both its ends is also zero. This fact implies that all the pressure supplied by the main fan is used in overcoming the resistances of the shafts, therefore:

$$P_T = R_{s_1-s_2}Q_T^2; \; Q_T' = \sqrt{\frac{P_T}{R_{s_1-s_2}}} = \sqrt{\frac{1500\,\text{Pa}}{0.816\,\frac{\text{N s}^2}{\text{m}^8}}} = 42.87\,\frac{\text{m}^3}{\text{s}}$$

The air quantity that circulates through gallery B after installing the booster fan (Q'_B) is the same as the total quantity circulating through the shafts (Q'_T). This is so as the quantity through gallery A is zero. In this way, as the resistance of gallery B is also known, we have:

$$P'_B = R_B Q_T'^2 = 2.22 \frac{\text{N s}^2}{\text{m}^8} \left(42.87 \frac{\text{m}^3}{\text{s}} \right)^2 = 4080 \, Pa$$

Exercise 6.2 The main fan of the mine shown in the figure communicates an airflow rate of 50 m³ s⁻¹ operating at a pressure of 6 kPa. The airflow rate which circulates through gallery A is 35 m³ s⁻¹ and through gallery B is 15 m³ s⁻¹. Determine the pressure at which a booster fan located in gallery B should operate to increase its airflow rate to 25 m³ s⁻¹. Assume that the pressure supplied by the main fan does not change after installing the booster fan.

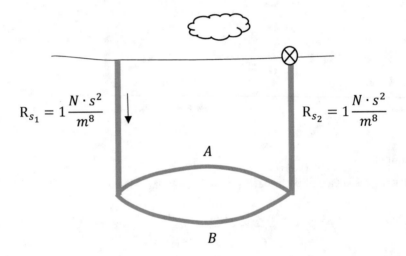

Solution

Prior to the installation of the booster fan, have for branch A:

$$6000 \, \text{Pa} = \left(R_{s_1} + R_{s_2} \right) Q_T^2 + R_A Q_A^2 = (1+1) \frac{\text{N s}^2}{\text{m}^8} \left(50 \frac{\text{m}^3}{\text{s}} \right)^2 + R_A \left(35 \frac{\text{m}^3}{\text{s}} \right)^2$$

Solving for R_A:

$$R_A = 0.816 \frac{\text{N s}^2}{\text{m}^8}$$

Similarly, for branch B:

$$6000 \, \text{Pa} = \left(R_{s_1} + R_{s_2} \right) Q_T^2 + R_B Q_B^2 = (1+1) \frac{\text{N s}^2}{\text{m}^8} \left(50 \frac{\text{m}^3}{\text{s}} \right)^2 + R_B \left(15 \frac{\text{m}^3}{\text{s}} \right)^2$$

So, solving for R_B:

$$R_B = 4.44 \, \frac{N \, s^2}{m^8}$$

After the installation of the booster fan in branch B, the total airflow rate and the airflow rates in branches A and B will vary. These new values are denoted as Q_T', Q_A' and Q_B', respectively. In this case, if the condition imposed in the statement of no pressure variation in the main fan is applied, we have:

$$6000 \, Pa = \left(R_{s_1} + R_{s_2} \right) Q_T'^2 + R_A Q_A'^2$$

Since: $Q_A'^2 = \left(Q_T' - Q_B' \right)^2$, we have:

$$6000 \, Pa = (1+1) \, \frac{N \, s^2}{m^8} Q_T'^2 + 0.816 \frac{N \, s^2}{m^8} \left(Q_T' - 25 \, \frac{m^3}{s} \right)^2$$

So: $Q_T' = 51.99 \, \frac{m^3}{s}$.
Then:

$$6000 \, Pa + P_B = \left(R_{s_1} + R_{s_2} \right) Q_T'^2 + R_B Q_B'^2$$

$$6000 \, Pa + P_B = (1+1) \, \frac{N \, s^2}{m^8} \left(51.99 \, \frac{m^3}{s} \right)^2 + 4.44 \, \frac{N \, s^2}{m^8} \left(25 \, \frac{m^3}{s} \right)^2$$

So, finally $P_B = 2180.92 \, Pa$

Auxiliary Fans

Auxiliary fans are the ones used to ventilate development headings. Altough it is more complex than this, they are usually employed for directing (boosting) the air to those areas where it is most needed and orienting the flow to the working face. Calculations involved in this type of ventilation are treated in Chap. 8.

6.2.4 Energy Consumed in Ventilation

Let there be a fan taking an airflow rate Q_1 at velocity v_1, static pressure P_1 and height z_1. The fan increases the airflow (Q_2) pressure up to a dynamic pressure given by v_2, static pressure P_2 at height z_2 (Fig. 6.11).
Using Eq. 1.12, the energy (E_v) supplied by the fan would be[9]:

$$\frac{1}{2} \rho v_1^2 + \rho g z_1 + P_1 + E_v = \frac{1}{2} \rho v_2^2 + \rho g z_2 + P_2$$

---

[9]A thoroughly demonstration is offered in Chap. 1.

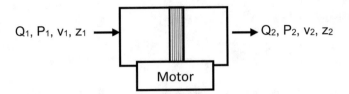

Fig. 6.11 Input and output variables on a control volume of air crossing a fan

Considering that fans are normally in the horizontal, $z_1 = z_2$, therefore:

$$\frac{1}{2}\rho v_1^2 + P_1 + E_v = \frac{1}{2}\rho v_2^2 + P_2$$

Consequently, the energy supplied by the fan is:

$$E_v = \left(\frac{1}{2}\rho v_2^2 + P_2\right) - \left(\frac{1}{2}\rho v_1^2 + P_1\right)$$

The above formula is usually understood in units of pressure; however, if these units are multiplied in numerator and denominator by metres, we have:

$$\left(\frac{N}{m^2}\right)\frac{m}{m} = \frac{J}{m^3}$$

Thus, another way of looking at the above expression is to consider that the fan supplies energy per unit volume ($J\,m^{-3}$) equal to the differences between the sum of the dynamic and static pressure in the fan outlet and inlet. So, according to the units analysis, if we multiply by the airflow rate, power units are obtained:

$$\frac{J}{m^3}\frac{m^3}{s} = \frac{J}{s}$$

Therefore, the energy needed to make the air pass through a gallery at a fixed pressure is equal to the total pressure[10] (FTP) communicated by the fan multiplied by the airflow that the fan moves, which is mathematically expressed as[11] (Eq. 6.1):

$$Pw_u = \text{FTP}\,Q \tag{6.1}$$

---

[10]We will go deeper into this concept in the next section.
[11]Note that where pressure and power terms coexist, the power is denoted as P_w and the pressure as P.

6.2.5 Fan Pressures

In Fig. 6.12, it can be seen how starting from a zero pressure at the beginning of the inlet duct, the fan delivers static and dynamic pressure to the air mass. The total pressure increase is the sum of the highest generated depression (fan inlet) and the maximum overpressure (fan outlet). Both inlet and outlet ducts are affected by the consequent friction pressure losses.

The *Fan Total Pressure* (*FTP*) represents the total energy per unit volume communicated to the fluid (Cermak and Murphy 2011). This parameter is defined as (Eq. 6.2):

$$FTP = TP_i - TP_o \tag{6.2}$$

where TP_i and TP_o are the total inlet (i) and outlet (o) pressures, respectively. Then (Eq. 6.3) is:

$$FTP = (SP_o + VP_o) - (SP_i + VP_i) \tag{6.3}$$

where SP and VP represent the static velocity pressures at the inlet (i) and outlet (o), respectively.

If the size of the fan inlet and outlet cross section match, then the velocities are also the same and with them the VP. In such a case (Eq. 6.4):

$$FTP = (SP_o) - (SP_i) \tag{6.4}$$

All static pressure at the outlet of the fan duct is lost due to the expansion of the airflow until atmospheric pressure is reached. There are two possibilities for converting this static pressure into dynamic pressure at the outlet: the nozzle and the grille. The idea in both cases is to reduce the duct cross section so that a greater

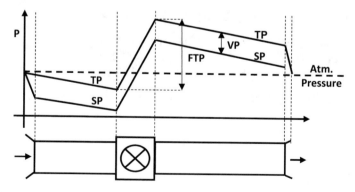

Fig. 6.12 Evolution of static (SP) and total (TP) pressures for a booster system (ducts coupled to the inlet and the outlet of the fan). Modified from ATECYR (2012)

amount of static pressure is transformed into velocity pressure. In addition, both give the airflow the desired exit direction.

Moreover, it is also common to transform a part of the dynamic pressure at the fan outlet into static pressure by means of a gradual expansion. This is achieved with devices called diffusers in the case of blower fans, which also help to couple the fan to the duct and *evasés* in exhaust fans, which discharge directly into the atmosphere.

Question 6.1 Pressure measurements have been made on the exhausting system shown in the figure.

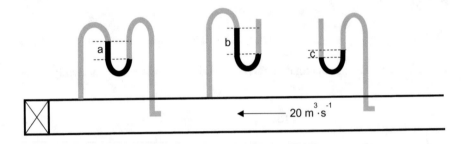

They provide the following results:

(a) 3.2 cmwc,
(b) 4.6 cmwc, and
(c) 1.4 cmwc.

You are asked to determine the:

(a) Static, dynamic and total pressure SI units and with the corresponding sign with regard to the atmospheric pressure; and
(b) Fan power.

Answer

(a) With the equivalence 1 Pa = 0.010197 cmwc; 1 cmwc = 98.068 Pa

- Dynamic pressure (a): $3.2 \cdot 98.068$ Pa = 313.8 Pa
- Static pressure (b): $-4.6 \cdot 98.068$ Pa = -451.1 Pa
- Total pressure (c): $-1.4 \cdot 98.068$ Pa = -137.3 Pa

(b) For an exhausting system, the pressure grade lines are similar to those in the figure.

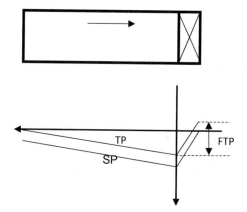

In order to be able to perform these calculations, information should be available on the dynamic pressure at the fan discharge. In the absence of data, this can be assumed the same as that existing in the fan inlet. Therefore:

$$Pw_u = \text{FTP } Q = (127.5 + 313.8) \, \text{Pa} \cdot 20 \, \frac{\text{m}^3}{\text{s}} = 8826 \, \text{W}$$

Exercise 6.3 Determine the useful power of a forcing system if it is capable of generating an airflow of 20 m^3 s^{-1} in a duct of 2.3 m^2 of cross section and the static losses between the fan outlet and the duct end are 1000 Pa. Assume no shock losses occur at the entrance.

Solution

A forcing system follows the layout of the figure.

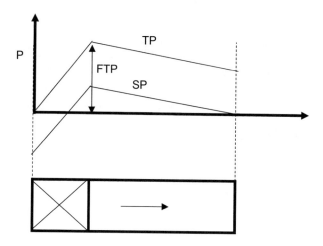

The average velocity inside the duct can be calculated from:

$$V = \frac{Q}{S} = \frac{20 \, \frac{m^3}{s}}{2.3 \, m^2} = 8.69 \, \frac{m}{s}$$

Therefore, the dynamic pressure is:

$$VP = \frac{1}{2}\rho_a V^2 = \frac{1}{2}1.2 \, \frac{kg}{m^3}\left(8.69 \, \frac{m}{s}\right)^2 = 45.31 \, \text{Pa}$$

$$FTP = FSP + VP = 1045.31 \, \text{Pa}$$

Then, the total power needed to move that air quantity (useful power supplied to the fluid) is:

$$Pw_u = FTP \, Q = 1045.31 \, \text{Pa} \cdot 20 \, \frac{m^3}{s} = 20906.2 \, W = 20.9 \, kW$$

Exercise 6.4 The static (SP) and dynamic (VP) pressures have been measured for the fans in the figures, their values being those given in the tables.

Case 1: Booster system

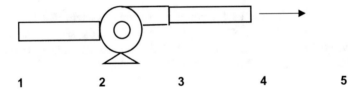

		Pressure	
		SP	VP
Point	1	0	0
	2	−400	80
	3	400	80
	4	0	80
	5	0	0

Case 2: Exhausting system

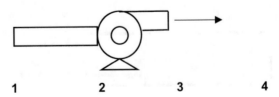

		Pressure	
		SP	VP
Point	1	0	0
	2	−400	80
	3	0	80
	4	0	0

Case 3: Forcing system

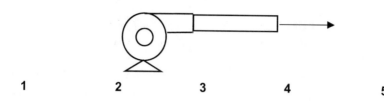

		Pressure	
		SP	VP
Point	1	0	0
	2[a]	−80	80
	3	400	80
	4	0	120
	5	0	0

[a]The static and dynamic pressures have the same absolute value but with contrary signs. Therefore, the total pressure is zero

You are requested to calculate the FTP and the FSP for each case.

Solution

Case 1:

		Pressure		
		SP	VP	TP
Point	1	0	0	0
	2	−400	80	−320
	3	400	80	480
	4	0	80	80
	5	0	0	0

$$FTP = TP_3 - TP_2 = 480 - (-320) = 800$$
$$FSP = FTP - VP_3 = 800 - 80 = 720$$

Note the negative total pressure values in the suction duct.

Case 2:

		Pressure		
		SP	VP	TP
Point	1	0	0	0
	2	−400	80	−320
	3	0	80	80
	4	0	0	0

$$\text{FTP} = \text{TP}_3 - \text{TP}_2 = 80 - (-320) = 400$$
$$\text{FSP} = \text{FTP} - \text{VP}_3 = 400 - 80 = 320$$

Case 3:

		Pressure		
		SP	VP	TP
Point	1	0	0	0
	2	−80	80	0
	3	400	80	480
	4	0	120	120
	5	0	0	0

$$\text{FTP} = \text{TP}_3 - \text{TP}_2 = 480 - 0 = 480$$
$$\text{FSP} = \text{FTP} - \text{VP}_3 = 480 - 80 = 400$$

6.2.6   Fan Losses

A fan is a machine that accumulates a series of operating losses. A general compilation of these is shown in Fig. 6.13 by means of a *Sankey diagram* (ATECYR 2012).

With all the above, the *total efficiency* (η_t) of the system can be defined as the quotient between the power received by the fluid (Pw_v) and that supplied to the fan shaft (Pw_{shaft}) (Eq. 6.5).

$$\eta_t = \frac{Pw_v}{Pw_{\text{shaft}}} \tag{6.5}$$

That is to say, the total efficiency of the system is the ratio of the final useful energy to the total energy supplied to the fan shaft, the formulation of which is (Eq. 6.6):

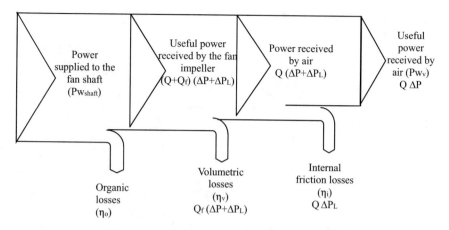

Fig. 6.13 Sankey diagram for a fan. Modified from ATECYR (2012)

$$\eta_t = \frac{Q\,\Delta P}{Pw_{\text{shaft}}} \tag{6.6}$$

Multiplying by the following unit factors:

$$\frac{(Q+Q_f)(\Delta P + \Delta P_L)}{(Q+Q_f)(\Delta P + \Delta P_L)}; \quad \frac{Q(\Delta P + \Delta P_L)}{Q(\Delta P + \Delta P_L)}$$

We obtain, after rearranging terms, Eq. 6.7, which is also deductible from the Sankey diagram in Fig. 6.12:

$$\eta_t = \left(\frac{(Q+Q_f)(\Delta P + \Delta P_L)}{Pw_{\text{shaft}}}\right)\left(\frac{Q(\Delta P + \Delta P_L)}{(Q+Q_f)(\Delta P + \Delta P_L)}\right)\left(\frac{Q\Delta P}{Q(\Delta P + \Delta P_L)}\right) \tag{6.7}$$

In other words, the total efficiency (η_t) is equal to the product of the *organic* (η_o), *volumetric* (η_v) and *friction* (η_f) *efficiencies* (Eq. 6.8) corresponding to the three respective parentheses of the above equation.

$$\eta_t = \eta_o \eta_v \eta_f \tag{6.8}$$

6.2.7 Electric Motor Losses

The *efficiency of an electric motor* (η_m) is the quotient between the useful power the motor provides (also called *shaft power*, Pw_{shaft}) and the power it absorbs (Pw) from the electrical grid (Eq. 6.9). Figure 6.14 schematizes this concept.

Fig. 6.14 Simplified Sankey
diagram for a motor

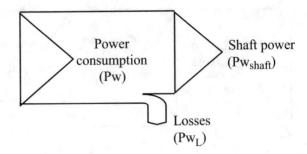

$$\eta_m = \frac{Pw_{\text{shaft}}}{Pw} \tag{6.9}$$

The *shaft power* (Pw_{shaft}) is the mechanical power output communicated by the motor to the fan shaft. This value can often be approximated by the *nominal power*, which is indicated on the nameplate, although divergences are possible due to fluctuations in the current, variations in the environmental working conditions or in the driving load of the motor. The idea is that the nominal power of the motor should not be exceeded except for very short periods as this results in shortening its useful life.

The *actual power, active power or real power* (Pw) is the power consumed and transformed into work (torque, motion...) and heat. It is, therefore, the power that is used to determine electricity demand. It is calculated, for a three-phase system, as (Eq. 6.10):

$$Pw = \sqrt{3}UI \cos \Phi \tag{6.10}$$

where

- U: Line voltage (V).
- I: Line current (A).
- Φ: Phase of the voltage relative to the current.
- $\sqrt{3}$: Constant for three-phase systems. Balanced phases are assumed, which is normal in motors and lines properly maintained (without wear and false contacts). It is measured in watts (W).

The *reactive power* (Q) is a type of wasted energy, but necessary for the operation of electrical equipment. This power bounces back and forth between the electrical grid and the motor. It is used in the generation of magnetic and electric fields in coils and capacitors. In the analogy of Fig. 6.15, it is the effort in the vertical component (perpendicular to the displacement) of an operator who drags an object and this effort does not result in the generation of work. The reactive power circulates through the networks, needing to make them with a higher dimension, but does not contribute to any particular work. Its expression for a three-phase system is (Eq. 6.11):

$$Q = \sqrt{3}UI \, sen \, \Phi \tag{6.11}$$

Fig. 6.15 Mechanical analogy for the electric power triangle

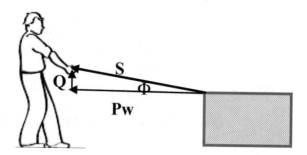

It is measured in volt-amperes reactive (var).

Aparent power (S) is the total power which includes both the power consumed, which is then transformed into heat and work (P), and the consumed and then returned (Q). It is calculated for a three-phase system by means of (Eq. 6.12):

$$S = \sqrt{3}UI \tag{6.12}$$

It is measured in volt-amperes (VA).

Exercise 6.5 The asynchronous motor of a mine fan has the following motor nameplate:

- Power = 3.2 kW,
- Voltage = 6 kV,
- Frequency = 50 Hz,
- Efficiency = 86%, and
- Cos φ = 0.85.

Determine:

(a) Power developed by the motor on its shaft,
(b) Power absorbed,
(c) Intensity absorbed, and
(d) Losses of power in the motor.

Solution

(a) The power developed in the shaft can be approximated by means of the nominal power, whose value is 3.2 kW.
(b) The actual or absorbed power (*Pw*) is calculated from the efficiency:

$$Pw = \frac{Pw_{shaft}}{\eta} = \frac{3200 \text{ W}}{0.86} = 3720.93 \text{ W}$$

(c) The absorbed intensity can be obtained from:

$$I = \frac{Pw}{\sqrt{3}\,U\cos\varPhi} = \frac{3720.93\ \text{W}}{\sqrt{3}\cdot 6000\ \text{V}\cdot 0.85} = 0.447\ \text{A}$$

(d) The lost power (Pw_L) is, therefore:

$$Pw_L = Pw - Pw_{\text{shaft}} = 3720.93\ \text{W} - 3200\ \text{W} = 520.93\ \text{W}$$

6.2.8 Affinity Laws in Fans

For these laws to be applied, it is required that the two fans are dynamically similar, which means that both are constructed geometrically similar and the velocity diagrams in their homologous points are also similar.

Rotation Speed Variation

In the case of a fan, whose speed varies from N_0 to N (Fig. 6.10), Eqs. 6.13, 6.14 and 6.15 are applicable (Fig. 6.16).

Flow rate (Q):

$$Q = Q_0 \frac{N}{N_0} \tag{6.13}$$

Pressure (P):

$$P = P_0 \left(\frac{N}{N_0}\right)^2 \tag{6.14}$$

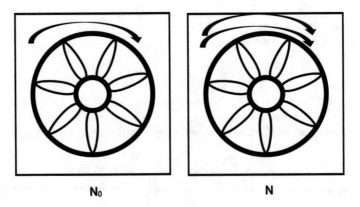

No   **N**

Fig. 6.16 Diagram of a fan rotating at a speed N_0 and $N > N_0$

Power (Pw):

$$Pw = Pw_0 \left(\frac{N}{N_0} \right)^3 \tag{6.15}$$

Diameter Variation

For two dynamically similar fans of diameter D_0 and D (Fig. 6.17) rotating at the same speed, Eqs. 6.16, 6.17 and 6.18 can be employed.

Flow rate (Q):

$$Q = Q_0 \left(\frac{D}{D_0} \right)^3 \tag{6.16}$$

Pressure (P):

$$P = P_0 \left(\frac{D}{D_0} \right)^2 \tag{6.17}$$

Power (Pw):

$$Pw = Pw_0 \left(\frac{D}{D_0} \right)^5 \tag{6.18}$$

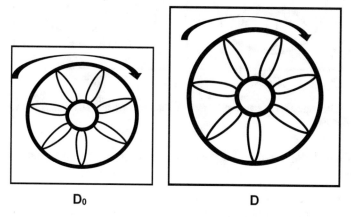

D_0 D

Fig. 6.17 Schematization of two similar fans of diameters D_0 and D, with the same rotational speed (N)

Simultaneous Variation of Several Parameters

For two similar fans of diameter D_0 and D (Fig. 6.18) rotating at speeds N_0 and N, respectively, Eqs. 6.19, 6.20 and 6.21 are fulfilled.

Flow rate (Q):

$$Q = Q_0 \left(\frac{D}{D_0} \right)^3 \frac{N}{N_0} \tag{6.19}$$

Pressure (P):

$$P = P_0 \left(\frac{D}{D_0} \right)^2 \left(\frac{N}{N_0} \right)^2 \tag{6.20}$$

Power (Pw):

$$Pw = Pw_0 \left(\frac{D}{D_0} \right)^5 \left(\frac{N}{N_0} \right)^3 \tag{6.21}$$

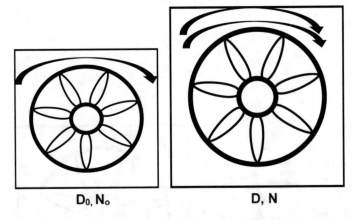

Do, No   **D, N**

Fig. 6.18 Schematization of two similar fans of diameters D_0 and D, rotating at revolutions N_0 and N, respectively

Exercise 6.6 A fan, rotating at 500 rpm, supplies 70 m³ s⁻¹ at a pressure of 600 Pa. Determine:

(a) The system impedance curve; and
(b) The pressure it would be able to deliver at those revolutions, if it provides an airflow rate of 140 m³ s⁻¹.

Solution

(a) Starting from the characteristic curve of the circuit:

$$P = R Q^2$$

Then solving for R:

$$R = \frac{600\,\text{Pa}}{\left(70\,\frac{\text{m}^3}{\text{s}}\right)^2} = 0.122\,\frac{\text{N s}^2}{\text{m}^8}$$

Therefore, the system impedance curve is the parabola:

$$P = 0.122\,Q^2$$

(b) As the resistance (R) is a characteristic of the circuit and remains invariable as long as it is not modified, we have:

$$R = \frac{P_1}{Q_1^2}$$

$$R = \frac{P_2}{Q_2^2}$$

So:

$$\frac{P_1}{Q_1^2} = \frac{P_2}{Q_2^2}$$

Therefore:

$$P_2 = P_1\frac{Q_2^2}{Q_1^2} = 600\,\text{Pa}\frac{\left(140\,\frac{\text{m}^3}{\text{s}}\right)^2}{\left(70\,\frac{\text{m}^3}{\text{s}}\right)^2} = 2400\,\text{Pa}$$

Exercise 6.7 A fan supplies 100 m^3 s^{-1} at a pressure of 1000 Pa. Under these conditions, its rotating speed is 500 rpm and receives 200 kW of power in the shaft. Determine:

(a) The speed at which it has to rotate for the flow to increase by 25%,
(b) The pressure that is capable of supplying at this rotation speed, and
(c) The total efficiency of the fan.

Solution

(a) The conditions indicated in the statement are:

$$Q_1 = 100\,\frac{\text{m}^3}{\text{s}}$$

$$Q_2 = 1.25 \cdot 100 \, \frac{m^3}{s} = 125 \, \frac{m^3}{s}$$

Applying the appropriated affinity law, we have:

$$N_2 = N_1 \left(\frac{Q_2}{Q_1} \right) = 500 \, \text{min}^{-1} \left(\frac{125 \, \frac{m^3}{s}}{100 \, \frac{m^3}{s}} \right)$$

$$N_2 = 625 \, \text{min}^{-1}$$

(b) Similarly, using the corresponding affinity law:

$$P_2 = P_1 \left(\frac{N_2}{N_1} \right)^2 ; \quad P_2 = 1000 \, \text{Pa} \cdot \left(\frac{625 \, \text{min}^{-1}}{500 \, \text{min}^{-1}} \right)^2$$

$$P_2 = 1562.5 \, \text{Pa}$$

(c) The useful energy supplied to the fluid (Pw_v) is:

$$Pw_v = P_T \, Q = 1000 \, \text{Pa} \cdot 100 \, \frac{m^3}{s} = 100 \, \text{kW}$$

The power received in the fan shaft (P_{shaft}) is 200 kW.
Therefore, the total efficiency of the fan is:

$$\eta_t = \frac{Pw_v}{Pw_{\text{shaft}}} = \frac{100 \, \text{kW}}{200 \, \text{kW}} = 0.5$$

6.2.9   Fan Characteristic Curves

The fan *characteristic curves* are the graphical representation of the interrelation between a number of parameters, namely, airflow rate, static or total pressure, fan speed, efficiency, and power requirements of a fan. Of them, the curves that represent the airflow rate provided by fan at a given total or static pressure are the most basic. In order to obtain these curves, airflow rates and pressures are analysed for a fan in a test rig, whose outlet orifice is progressively blocked (Fig. 6.19). In other words, the differential pressure to be overcome by the fan is increased and the flow rate it provides is measured. Thus, the fan curve represents the airflow rate provided by the fan at different system resistances.

It should be noted that when the system is in free delivery, the fan provides its maximum airflow. At this point, the static pressure is zero, and therefore, the total and dynamic pressures coincide. Conversely, when the fan outlet is completely sealed,

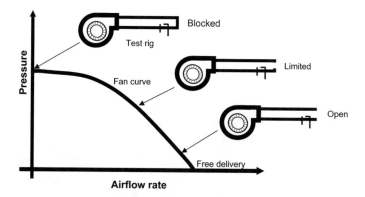

Fig. 6.19 Experimental procedure for obtaining the fan characteristic curve at constant rotational speed in a test rig. Modified from ASHRAE (2008)

Fig. 6.20 Obtaining of the operating point from the system characteristic curve and the total pressure supplied by the fan

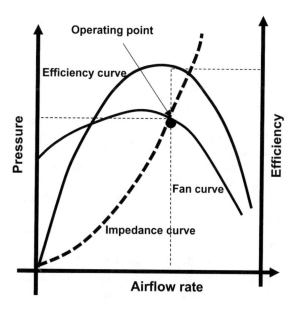

it does not supply any flow, so the dynamic pressure is zero. At this point, the total pressure and static pressure coincide.

The intersection between the characteristic, resistance or impedance curve of a given system[12] and the characteristic curve of the fan, obtained by the previous procedure, is the *duty point* or *operating point* of the fan (Fig. 6.20). This should be interpreted as the airflow rate that the fan is capable of supplying when operating against the resistance of a given circuit. It is important to bear in mind that this

---

[12] See Chap. 4 for the physical meaning of the system impedance curve.

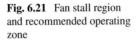
Fig. 6.21 Fan stall region and recommended operating zone

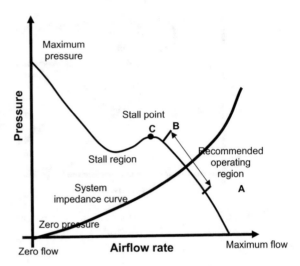

point does necessarily coincide with the *optimum efficiency point*, which is the one at which the efficiency is greatest.

The ideal working area for the fan lies in the region A-B of its static or total pressure curve (Fig. 6.21). The *stall point* (C) is an unstable operating point, which occurs when fans operate at low speed causing vibrations that can damage bearings and blades. This situation can easily be detected as the air moves forwards and stops, which results in oscillations of the pressure gauges in the airway, as well as in the ammeters attached to the fan's electric motor.

In the case of the main fans, the airflow rates and pressures caused by the natural draught of the mine should be taken into account. The combined characteristic curve is calculated, if both the main fan and natural draught flow in the same direction, by adding together the fan characteristic curve and the natural ventilation curve or by subtracting the natural ventilation curve from the fan curve, if they move in opposite directions (Fig. 6.22).

Once the operating point has been deduced, the power consumed can be obtained by means of graphs such as the one shown in Fig. 6.23. In this case, it is useful to draw a vertical line from the operating point until to where it cuts the corresponding power curve.[13] *Power values* can then be read on the right ordinate axis.

Fan characteristic curves can be obtained for the static pressure giving rise to the fan *static efficiency* (η_s) (Eq. 6.22):

$$\eta_s = \frac{P_s \, Q}{P w_{\text{shaft}}} \qquad (6.22)$$

where

---

[13]Note that for each characteristic curve, there is the corresponding power curve that is denoted by the same number.

Fig. 6.22 Effect of natural ventilation on the fan characteristic curve

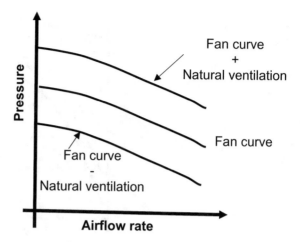

Fig. 6.23 Use of fan curve to determine power consumption

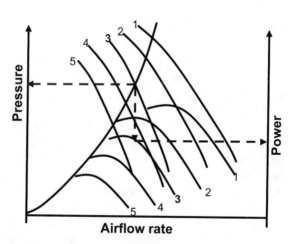

- P_s: Static pressure (Pa).
- Q: Airflow rate (m^3 s^{-1}).
- Pw_{shaft}: Power input to the fan shaft (W).

When selecting a fan, it must be borne in mind whether the manufacturer's data refer to the P_s or the P_T (Fig. 6.24). As a general criterion, one should choose the fan with the maximum efficiency closest to the intended operation point.

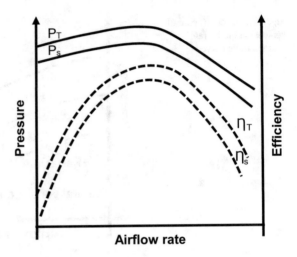

Fig. 6.24 Efficiency curves of the total and static pressures of a fan

6.2.10   Flow Control Strategies

The regulation of the flow supplied by a fan can be carried out using any of these fundamental mechanisms: *rotation speed variation, discharge dampers (throttle control[14]), inlet dampers* (throttle control), *inlet guide vanes,*[15] *blade pitch variation* and *bypass control.*

Variation of the rotation speed is achieved by utilizing variable *frequency drives*[16] acting on the electric motor, although in the past pulley mechanism were frequently used. Figure 6.25 depicts a family of fan characteristic curves generated reducing the rotation speed from n_3 to n_1. As a result, the cutting point with the system characteristic takes place in the curve at lower airflow rate and pressure. Variable frequency drives are probably the most efficient form of flow control. However, the system may not be economical when airflow rate variations are not frequent.

Dampers can be employed to throttle the air that leaves a fan system. The simplest method to throttle a fan is with a discharge damper.[17] In this case, the fan characteristic curve remains the same although the impedance curve of the system increases. Figure 6.26 shows the variation of the characteristic curve for the circuit with damper

---

[14]Creation of an obstruction in the system. Some authors use it, by extension, to refer to any element which reduces power or speed in a ventilation system.

[15]Their effect is to provide a pre-whirl to the air which modifies the fan performance curve, thus they should not be considered throttling devices.

[16]The *rotation speed* (ω) of a synchronous motor is given as the expression: $\omega = 60 \cdot f/P_p = 120 \cdot f/p$; where: f: frequency (50 Hz in Europe and 60 Hz in some other areas of the world); P_p: *number of pairs of poles* (N-S) and p: *number of poles* of the motor, which is a fixed design value and not accessible to the user. Thus, for a motor which has 2 pairs of poles and is connected to a 50-hertz grid, its synchronism speed is 1500 rpm. In the case of the asynchronous motor, the speed would be lower, about 1450 rpm, because of the slip.

[17]These elements provide an analogous effect to the choke valves on the pumps.

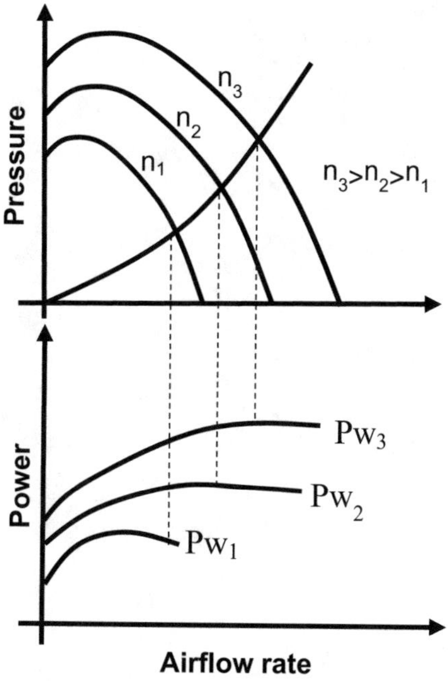

Fig. 6.25 Effect of increased rotation speed on the airflow rate and power of a fan. Modified from Jorgensen (1999) and Hensaw et al. (1999)

Fig. 6.26 Effect of the throttling of a fan with a discharge damper. Modified from Jorgensen (1999) and Hensaw et al. (1999)

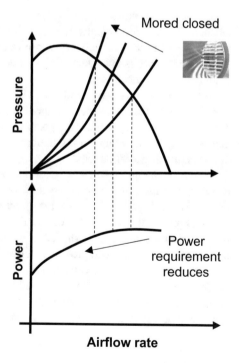

Fig. 6.27 Effect of free cross section variation utilizing inlet guide vanes. Modified from Jorgensen (1999) and Hensaw et al. (1999)

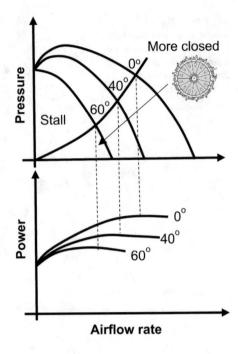

closure. This variation results in a reduction in airflow rate, an increase in the pressure provided by the fan and a decrease in power requirement, although the power requirement curve is not affected. The outlet damper system is very inefficient due to an increase in shock losses.

There are two main types of *inlet dampers*, namely *louver dampers* and *inlet variable guide vanes*. Louver dampers are located in the fan inlet box and their resistance is varied by turning the position of the slats. This system is cheap, simple and provides a wide-range of regulation; however, it results in high pressure losses and thus higher power demand.

Variable inlet guide vanes, also termed *radial dampers*, *vane dampers* or *vortex dampers*, are generally used at the inlet as they are more efficient from an energy point of view. Figure 6.27 shows the effect of increasing the angle of the inlet fan vanes to reduce the available cross section on the fan characteristic curve and the power requirement curve. In this case, the operating point moves downwards, and hence the airflow rates and pressures delivered by the fan decrease. Besides, power requirement curves move down also reducing power consumption. This is so because guide vanes deliver a pre-whirl to the airflow which improves the fan performance curve.[18] This system offers better control than traditional dampers. Excessive closure of the inlet guide vanes, however, may result in a *stall phenomenon*.

---

[18]For performance to improve, the pre-whirl direction must the same as the rotation direction of the fan.

There are guide vanes that are already integrated with the fan before purchase, while others are coupled later on. In the first case, the effect of the vane is frequently included in the fan characteristic curves supplied by the manufacturer; in the second case, the user must calculate their effect.

Variable-pitch control is possible for axial fans with adjustable propeller blades. In such cases, the airflow rate can be controlled by varying the angle of the impeller blades so that if the angle of attack increases, the flow rate increases. Varying the angle of attack can be performed with the fan at standstill or in motion. Present-day variable-pitch adjustment is obtained using a hydraulic actuator axially stationary relative to the main shaft. When the fan is in operation the system permits instant matching to the demand. Fan duty control employing variable-pitch blades provides an infinite range of airflow rates, has relatively high efficiency and can move air backward employing negative pitch angles. The greatest drawbacks of this system are its high cost and very high risk of stall if the pitch angle adjustment is not correct.

The last options for flow control would be either to expel the excess airflow rate to the atmosphere or to feedback this excess current to the suction side (*bypass control*). Bypass control acts by lowering down the system impedance curve while maintaining the fan curve. The system is only used in fans with steep fan curves, such as centrifugal backward-curved and axial fans. The method is not often used as it creates high losses.

Question 6.2 Analyse, using a graph showing pressures and flows, the effect that the increase in the total length of the ventilation circuit has on the airflow rate supplied by a fan. Make a synoptic table indicating the main differences between discharge damper control and vane control with regard to the system characteristic and fan characteristic.

Answer

Increasing the length of a circuit means increasing its aerodynamic resistance. Therefore, since the basic relationship $P = R \, Q^2$ is met, the increase in resistance results in an increase in the slope of the curve, with the same considerations as those made for Fig. 6.26.

The main differences between discharge damper control and inlet vane control are:

Control system	System characteristic	Fan characteristic
Discharge damper	Altered	Unaltered
Inlet vane control	Unaltered	Altered

It is key to remember that due to the pre-whirl in the direction of fan rotation created by the inlet vanes, the airflow, pressure and power are reduced, thus altering the fan performance curve.

6.2.11 Power–Flow and Pressure–Flow Curves for Axial and Centrifugal Fans

The shape of the characteristic curves of the fans differs depending on whether they are axial or centrifugal. The prototypical shape of the pressure and power curves of axial and centrifugal fans is shown in Fig. 6.28.

Fig. 6.28 Prototypical characteristic curves for axial and centrifugal fans. Modified from Osborne and Turner (1992, p. 133)

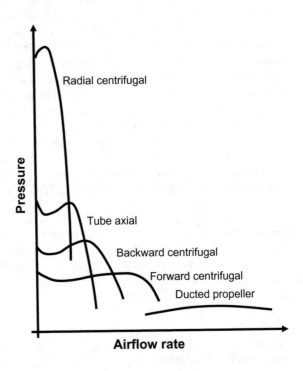

In short, it can be said that the main difference between the pressure–airflow rate curves of axial and centrifugal fans lies in the fact that the former are more horizontal. This results in the fact that axial fans provide relatively high airflow rates with regard to pressures rise, while centrifugal fans provide relatively high pressures with regard to airflow rates.

Moreover, the typical shape of the same curves for radial, forward-curved or backward-curved centrifugal fans is shown in Fig. 6.29a. The reader should take note of the fact that the previous graph is a simplification as the curves correspond to some specific models.

Perhaps, the most obvious characteristic of both axial and backward-curved fans is that they are *non-overloading fans* (Fig. 6.29b). This results in power curves peaking and then dropping off. As a result, a motor already selected to support a power peak,

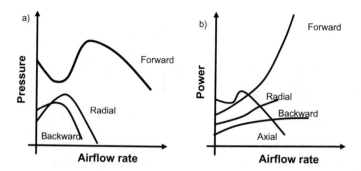

Fig. 6.29 **a** Pressure and **b** power curves for different types of centrifugal fans

is not overloaded as long as the speed of rotation remains constant despite variations in the resistance of the system.

6.2.12 Characteristic Curve for Series and Parallel Connection of Fans

In a series arrangement of identical fans without mutual interference, the total pressure is equal to the sum of the individual total pressures provided by each one of them (Fig. 6.30a, b). In this type of connection, if one of the fans is weaker than the other is, it can act by slowing down the most powerful one (*tandem effect*).

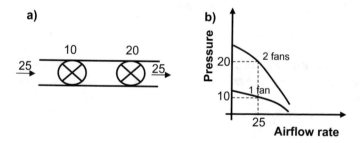

Fig. 6.30 **a** Connection diagram of fans in series; **b** characteristic curves of the series connection of two identical fans

Moreover, in a *parallel system* of fans, the total airflow rate is equal to the sum of the individual flows provided by each of them (Fig. 6.31a, b).

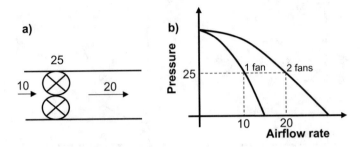

Fig. 6.31 a Parallel fan connection diagram; **b** characteristic curves of parallel connection of two equal fans

Question 6.3 You have determined the pressure and airflow rate required to ventilate a mine. The largest fan available supplied by the manufacturer is insufficient for the air quantity demand of the mine. One of the members of your team is inclined to place two fans in series, another to place two in parallel and a third comments on the possibility of splitting the air currents. Discuss which of the above decisions are correct and the price of each one of them.

Answer

To place two fans in parallel would allow us to increase the airflow rate supplied, while placing them in series would not. Splitting the airflow will decrease the total mine resistance, but may result in insufficient airflow rate in each split, so additional checks should be carried out. This decision, although it may be the cheapest from the point of view of energy consumption, requires earth movement for the execution of the additional galleries, thus greatly increasing costs.

Exercise 6.8 To undertake successfully the secondary ventilation of a development heading 45.8 m^3 s^{-1} of air are needed at the face. The leaking calculations made indicate that this can be carried out by means of a forcing system in which the fan blows 46.21 m^3 s^{-1} (Q_{fan}) at a pressure of 4861.7 Pa (P_{fan}). The following data is also available:

  System impedance curve:

Q_v	P
0	0
10	228
20	911
30	2049
40	3643
50	5693
55	6888

- 130 kW-fan characteristic curve: $P_{fan} = -1.0745Q^2 + 60.885Q + 1613.2$
- 150 kW-fan characteristic curve: $P_{fan} = -1.1906Q^2 + 54.583Q + 2424.9$

Determine the optimal series configuration of two equal fans which better meets the system needs.

Solution

The system impedance curve, obtained from the table, is:

$$P_{sist} = 2.277Q^2$$

The options of two units of 130 kW and of two units of 150 kW, in series, are analysed.

(a) Coupling in series 2 units of 130 kW.

The characteristic curve of the fans and the system is:

$$P_{fan} = -2.149Q^2 + 121.77Q + 3226.3$$
$$P_{sist} = 2.277Q^2$$

The common point is obtained by solving the system by equalization:

$$2.277Q^2 = -2.149Q^2 + 121.77Q + 3226.3$$

Therefore:

$$Q = 44.06 \, \text{m}^3 \, \text{s}^{-1} \qquad P = 4419.8 \, \text{Pa}$$

(b) Coupling in series 2 units of 150 kW.

The characteristic curve of the fans and the system is:

$$P_{fan} = -2.3812Q^2 + 109.166Q + 4849.8$$
$$P_{sist} = 2.277Q^2$$

Equaling both expressions:

$$2.277Q^2 = -2.3812Q^2 + 109.166Q + 4849.8$$

Therefore:

$$Q = 46.05 \, \text{m}^3 \, \text{s}^{-1} \qquad P = 4827.56 \, \text{Pa}$$

The solution that best suits the needs is that of two fans of 150 kW arranged in series. A comparative analysis of the different options is given in the following figure:

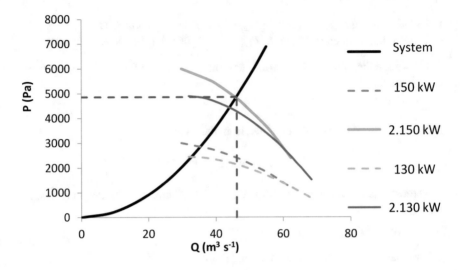

6.2.13   Fans Operation

When a three-phase motor is started with a direct-on-line motor starter, a peak of current consumption takes place. This peak can reach 4–10 times higher than the motor running current and may cause a voltage dips in the grid. For this reason, the following starting systems are usually adopted, namely:

- Star-delta starter,
- Auto-transformer starter, and
- Soft starter.

The manufacturer determines the type that best suits the motor for each model of starter.

As far as maintenance is concerned, modern fans have electronic devices to protect the motor. This includes thermistors to control temperatures, accelerometers to control vibrations and Petermann probes to detect the stall effect.

In addition, the following recommendations are established:

- In order to avoid bearing failure, the alignment between the motor and fan shafts must be checked at least once every six months.
- The oil tanks should be checked at the frequency indicated by the manufacturer. The same applies to the greasing of the fan parts.
- The blades must be checked annually to avoid material deposits and wear.
- A complete overhaul of the bearings is recommended every year.
- In general, action should be taken on the fan if it shows: abnormal vibrations, excessive noise, high-energy consumption or low performance.

6.3 Airflow Control Devices

Flow control elements are primarily used to block air passage, direct air to the points where it is most needed or reduce air short circuits. The use of these devices results in an increase in the resistance of the ventilation circuits and therefore in an increase in power requirements.

6.3.1 Deflector Brattices

Brattice is a general category that includes different types of air control devices used for the partition of airflow in underground mining. Their mission can be to direct the air into a dead-end gallery or a working place above or below the main airway. They are also extensively used in emergency events such as fires or landslides. If they have a permanent mission, they are usually made up of wood, brick, metal or concrete whether their character is temporary or they are made of cloth of fire-proofed canvas. Temporary brattices also receive the name of *cloths, brattice curtains* or *simply curtains*. Airtightness offered by temporary brattices is usually very low.

6.3.2 Overcasts and Undercasts

It is common for intake and return air currents to cross within the mine, especially in mines with horizontal or sub-horizontal seams. In these cases, overcasts and under-casts are used to prevent short circuits. *Overcasts* are bridge-type works that allow two airways excavated at the same level to pass over one another without contact between them while maintain the cross-sectional area (Fig. 6.32). *Undercasts*, also called *underpasses*, are inverted overcasts, that is to say, one airway descends to pass beneath the other. Undercasts are not usually used in mining due to their tendency to fill up with water and debris.

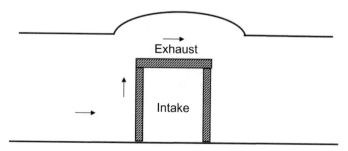

Fig. 6.32 Air overcast. Exhaust air usually flows through the overcast, as it is warmer and has a natural draught

6.3.3   Stoppings and Seals

This type of air control device is constructed to obstruct flow through a given section. The most commonly used materials are wood, concrete blocks, polyurethane slabs and so on, which are fit together to cover the entire gallery cross section (Fig. 6.33).

Fig. 6.33   Wood stopping

Stoppings can be temporary or permanent. *Temporary stoppings* are generally used in the most active parts of the mine, where ventilation requirements change frequently. Their airtightness is moderate, and the materials used in them are reusable.

Permanent stoppings are also called bulkheads[19] in metal mines and seals in coal mines where a long-term airflow closure is required. This includes main intakes and returns, belt entries, vent raises, booster fan zones, abandoned goafs and burning or self-heating zones. These closures were made in the past with stonewalls filled with sand from the workings, bricks and nowadays with concrete blocks, sprayed concrete and metallic panels. Modern configurations of the later include man doors, louver dampers and booster fan mounts.

6.3.4   Doors

These are devices are used for closing a gallery completely and also admit the passage of the people and equipment. This system usually has a small sliding panel in the middle or a shutter to facilitate flow regulation (*regulating doors*) (Fig. 6.34).

---

[19]This term is also used for the walls engineered to retain water inside the mine workings.

Fig. 6.34 Ventilation door

The door hinges are not perfectly vertical so that the doors have a certain tendency to close by their own weight and by the action of the preferential direction of the air.[20] As a basic principle, doors should always remain closed, so it is necessary to eliminate those which are no longer needed.

6.3.5 Airlocks

Locks are devices that are generally associated with the isolation of the air inside the mines from the air outside, although they can also be present in galleries. The lock system consists of two doors separated from each other by a few metres (as many as the length of the longest vehicle), so that during the passage of goods and personnel, at least one of the two remains closed, permitting the change of pressure without allowing the passage of air. Nowadays, it is common that the co-ordination between both doors is done by means of electromechanical devices which facilitate that one of them remains always closed.

When their purpose is to isolate the shaft from the outside, at the base of the headgear, they are named surface airlocks. A widely used system is the *Briart airlock* (Fig. 6.35). This system has an *auxiliary valve* in the centre intended for the passage of the cage cable. In a certain part of the cable, immediately above the cage, there are some protrusions which open the auxiliary valve with the upward movement of the cage. The purpose of the auxiliary lock is to avoid half a tonne that is added to the weight of the main door due to the depression inside the shaft.

Once the auxiliary lock is opened, the main one opens by direct pushing of the ascending cage. The tightness of the shaft remains assured because the lower part of the cage (specially sealed) acts as a seal together with the main door, now in a vertical position. When the auxiliary lock closes as the cable is lowered, the combined action of the vacuum and its weight closes the main door again.

---

[20]Doors open against the air direction.

Fig. 6.35 Scheme of a
Briart airlock

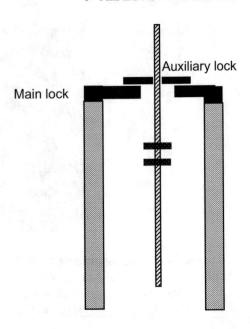

6.3.6   *Regulators*

A *regulator* is an obstacle to the flow of a current which consists of an opening whose section can vary offering different resistances to the passage of air. The materials from which the regulators are manufactured depend on whether they are temporary or permanent, the pressure differences on both sides of the gallery, the expected effects of the blasting waves and so on. There are different types of regulators, the characteristics of which are described in detail below.

- *Mine ventilation louvers*: These consist of folding panels similar to Venetian blinds, but generally made of metal. All these panels, in addition to rotating around their respective horizontal axes to open and close individually, can rotate as a whole on a vertical axis. This axle supports the assembly and, as if it were a revolving door, allows for the passage of personnel.
- *Inflatable regulators*: These inflatable devices cover the entire cross section of the gallery. The initial idea was for this type of regulator to be of a temporary nature, but current commercial catalogues indicate that they may be permanent. Among their main advantages are their transportability, lightweight and simple installation. In their most modern versions, they incorporate doors for the passage of the personnel.

6.4 Calculation by the Simplified Expression of the Regulator Area

In order to solve the problems of regulator area selection, the expression previously deduced for the equivalent orifice (Eq. 6.23) (Murgue 1873) is often used:

$$A = \frac{1.2}{\sqrt{\Delta P}} Q \qquad (6.23)$$

Different coefficients are applicable depending on the geometry involved. The above expression with coefficient 1.2 is the most popular, although it is more accurately adapted to box-type regulators.[21]

Exercise 6.9 Determine the approximate area of a sliding panel regulator which is capable of letting a flow rate of 12 m³ s⁻¹ pass when the static pressures windward and leeward of it are 530 Pa and 340 Pa, respectively.

Solution

Applying Eq. 6.24, we have:

$$A = \frac{1.2}{\sqrt{\Delta P}} Q = \frac{1.2}{\sqrt{(530 - 340)\,\text{Pa}}} \cdot 12\, \frac{\text{m}^3}{\text{s}} = 1.04\,\text{m}^2$$

It should be noted that shock losses have been neglected in this expression, leading to the consequent error.

Exercise 6.10 The mining district in the figure is traversed by an airflow rate of 32 m³ s⁻¹ when the pressure difference between the intake and the return airways is 250 Pa. This airflow rate is to be reduced to 27 m³ s⁻¹ by means of a regulator installed in the return airway. Determine the area of the regulator.

---

[21] Sliding panel ventilation door.

Solution

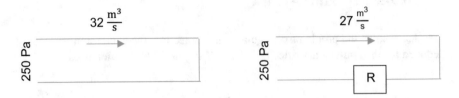

First, we calculate the total resistance of the district without regulator (R_{dist}):

$$R_{\text{dist}} = \frac{\Delta P}{Q_{\text{dist}}^2} = \frac{250\,\text{Pa}}{\left(32\,\frac{\text{m}^3}{\text{s}}\right)^2} = 0.244\,\frac{\text{N}\,\text{s}^2}{\text{m}^8}$$

For airflow rate to be $27\,\text{m}^3\,\text{s}^{-1}$, the total resistance in the circuit with regulator must be:

$$R_{\text{eq}} = \frac{\Delta P}{Q_{\text{reg}}^2} = \frac{250\,\text{Pa}}{\left(27\,\frac{\text{m}^3}{\text{s}}\right)^2} = 0.343\,\frac{\text{N}\,\text{s}^2}{\text{m}^8}$$

The regulator must provide the difference in resistance, thus:

$$R_{\text{reg}} = R_{\text{eq}} - R_{\text{dist}} = 0.099\,\frac{\text{N}\,\text{s}^2}{\text{m}^8}$$

The area of the regulator can be obtained as:

$$A = \frac{1.2}{\sqrt{\Delta P}}\,Q = \frac{1.2}{\sqrt{R_{\text{reg}}\,Q^2}}\,Q = \frac{1.2}{\sqrt{R_{\text{reg}}}}$$

And therefore:

$$A = \frac{1.2}{\sqrt{R_{\text{reg}}}} = \frac{1.2}{\sqrt{0.099\,\frac{\text{N}\,\text{s}^2}{\text{m}^8}}} = 3.81\,\text{m}^2$$

Question 6.4 The system a of the figure represents a water tank, which is freely discharged through a network of hoses. System a' is identical to system a except for the fact that one of the hoses is constricted by means of a clamp. You are asked to determine if the flow rate (Q) exiting the system a is greater or equal to that exiting system a' (Q'). Use the mathematical expressions studied in this section in your reply.

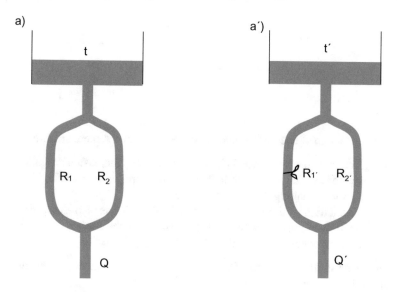

Modified from Passaro et al. (1994)

Answer

Pressure variation between the top and the bottom of each system is the same (ΔP). When the clamp constricts the hose in system a', the resistance of the branch increases ($R_1' > R_1$), thus raising the equivalent resistance of the parallel hoses ($R_{(1,2)}' > R_{(1,2)}$). Given that in system a:

$$Q = \sqrt{\frac{\Delta P}{R_{(1,2)}}}$$

And in system a':

$$Q' = \sqrt{\frac{\Delta P}{R_{(1,2)'}}}$$

Then, $Q > Q'$.

Exercise 6.11 There is a need to establish permanent machinery (M) in gallery 2 of the figure. This machinery raises its resistance up to 4 Ns^2 m^{-8} reducing the air quantity down to 30 m^3 s^{-1}. When these conditions are met, the total quantity into the system is 100 m^3 s^{-1}. You are asked to:

(a) Propose three solutions so that air quantities that circulate through both galleries are the same (are compensated).
(b) Estimate the area of the regulator that allows the airflow rates of both branches to be compensated at 45 m^3 s^{-1}.

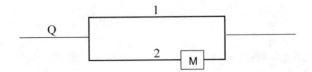

Solution

(a) It is a problem of decompensated ventilation; to solve it, there are three options:

(1) Reducing the resistance R_2 in some way, e.g. by increasing the gallery cross section, which is an extremely expensive option, or reducing its friction factor by making the walls more even.

(2) Installation of a booster fan in gallery 2, which is also expensive, as the equipment must be purchased and also consumes energy.

(3) Placing a regulator in gallery 1.

(b) In this case, for conditions of resistance raised by the machinery, we have, for a parallel system:

$$\Delta P_1 = \Delta P_2 \quad \rightarrow \quad R_1 \, Q_1^2 = R_2 \, Q_2^2$$

Then, operating and substituting:

$$\frac{70}{30} = \sqrt{\frac{4}{R_1}}$$

Solving for R_1:

$$R_1 = 0.735 \, \mathrm{Ns^2 \, m^{-8}}$$

After installing the regulator, the total flow is reduced, as seen in the previous exercise. In this case, we want it to be equal to 45 m^3 s^{-1} in both branches, therefore:

$$\frac{45}{45} = 1 = \sqrt{\frac{4}{R_1'}}$$

Thus, solving for the total resistance of line 1 after installing the regulator (R_1'), we have $R_1' = 4 \, \mathrm{Ns^2 \, m^{-8}}$, from which it can be deduced that the flow is the same if the resistances coincide.

Moreover, the resistance of the regulator is:

$$R_\mathrm{reg} = R_1' - R_1 = 3.265 \, \mathrm{Ns^2 \, m^{-8}}$$

Applying $\Delta P = R_\mathrm{reg} Q^2 = 3.265 \, \mathrm{Ns^2 \, m^{-8}} \cdot \left(45 \, \mathrm{m^3 \, s^{-1}}\right)^2 = 6611.63 \, \mathrm{Pa}$

Finally, the area of the regulator can be estimated by applying Eq. 6.24:

$$A = \frac{1.2}{\sqrt{\Delta P}} Q \frac{1.2}{\sqrt{6611.63 \, \text{Pa}}} \cdot 45 \frac{\text{m}^3}{\text{s}} = 0.66 \, \text{m}^2$$

6.5 Precise Calculation of the Regulator Area

At present, formulations that are more precise than the preceding one tend to be used (e.g. Kingery 1960; Hartman 2012). In order to define them, one can start from the abrupt contraction in the airway of Fig. 6.36. In it, the following geometric contraction coefficients can be defined:

$$c = \frac{A_c}{A_0}$$

$$N_c = \frac{A_0}{A_a}$$

$$N_e = \frac{A_0}{A_e}$$

We can also apply the expression that gives the shock loss factor (X_e) with regard to the area of the orifice (A_e) (Eq. 6.24):

$$X_e = \left(\frac{\frac{1}{c} - N_e}{N_e} \right)^2 \tag{6.24}$$

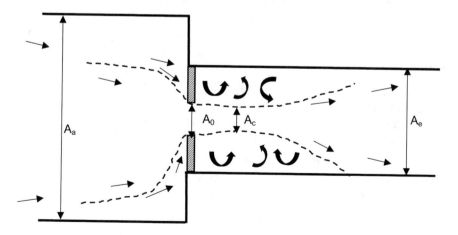

Fig. 6.36 Flow conditions through a mine regulator orifice

The coefficient c can be calculated by means of:

$$c = \frac{1}{\sqrt{Z - ZN_c^2 + N_e^2}}$$

If the above equation is particularized for the case in which the airways located at the inlet of the regulator and that located at the outlet have the same dimension ($A_a = A_e$). Then, $N_c = N_e = N$ and simplifies to Eq. 6.25:

$$c = \frac{1}{\sqrt{Z - ZN^2 + N^2}} \tag{6.25}$$

Combining Eqs. 6.24 and 6.25, we obtain Eq. 6.26:

$$N = \sqrt{\frac{Z}{X + Z + 2\sqrt{X}}} \tag{6.26}$$

where

- Z: Empirical factor which depends on the shape of the regulator (usually between 1 and 3.8). It is more common to use 2.5 for rectangular ducts.
- X: Shock loss factor,

$$X = \frac{P_x}{P_v}.$$

- P_x: Pressure loss caussed by the regulator (Pa).
- P_v: Dynamic pressure at the inlet (Pa).

In practice, N is usually smaller than 0.2.
So finally, the area of the regulator (A_0) is (Eq. 6.27):

$$A_0 = N A_a \tag{6.27}$$

where

- A_a: Cross-sectional area of the gallery (m^2),
- A_0: Cross-sectional area of the regulator (m^2),
- A_c: Cross-sectional area of the *vena contracta* (m^2), and
- A_e: Cross-sectional area of the smallest-diameter airway (m^2).

A comparison with the sizing of regulators by means of the equivalent orifice formula indicates that this equation yields values 10% higher for $N = 0.1$ and 30% higher for $N = 0.3$ (Brackett and McElroy 1941).

Exercise 6.12 A regulator installed in a mining gallery of 4.3 m^2 of cross section lets an airflow rate of 4.1 m^3 s^{-1} pass when the pressure difference on both sides of it is 60 Pa. Estimate the area of the regulator: (a) by the expression of the equivalent orifice, (b) by the precise method and (c) determine the deviation between both expressions.

Solution

(a) Directly applying Eq. 6.24:

$$A = \frac{1.2}{\sqrt{\Delta P}} Q = \frac{1.2}{\sqrt{60 \, Pa}} \cdot 4.1 \frac{m^3}{s} = 0.63 \, m^2$$

(b) The velocity is calculated as:

$$V = \frac{Q}{S} = \frac{4.1 \frac{m^3}{s}}{4.3 \, m^2} = 0.95 \frac{m}{s}$$

The dynamic pressure is, therefore:

$$P_v = \frac{1}{2} \rho V^2 = \frac{1}{2} 1.2 \frac{kg}{m^3} \left(0.95 \frac{m}{s} \right)^2 = 0.54 \, Pa$$

The shock loss factor can then be calculated as:

$$X = \frac{P_x}{P_v} = \frac{60 \, Pa}{0.54 \, Pa} = 111.11$$

Then, the coefficient N is:

$$N = \sqrt{\frac{Z}{X + Z + 2\sqrt{X}}} = \sqrt{\frac{2.5}{111.11 + 2.5 + 2\sqrt{111.11}}} = 0.136 \, (< 0.2)$$

Therefore, the area of the regulator is:

$$A_0 = N A_a = 0.136 \cdot 4.3 \, m^2 = 0.585 \, m^2$$

(c) The difference between the two approximations, expressed as a relative error (E_r), is:

$$E_r = \frac{(0.63 - 0.585)}{0.585} \cdot 100 = 7.7\%$$

Question 6.5 (a) Indicate the main consequences of an air velocity outside the normal range for a gallery. (b) Specify the measures you would take if the mine you are working in has a gallery with an excessive airflow speed.

Answer

(a) If the speed is high:

– Increased fan consumption,
– Pressure rise,
– Greater dust suspension, and
– Lack of comfort for workers due to less thermal sensation.

   If the speed is low:

– Reduced turbulence and less capacity to mix and evacuate gases.

(b) In order to slow the speed down, it would be possible to:

– Split the airstream; and
– Diminish the flow of air accessing to the gallery, for example, by means of a regulator. In order to avoid inappropriate speeds, galleries must be designed so that they have the correct section.

6.6   Conventional Signs on the Ventilation Maps

Mining maps symbols are not standardized. The following is a suggestion of simplified symbols for some common ventilation devices (Fig. 6.37).

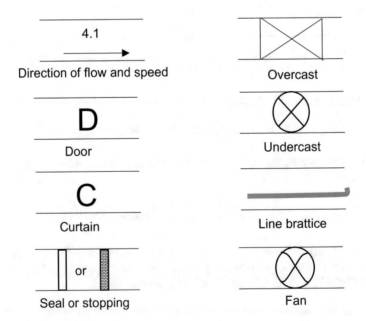

Fig. 6.37 Simplified symbols used in ventilation schematics

Question 6.6 Design a bi-directional ventilation system, which carries air to all faces, for room and pillars development in the figure. You have for it:

- 8 seals,
- 3 curtains,
- 1 regulator, and
- 5 line brattices.

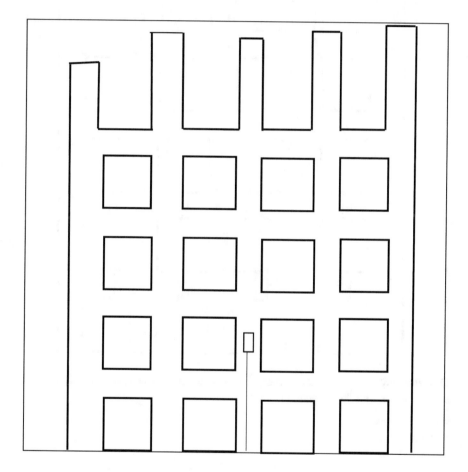

Solution

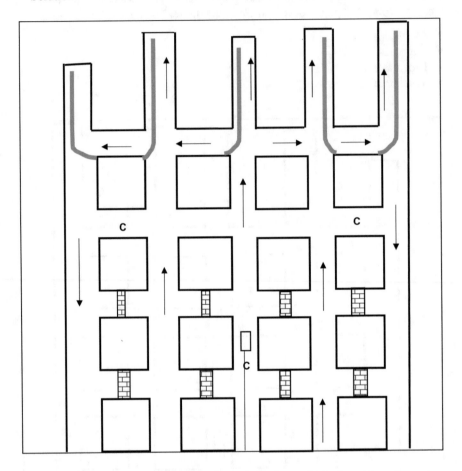

Question 6.7 The room and pillar mine of the figure is designed according to a unidirectional ventilation scheme. This system consists of:

- 5 stoppings,
- 8 curtains,
- 1 regulator,
- 2 doors (equivalent to an airlock) for exiting vehicles, and
- 5 line brattices.

According to the regulations, in order to properly isolate the conveyor belt located in the central area of the development, you must use permanent insulation, a fireproof curtain and a regulator.

Modified from Collet (2006).

Solution

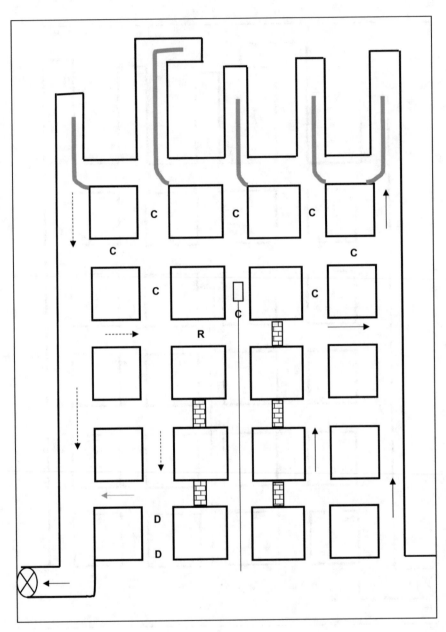

References

ASHRAE (2008). ASHRAE Handbook-Fundamentals, American Society of Heating. Refrigeration, and Air-Conditioning Engineers, Inc., Atlanta.

ATECYR (2012). *Selección de equipos de transporte de fluidos. Instituto para la Diversificación y Ahorro de la Energía.* España, Madrid: Ministerio de Industria Energía y Turismo.

Brackett, F., & McElroy, G. E. (1941). Mine ventilation (Section 14). In R. Peele (Ed.), *Mining engineers' handbook.* Wiley.

Cermak, J., & Murphy, J. (2011). Select fans using fan total pressure to save energy. *ASHRAE Journal, 53*(7), 44.

Collett, W. (2006). *Unit 7—Part 2. Ventilating a mine. Mine foreman training ventilation.* Office of Mine Safety & Licensing.

Hartman, H. L., Mutmansky, J. M., Ramani, R. V., & Wang, Y. J. (2012). *Mine ventilation and air conditioning.* Wiley.

Henshaw, T., Karassik, H., Bowman, J., Dayton, B., & Jorgensen, R. (1999). Fans, pumps and compressors. In *Mark's standard handbook for mechanical engineering.*

Jorgensen, R. (1999). *Fan engineering: An engineer's handbook on fans and their applications.* Howden Buffalo.

Kingery, D. (1960). *Introduction to mine ventilating principles and practices.* US Department of the Interior, Bureau of Mines.

Murgue, D. (1873). *Theories and practices of centrifugal ventilating machines.* London.

Osborne, W. C., & Turner, C. G. (1992). *Woods practical guide to fan engineering* (6th ed.). Woods of Colchester Limited.

Passaro, P. D., Cole, H. P., & Wala, A. M. (1994). Flow distribution changes in complex circuits: Implications for mine explosions. *Human Factors, 36*(4), 745–756.

Ramani, R. V. (1992). Mine ventilation. In H. L. Hartman (Ed.), *SME mining engineering handbook* (Vol. 2). Denver: Society for Mining, Metallurgy, & Exploration (SME).

Bibliography

Cory, W. (2010). *Fans and ventilation: A practical guide.* Elsevier.

McDermott, H. J. (2001). *Handbook of ventilation for contaminant control* (3rd ed.). ACGIH.

Salvador Escoda. (Ed.). (2013). *Manual Práctico de Ventilación.* Barcelona: Salvador Escoda, S. A.

Chapter 7
The Role of Ventilation in Fires and Explosions

7.1 Introduction

Mine fires and explosions can take place in both working and inactive mines. The latter case relates mostly to spontaneous self-ignition of coal seams, which prove impossible to extinguish and may remain active for decades. Coal-seam fires are a global problem because of the release of tons of CO, CO_2, CH_4, NO_x, SO_x, Hg and ashes with harmful effects on both soil and water quality. They are so extensive that an estimated 2% to 3% of global CO_2 emissions from fossil fuels come from unextinguished fires in coal mines (Zhang et al., 2004a; Kuenzer et al., 2007). In addition to the above effects, if fires arise in coal seams of any great thickness, subsidence phenomena may occur, with unfortunate consequences for surface structures above them. According to the Office of Surface Mining Reclamation and Enforcement (OSMRE), in the U.S.A., an estimated 98 coal-seam fires are currently active. This chapter will focus on the role that mine ventilation can play in controlling and extinguishing active mine fires.

7.2 Conditions for the Occurrence of Fire

The *fire triangle* (Fig. 7.1) shows the three elements necessary for a fire to take place: *activating energy, combustible material* and *oxygen (oxidizer)*.

Ventilation plays a major role with regard to the oxygen element as it is responsible for supplying this gas. Furthermore, if there are flammable or combustible[1] gases or vapours their concentration can be reduced with appropriate ventilation. Moreover, mine ventilation equipment, such as fans, must be specifically designed for operation in *explosive atmospheres* when relevant, so that they will not act as an ignition source (EN 1127-2:2014).

---

[1]Flammable materials refers to those combustible materials that can be easily ignited at room temperatures.

C. Sierra, *Mine Ventilation*, https://doi.org/10.1007/978-3-030-49803-0_7

Fig. 7.1 Fire triangle

The *activation energy sources* for fires are included in European Standard EN 1127-1:2012, and they are:

1. Hot surfaces;
2. Flames and hot gases;
3. Mechanical sparks;
4. Electrical material;
5. Eddy currents, cathodic corrosion protection;
6. Static electricity;
7. Lightning strikes–some of them can be directly transmited into underground mines or cause fires in the exterior that may end up in the interior;
8. Electromagnetic fields in the range 9 kHz–300 GHz;
9. Electromagnetic radiation in the range 300 GHz to 3×10^{16} Hz or wavelengths from 1000 to 0.1 μm (optical spectrum range);
10. Ionizing radiation;
11. Ultrasounds;
12. Adiabatic compression, shock waves, circulating gases; and
13. Chemical reactions.

Of the above, historically the commonest sources of fires in mines have been (Stracher and Taylor 2004):

- Soldering and welding,
- Cutting works,
- Electrical apparatus, and
- Explosives.

With regard to *combustible materials (fuels)*, they can be found in:

- Gas accumulation zones;
- Explosive dust accumulation areas;
- Rubber conveyor belts;[2]
- Wooden supports for mine slopes and access shafts, boarded areas and underground pumping stations;
- Diesel engines;
- Places where spontaneous combustion of coal may take place; and
- Oxidation zones of pyrites.

An important reference document for the selection of suitable equipment in explosive atmospheres is Directive 2014/EU-ATEX.

7.3 Gas Movements During Fires

Fires trigger a reduction in the density of the air, which, together with the combustion products, rises up and impinges on the roof of the gallery. This is termed the *buoyancy effect*, *natural draught*[3] or *chimney effect*. On the assumption, there is a slightly inclined airway with downwind circulation, one part of the gases will travel along the gallery roof, following the general circulation pattern (Fig. 7.2a). However, other combustion products will have enough energy to run counter to the wind for some time, in a phenomenon known as *rollback*, *backlayering* or *smoke reversal* (Fig. 7.2b). The combustion products moving upwind cool down, lose their capacity to run against the airflow and are dragged back towards the fire (Fig. 7.2c). The possibility of a rollback effect must be taken into account by rescue teams (see, for example, LOM 2013; Zhou and Wang 2005). The rollback effect can be countered by increasing the airflow in the gallery –reducing its cross section–. The *critical ventilation velocity* is the minimum airflow speed that prevents the rollback effect. Another option would

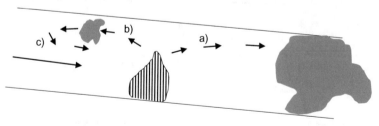

Fig. 7.2 Buoyancy effects in a mine fire: **a** downstream movement, **b** backlayering (rollback), **c** air return due to cooling

---

[2]In modern mines, both the conveyor belts and their coatings are made of fire-resistant materials.

[3]The *draught* (American English: *draft*) is the flow rate induced by the (chimney) *stack effect*.

be to direct fogging sprays towards the roof so that an air induction effect will aid downwards circulation.

Mine fires cause an increase in the volume of gases downstream of the fire. This is the consequence of both gas expansion and the incorporation of combustion products –gases and water– into the airflow. This gas expansion increases the speed of the dowstream air, which, in turn, is responsible for an increase in pressure shock losses as they depend on the square of the air speed. In other words, what takes place is a local increase in the resistance of the airway. This phenomenon is termed the *choke* or *throttling effect*. Airflow reduction due to this effect can range from around 10–25% relative to conditions before the fire started (Gillies et al. 2004).

The ascending or descending direction of the air in mine stopes must be taken into account in fire management, as the evolution of the fire is different in the two cases. Thus, with *ascending ventilation* (Fig. 7.3a), there is an increase in the driving force as the gases get closer to point A_2. This is so because they are warmer when they reach the vertical column, which is the only place where they can ascend. At A_3 the gases are still hot enough to have some driving force, whereas at A_4 their driving force is zero, although they do still retain a capacity to reduce the flow of air by throttling.

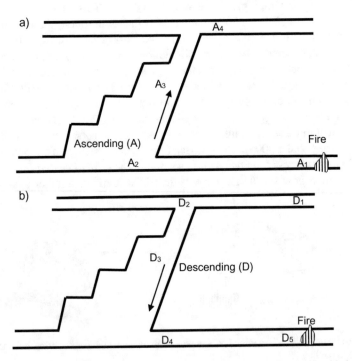

Fig. 7.3 Evolution of a fire in a stope or inclined excavation **a** with ascending and **b** with descending ventilation. Modified from Luque (1988)

In the case of *descending ventilation* (Fig. 7.3b), at point D_1 the air cross section is reduced by throttling. From D_1 to D_2 rollback increases, until it reaches a maximum at D_2, where interruptions of the airflow (stagnation) are possible. At D_4 it undergoes rollback similar to the previous point. Finally, at D_5 throttling takes place once again, with consequences similar to D_1.

It should be noted that the severity of a fire, as long as no action is taken against it, is conditioned by the point in the mine where it originates. Thus, the closer it occurs to the downcast shaft, the worse it will be, as fumes and gases will affect more areas of the mine. Conversely, a fire in the vicinity of the upcast shaft would have a lesser capacity to spread gases through the mine, so its effects will be more limited. If the fire takes place in a stope or steeply inclined excavation, it can be much more problematic because, as a result of the type of work carried out in such areas, it can be difficult to put fire barriers in place. In such cases, appropriate channelling of ventilation flows through them is of vital importance.

7.4 Estimating Fire Pressures

It is possible to start from the expression of the stack or chimney effect, well known in the architecture (Eq. 7.1):

$$H_f = (\gamma_a - \gamma_h)\Delta z = \gamma_a \Delta z \left(1 - \frac{\gamma_h}{\gamma_a}\right) \qquad (7.1)$$

where

- H_f: Pressure generated by the fire (kPa m^{-2}).
- γ_a: Air density (kg m^{-3}).
- γ_h: Hot gas density (kg m^{-3}).
- Δz: Active height difference (m). With ascending ventilation this is the difference in height traveled by the hot gases. If the ventilation is descending, it is the difference between the level of the fire and the lowest level reached by the hot gases.

If the following approximation[4] is made:

$$\frac{\gamma_h}{\gamma_a} \approx \frac{T_a}{T_h}$$

where

- T_a: Air temperature before the fire (K), and
- T_h: Smoke temperature (K).

---

[4]The approximation is based on Charles' law: "At constant pressure, the volume V of a gas is directly proportional to its absolute temperature T". Therefore, the densities can be considered inversely proportional to the absolute temperatures.

From this, Eq. 7.2 emerges:

$$H_f = \gamma_a \Delta z \left(1 - \frac{T_a}{T_h} \right) \tag{7.2}$$

Working on the basis that $T_h = T_a + \Delta T$, Eq. 7.3 yields:

$$H_f = \gamma_a \Delta z \left(1 - \frac{T_a}{T_a + \Delta T} \right) \tag{7.3}$$

Thereafter, on the basis of the initial approximation, relating densities to temperatures, Eq. 7.4 is obtained:

$$H_f \approx \gamma_a \Delta z \left(\frac{\Delta T}{T_a + \Delta T} \right) \tag{7.4}$$

Finally, if T_a approaches 300 K (27 °C), the outcome is Eq. 7.5:

$$H_f \approx 1.2 \left[\frac{kg}{m^3} \right] \left(\frac{\Delta T}{300 + \Delta T} \right) \Delta z \tag{7.5}$$

Temperature variations can be obtained from the cooling curve, with certain limitations in the case of descending ventilation. Readers interested in more in-depth information may consult Trutwin (1972).

7.5   Cooling Curves

Cooling curves are functions that approximate temperatures observable at different distances from the fire. They usually correspond to the family of expressions noted in Budryk (1956) and Surkov (1975) (Eq. 7.6):

$$T_x - T_a = (T_f - T_a)e^{-Sx} \tag{7.6}$$

where

- x: Distance between the point under consideration and the fire (m).
- T_x: Temperature of the flow of gases and smoke at distance x from the fire (°C).
- T_f: Temperature of the fire (°C).
- T_a: Ambient temperature before the fire, or rock temperature (°C).
- S: Co-efficient (per metre). If it is assumed that S is essentially dependent on air speed, then $S = \frac{0.0175}{v^{0.64}}$ (Simode 1976).
- v: Average speed of the fire-driven gas flow (m s^{-1}).

These curves give an approximate model for the phenomenon of temperature decay, as they do not incorporate various phenomena, such as heat released during the condensation of water at temperatures below 100 °C.

More adequate curves are obtained using an average value that characterizes the entire plume of heated air as a whole. To obtain it, the integral mean value for the above family of functions (Eq. 7.7) can be used:

$$\Delta T_M(x) = \left(\frac{1}{x-o}\right) \int_0^x \Delta T_f \, e^{-\left(\frac{0.0175}{v^{0.64}}\right)x} dx \tag{7.7}$$

where ΔT_f is the increase in temperature –with respect to that previous to the fire– caused by the fire in the exact location where it takes place ($x = 0$), and $\Delta T_M(x)$ is the average increase in temperature –with respect to that previous to the fire– of the current up to a distance of x from the focus.

Integrating Eq. 7.7 results in Eq. 7.8:

$$\Delta T_M(x) = -\left(\frac{1}{x}\right)\left(\frac{\Delta T_f v^{0.64}}{0.0175}\right)\left(e^{-\frac{0.0175}{v^{0.64}}x} - 1\right) \tag{7.8}$$

If curves are plotted for different speed values by taking[5] $\Delta T_f \sim 800$ °C, the family of curves corresponds to Eq. 7.9 (Schmidt et al. 1973):

$$\Delta T_M(x) \sim \frac{45,700 \, v^{0.64}}{x}\left(1 - e^{-\frac{0.0175 \cdot x}{v^{0.64}}}\right) \tag{7.9}$$

Exercise 7.1 A fire has been detected in a mining stope. It is necessary to:

(a) Determine the mean temperature variation (ΔT_M) experienced by the fumes up to a distance of 500 m from the focus when the air velocity dragging them is 2 ms^{-1}.

(b) What would the average temperature variation up to 500 m from the focus be if the air speed is tripled?

(c) Why is it that the higher the air velocity, the greater the average temperature increase of gases up to a distance of x?

(d) Construct cooling curves for speeds of 2, 4, 6, 8 and 10 m s^{-1}.

Solution

(a) Substitution with Eq. 7.9 yields the result that at 500 m from the fire there is an increase of 142 °C above the ambient temperature prior to the fire.

(b) Similarly, at the same distance, with a tripling of speed to 6 m s^{-1}, there is an increase of 270 °C relative to the temperature prior to the fire.

---

[5]The temperature in the fire zone is taken to be approximately 1073 K.

(c) The higher the velocity of the gases, the more difficult it will be for them to exchange their heat with the gallery walls. This causes the gases to remain hot while transported over longer distances.

(d) The table and plot show the values, obtained from Eq. 7.9, of ΔT (°C) for different distances (x) to the fire, depending on the air speed (v).

		Speed (m s^{-1})				
		2	4	6	8	10
Distance to the fire (x) (m)	0.5	798	798	799	799	799
	1	795	797	798	798	798
	10	756	772	778	782	784
	30	679	719	737	747	754
	50	612	672	698	714	725
	80	528	608	646	668	684
	100	480	570	613	640	659
	150	387	489	542	577	601
	250	268	371	432	474	505
	500	142	216	270	312	345
	750	95	147	189	223	253
	1000	71	111	143	171	196
	1500	47	74	96	115	133
	2000	36	55	72	86	100

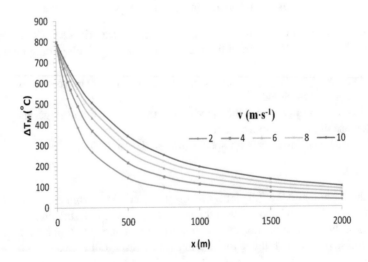

If, on the contrary, Eq. 7.6 were to be used, the following table would emerge. Considerable differences relative to Eq. 7.9 may be observed, particularly for longer distances away from the focus. The reader is invited to calculate the limit of both functions when $x \to \infty$ for different air speeds.

Distance to the fire (x) (m)		Speed (m s^{-1})				
		2	4	6	8	10
	0.5	796	797	798	798	798
	1	791	794	796	796	797
	10	715	744	757	764	769
	30	571	644	677	696	709
	50	456	558	606	635	655
	80	326	449	513	553	581
	100	260	389	459	504	536
	150	148	271	347	400	438
	250	48	132	199	252	294
	500	3	22	50	79	108
	750	0	4	12	25	40
	1000	0	1	3	8	15
	1500	0	0	0	1	2
	2000	0	0	0	0	0

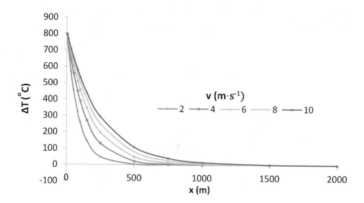

7.6 Average Temperatures at Junctions and Forks

At junctions, several airflows, including fumes, come together and mix. At bifurcations, or forks, a flow divides into several different streams. Here the focus will be on two processes:

(a) Two Airways Forking into One

When a mass flow rate of fumes (G_h) at temperature (T_h) converges with a mass flow rate of air (G_a) at temperature (T_a) at a point configured as shown in Fig. 7.4,

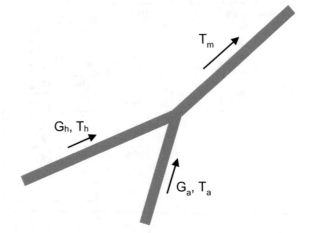

Fig. 7.4 Estimate of average temperatures of gases where air and smoke streams converge

the average temperature (T_m) of the resulting current is calculated by means of Eq. 7.10 (Simode 1976). This expression corresponds to the weighted average of the temperatures, based on the main variable, that is, the mass flow rate, with the simplification of taking the specific heat of the different gases that compose each flow to be the same:

$$T_m = \frac{G_h T_h + G_a T_a}{G_h + G_a} \tag{7.10}$$

where

$$G = Q\gamma$$

(b) Concatenated Forks

In Fig. 7.5, speeds are different in the AB branch from those in the BC branch, because of changes in cross section. In this case, the temperature increase in C due to fumes is found by following the scheme in Fig. 7.6. The procedure is:

(1) First, determine the temperature of the fumes in B (T_B) on the curve V_A.
(2) Once this is found, take the fumes to have approximately the same temperature at the end of AB as at the beginning of BC. For this purpose, read the imaginary abscissa or first coordinate X_2 from the V_B curve for the previous temperature.
(3) From the previous abscissa, mark the distance BC and determine from the V_B curve the temperature of the fumes at C (T_c).

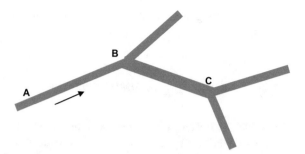

Fig. 7.5 Diagram of a point separated from the focus of a fire by several forks

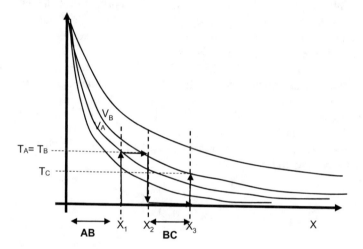

Fig. 7.6 Obtaining temperatures at a point separated from the focus of a fire by several bifurcations

The explanation given above was chosen for its conceptual simplicity. However, the current trend in the study of fires inside mines involves the use of Computational Fluid Dynamics (CFD). This type of work goes beyond the purposes of this text. Readers interested in further information can consult Surkov (1975) and Hansen (2010), which provide an exhaustive bibliographical survey.

7.7 Calculation of Critical Speed

In the event of a fire inside a gallery or tunnel, it is necessary to ensure that fumes move in the opposite direction to that of evacuation. If conditions are such that backlayering occurs then evacuation may be compromised.[6] In order to supply sufficient fresh

---

[6]Normally, the direction of evacuation opposes that of the fresh airflow.

airflow to overcome backlayering, the speed of the air supplied must be greater than the so-called *critical speed* (V_c).

7.7.1   Critical Speed According to NFPA 502: 2017

In Annex D of NFPA 502:2017, Kennedy's (1996) analytical model is recommended as a method for calculating critical speed in tunnels. This model employs Thomas' equation (1970), as a basis, using the average temperature of the hot gases in the fire zone. The model introduces a dimensionless constant K_1 as a function of the Froude number (F_r) and a dimensionless factor K_g in order to calculate the critical speed in the case of tunnels with a certain slope. It is interesting to note here that its application is limited to slopes of $\pm 6\%$, and this gradient is exceeded in most mines.

The critical speed is determined by solving Eqs. 7.11 and 7.13 iteratively, as described in the American standard NFPA 502:2017. The first step $(i = 1)$ is to obtain an initial value for the critical speed V_c (Eq 7.13) using a temperature corresponding to that of the fresh air, $T_{f\,(i=1)}$. Taking this calculated value of $V_{c\,(i=1)}$, we can then obtain a new temperature $T_{f\,(i=2)}$, which then gives a second iteration value for $V_{c\,(i=2)}$ and so on until the system is stabilized.

The average temperature of the gases at the site of the fire (T_f) in degrees K is calculated according to Eq. 7.11:

$$T_f = \left(\frac{Q}{\rho_a \, C_p \, A_x \, V_c} \right) + T_a \qquad (7.11)$$

where

- Q: Heat Release Rate (HRR) directly into the air at the location of the fire (kW).
- ρ_a: Average density of the fresh air (kg m^{-3}) which is taken to be 1.204 kg m^{-3} for dry air at 20 °C and at atmospheric pressure. For other temperatures Eq. 7.12 can be used:

$$\rho_a = 1.204 \frac{273.15 + 20}{273.15 + T} \qquad (7.12)$$

where T is the temperature in °C.

- C_p: Specific heat of the air [kJ (kg K)$^{-1}$] which is taken to be 1.005 kJ (kg K)$^{-1}$ for dry air at 20 °C and at atmospheric pressure.
- A_x: Cross section of the gallery (m^2).
- V_c: Critical speed (m s^{-1}).
- T_a: Fresh air temperature (K).

The velocity (V_c) can be calculated using Eq. 7.13:

Table 7.1 K_1 values for different heat release ratios (adapted from NFPA 502:2017)

Q (kW)	K_1
<10,000	0.87
10,000	0.87
30,000	0.74
50,000	0.68
70,000	0.64
90,000	0.62
100,000	0.606
>100,000	0.606

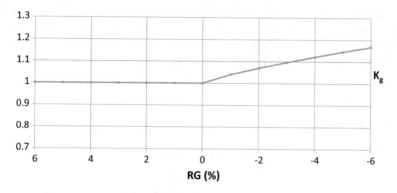

Fig. 7.7 Grade factor (K_g) as a function of roadway grade (RG, %)

$$V_c = K_1 \, K_g \left(\frac{g \, H \, Q}{\rho_a \, C_p \, A_x \, T_f} \right)^{\frac{1}{3}} \tag{7.13}$$

where

- K_1: A factor that is related to the Froude number (Fr) as: $K_1 = \mathrm{Fr}^{-1/3}$. Experimental models indicate that Fr < 4.5 is required to prevent smoke backlayering (e.g. Lee et al. 1979). When Fr = 4.5, then the value of $K_1 = 0.606$. Alternatively, this dimensionless value can be obtained for different Heat Release Ratios (HRR) from NFPA 502:2017 (see Table 7.1).
- g: Acceleration due to gravity, 9.81 (m s^{-2}).
- H: Height of the tunnel or gallery at the location of the fire (m).
- K_g: Grade factor, a dimensionless value that can be obtained from Fig. 7.7 (see NFPA 502:2017) as a function of the road grade[7].

An airway slope is:

---

[7]Ratio of uphill height to horizontal length (slope).

(a) Downgrade (negative): The entrance of fresh air is at a higher elevation than the fire source. In this situation, the flowrate must be sufficiently high to overcome the draught caused by the fire. Calculations indicate that V_c will be greatest in this kind of situation.
(b) Upgrade (positive): Clean air flows from bottom to top coinciding with the draught caused by the fire.

In the case of downgrade slopes, an alternative expression is frequently used to describe variation in K_g (Eq. 7.14):

$$K_g = 1 + 0.04[-RG(\%)]^{0.8} \tag{7.14}$$

Exercise 7.2 A fire with Heat Release Rate of 6 MW has started in a 3.5 m high gallery with a free passage cross section of 10 m². The gallery is on a downhill slope of 4%. Calculate the critical speed at which the ventilation air must be supplied to ensure that there is no backlayering. The average temperature of the ambient air is 26 °C.

Solution

For the first iteration:

$$T_{f(i=1)} = 273.15 + 26 = 299.15 K$$

$$V_c = K_1 K_g \left(\frac{g H Q}{\rho_a C_p A_x T_f} \right)^{\frac{1}{3}}$$

$$V_c(i = 1) = 0.87 \cdot 1.1188 \left(\frac{9.81 \cdot 3.5 \cdot 6000}{1.204 \cdot 1.005 \cdot 10 \cdot (273.15 + 26)} \right)^{\frac{1}{3}} = 3.76945$$

$$T_f = \left(\frac{Q}{\rho_a C_p A_x V_c} \right) + T_a$$

For the second iteration, we have that:

$$T_f(i = 2) = \left(\frac{6000}{1.204 \cdot 1.005 \cdot 10 \cdot 3.76945} \right) + (273.16 + 26) = 160.24$$

The following table shows the results of iterations 1–8:

Iteration (i)	T_f (°C)	V_c (m s^{-1})
1	26.00	3.76945
2	160.24	3.33131
3	177.89	3.28727

(continued)

(continued)

Iteration (i)	T_f (°C)	V_c (m s^{-1})
4	179.93	3.28234
5	180.16	3.28178
6	180.19	3.28171
7	180.19	3.28171
8	180.19	3.28171

The calculation is stabilized by iteration 6 indicating a solution of:

$$V_c = 3.28 \, \text{m s}^{-1} \quad T_f = 180.2 \, °C \, (453.15 \, K)$$

Ventilation air introduced into the fire gallery must have a velocity greater than 3.28 m s^{-1}.

7.7.2 Updated Calculation According to NFPA 502:2020

NFPA 502 has been updated for the year 2020 by including a new method for calculating critical speed (V_c) (contained in a revised Annex D[8]). As with the original calculation method, its validity is limited to slopes of ±6%.

According to these new recommendations, parameters A, B and C must be defined. These are calculated according to Eqs. 7.15, 7.16 and 7.17, respectively:

$$A = \left(\frac{Q}{\rho_a \, C_p \, T_a \, g^{\frac{1}{2}} \, H^{\frac{5}{2}}} \right) \tag{7.15}$$

$$B = 0.15 \left(\frac{H}{W} \right)^{-\frac{1}{4}} \tag{7.16}$$

$$C = \sqrt{g \, H} \tag{7.17}$$

The critical speed is then calculated according to Eqs. 7.18 and 7.19:

$$\text{If } A \leq B \quad V_c = K_g \cdot 0.81 \cdot C \cdot A^{\frac{1}{3}} \left(\frac{H}{W} \right)^{\frac{1}{12}} e^{\left(-\frac{L_b}{18.5 \, H} \right)} \tag{7.18}$$

$$\text{If } A > B \quad V_c = K_g \cdot 0.43 \cdot C \cdot e^{\left(-\frac{L_b}{18.5 \, H} \right)} \tag{7.19}$$

---

[8] Note that, as stated in the document itself: "This annex is not part of the requirements of this NFPA document and is included only for information purposes".

where

- L_b: Backlayering or distance traveled (m) by the fumes in the opposite direction to the ventilation airflow. The critical speed (V_c) corresponds to the value of L_b = 0, i.e. no backlayering.
- W: Width of the airway (m).

All other parameters and variables remain the same as before, with the same definitions, values and units.

Exercise 7.3 Calculate the critical speed for the situation described previously in Exercise 7.2 using the revised model proposed in NFPA 502:2020. Assume that the cross section of the gallery resembles an ellipse.

Solution

Where NFPA 502:2020 uses the height (H) and width (W) of the tunnel, the NFPA 502:2017 uses the height (H) and free area of the cross section (A_x). For the purpose of comparative calculations, the geometric figure of the area is considered to be an ellipse. The formula applied to calculate the area is, therefore:

$$A_x = \pi \frac{H}{2} \frac{W}{2}$$

where solving for W we have that:

$$10 = \pi \frac{3.5}{2} \frac{W}{2}$$

$$W = 3.64 \, \text{m}$$

Parameter A is calculated as:

$$A = \left(\frac{Q}{\rho_a \, C_p \, T_a \, g^{\frac{1}{2}} \, H^{\frac{5}{2}}} \right)$$

$$A = \left(\frac{6000}{1.204 \cdot 1.005 \cdot (273.15 + 26) \cdot 9.81^{\frac{1}{2}} \cdot 3.5^{\frac{5}{2}}} \right) = 0.23565$$

Parameter B is obtained from:

$$B = 0.15 \left(\frac{H}{W} \right)^{-\frac{1}{4}}$$

$$B = 0.15 \left(\frac{3.5}{3.64} \right)^{-\frac{1}{4}} = 0.15146$$

Finally, parameter C is:

$$C = \sqrt{g\,H}$$

$$C = \sqrt{8.81 \cdot 3.5} = 5.8596$$

The condition of $A > B$ is met, thus we can use Eq. 7.19:

$$V_c = K_g \cdot 0.43 \cdot C\,e^{\left(-\frac{L_b}{18.5\,H}\right)} = 1.1188 \cdot 0.43 \cdot 5.8596\,e^{\left(-\frac{0.0}{18.5\cdot3.5}\right)} = 2.82\frac{m}{s}$$

Therefore, the ventilating air blown into the fire gallery must have a velocity greater than 2.82 m s^{-1}.

Exercise 7.4 For a fire with Heat Release Rate of 35,000 kW, calculate the critical speed for a gallery with maximum height and width of 6 and 8 m, respectively, and a cross-sectional area of 37.7 m^2. The gallery has a 4% downhill slope and clean air must be introduced downwards, to drag the fumes in the opposite direction to the fire draught. The ambient temperature is 26 °C. In addition, plot a graph of the critical speeds for values of Heat Release Rate between 0 and 100 MW. In all cases, compare values obtained using the original and updated NFPA methods.

Solution

- Heat Release Rate (HRR): 35,000 kW;
- NFPA 502:2017; $V_c = 3.72$ m s^{-1}, $T_f = 236.7$ °C; and
- NFPA 502:2020; $V_c = 3.69$ m s^{-1}.

Given the gallery cross-section of 37.7 m^2, the critical speeds for different values of the Heat Release Rate for both methods are represented in the graph below:

Parameters: H = 6 m, W = 8 m, A = 37.70 m^2, RG = −4 %, T$_a$ = 26 °C.

Q (HRR) MW	V_c (2017) m s^{-1}	V_c (2020) m s^{-1}
0	0.00	0.00
1	1.57	1.47
5	2.61	2.52
10	3.20	3.17
30	3.59	3.69
50	3.63	3.69
70	3.58	3.69
90	3.58	3.69
100	3.53	3.69
120	3.62	3.69

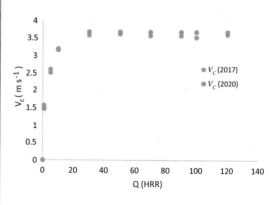

Ten iterations were used to calculate V_c using the NFPA 502:2017 method. In the HRR table, we have not performed an interpolation. Our criteria has been to consider between the two extreme values, the one that provides the most critical speed.

Considering calculations made using the NFPA 502: 2020 method, for values of HRR up to 15 MW the condition that A $\leq$ B holds, thus V_c was calculated according to Eq. 7.18, that is to say, V_c is a function of Q. From 16 MW the condition $A > B$ holds and calculations were performed with Eq. 7.19, which does not depend on Q. In contrast, for calculations made following NFPA 502: 2017, V_c is always a function of Q and the accuracy of the value obtained is determined by the number of iterations performed.

7.8   Explosiveness of Mine Gases

Of all coal-mine gases, methane (CH_4) is perhaps the most feared. However, others such as carbon monoxide (CO), hydrogen (H_2) and hydrogen sulphide (H_2S) can also play a major role.[9] The explosibility, explosiveness, or explosive nature of all of these gases is a function, not only of their concentration, but also of the presence of other gases with which they interact.

To determine whether a mixture is explosive or not, certain limits can be used for a first approximation. The *Lower Explosive Limit (LEL)* is the minimum concentration of a combustible gas or vapour required for combustion in air to take place. The *Upper Explosive Limit (UEL)* is the maximum concentration of a gas or vapour that will burn in the presence of air. The range between the LEL and the UEL is known as the explosive range (Table 7.2).

Table 7.2 Lower (LEL) and upper (UEL) explosive limits for various mine gases (Zabetakis et al. 1959; Zabetakis 1965; Yaws and Braker 2001)

Gas	LEL (%)	UEL (%)
CH_4	5	15
CO	12.5	74
H_2	4	75
H_2S	4	44

Question 7.1 Indicate the hazards of hydrogen sulphide and methane.

Answer

Hydrogen sulphide is an extremely poisonous gas, so much so that concentrations of 0.07% can lead to death within an hour. It must thus be monitored at all times if its presence is suspected.

---

[9]NH_3 in the presence of air at 16–27%$_v$ can form explosive mixtures, but as this is a very high concentration it is generally considered to be non-explosive (Geadah 1985).

Methane is a combustible gas which when mixed with air forms an explosive mixture. When the methane concentration is between 5 and 15%, the mixture is extremely explosive. From a toxicological point of view, this gas displaces oxygen.

Question 7.2 Indicate with an x the gases meeting the characteristics in the attached table. Also indicate the range of concentrations in which they are explosive.

Gas	Flammable	Explosive	Toxic	Asphyxiating	Concentration range at which it is explosive (%)
CO					
CO_2					
CH_4					
SO					
N_2					
NO_x					
H_2S					
H_2					
Benzene					

Answer

Gas	Flammable	Explosive	Toxic	Asphyxiating	Explosive limits (%)
CO	x	x	x	x	12.5–73
CO_2				x	
CH_4	x	x	x	x	5–15
SO			x		
N_2				x	
NO_x			x		
H_2S	x	x	x	x	4.3–46
H_2	x	x		x	4.1–74
Benzene	x	x			1.3–7.9

7.8.1 Explosive Limits of Mixtures of Gases or Vapours

When several substances coexist in a gaseous mixture, the explosive limits of the mixture do not coincide with the limits of these substances separately. Even the most conservative approach cannot be used, as the mixture can be more dangerous than each gas separately. With this caveat, it is possible to use Chatelier's principle (equilibrium law) as a first approximation for the limits of explosiveness (Eqs. 7.20 and 7.21):

$$LEL(\%_v) = \frac{100}{\frac{c_1}{LEL_1} + \frac{c_2}{LEL_2} + \frac{c_3}{LEL_3} + \cdots + \frac{c_i}{LEL_i}} \tag{7.20}$$

$$UEL(\%_v) = \frac{100}{\frac{c_1}{UEL_1} + \frac{c_2}{UEL_2} + \frac{c_3}{UEL_3} + \cdots + \frac{c_i}{UEL_i}} \tag{7.21}$$

where

- c_i: Concentration of each fuel gas relative to all fuels, expressed as a percentage of volume.
- LEL_i: Lower Explosive Limit of gas i.
- UEL_i: Upper Explosive Limit of gas i.

Exercise 7.5 Estimate the Lower Explosive Limit for a mixture of 1% hexane and 3% methane in air.

Data:

Gas	LEL (%)
Hexane	1.1
Methane	5

Solution

There is a total of $5 + 1.1 = 6.1\%$ of flammable gases, and therefore, 93.9% of air. Hence, the concentration of each of the two gases over the total of combustible gases (6.1%) is as follows:

$$c_{hexane} = 100 \left(\frac{1.1}{6.1} \right) = 18.03$$

$$c_{methane} = 100 \left(\frac{5}{6.1} \right) = 81.97$$

Therefore:

$$LEL(\%_v) = \frac{100}{\frac{18.03}{1.1} + \frac{81.97}{5}} = 3.05\%$$

Since the mixture is $1 + 3 = 4\%$, higher than the determined 3.05%, the mixture must be considered dangerous.

7.8.2 Explosiveness Determinations with the USBM Method

Of all the existing explosiveness diagrams, the graphic produced by the USBM (Zabetakis et al. 1959) is illustrated here.

In using this diagram, a point must first be represented on the x and y axes of Fig. 7.8.

The x-axis constitutes the percentage of inert gases, in accordance with Eq. 7.22:

$$X(\%) = N_{ex}(\%) + 1.5\,CO_2(\%) \tag{7.22}$$

Excess nitrogen (N_{ex}) is calculated as:

$$N_{ex}(\%) = N_2(\%) - O_2(\%)\,\frac{79.04}{20.93}$$

Hence Eq. 7.23

$$N_{ex}(\%) = N_2(\%) - 3.8\,O_2(\%) \tag{7.23}$$

Equation 7.24 ensues:

$$X(\%) = N_2(\%) - 3.8\,O_2(\%) + 1.5\,CO_2(\%) \tag{7.24}$$

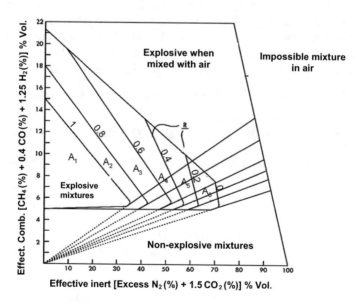

Fig. 7.8 Explosiveness diagram for mixtures of N_2, CO_2, CH_4, H_2 and CO at 26.67 °C and atmospheric pressure. Modified from Zabetakis et al. (1959)

Moreover, the y-axis represents the percentage of effective combustible gases, according to the expression shown in Eq. 7.25:

$$Y(\%) = CH_4(\%) + 0.4\,CO(\%) + 1.25\,H_2(\%) \tag{7.25}$$

The expression given above is the classic version, but nowadays it is expanded to take into account other gases, yielding Eq. 7.26:

$$Y(\%) = CH_4(\%) + 0.4\,CO(\%) + 1.25\,H_2(\%) + 0.6\,C_2H_6(\%) + 0.54\,C_2H_4(\%) \tag{7.26}$$

Finally, when a curve is being selected from the diagram, the ratio of CH_4 to total combustible gases (R) as in Eq. 7.27 must be used:

$$R = \frac{CH_4(\%)}{CH_4(\%) + H_2(\%) + CO(\%)} \tag{7.27}$$

In the case of more recent publications, R is expanded to include further gases (Eq. 7.28):

$$R = \frac{CH_4(\%)}{CH_4(\%) + H_2(\%) + CO(\%) + 0.6\,C_2H_6(\%) + 0.54\,C_2H_4(\%)} \tag{7.28}$$

The mixtures that fall within each of the areas A_1, A_2, A_3 and so on are explosive. Each R also defines the oblique line that joins the extreme right of each area with the origin of the coordinates. This line marks the boundary between explosive mixtures and those that are explosive when mixed with air. An important note is that the method is not valid if CO (%) > 3.0 and H_2 (%) > 5 (Zabetakis et al. 1959).

Exercise 7.6 Determine the explosiveness of the following gas mixtures using the USBM graph:

(a) $O_2 = 8.7\%$, CO $= 0.94\%$, $CH_4 = 1.8\%$, $CO_2 = 11.2\%$, $H_2 = 3.3\%$, $N_2 = 74.06\%$.
(b) $O_2 = 2.0\%$, CO $= 1.3\%$, $CH_4 = 3.0\%$, $CO_2 = 26.0\%$, $H_2 = 1.2\%$, $N_2 = 66.5\%$.

Solution

(a)
First, calculate the percentage of N_2 excess:
N_2 excess (%) $= N_2$ sample (%) $- N_2$ normal (%) $= 74.06\text{-}8.7 \cdot (79.04 \div 20.93)$
$= 41.2$
Second, calculate the effective inert gases:
X (%) $= N_2$ excess (%) $+ 1.5\,CO_2$ (%) $= 41.2 + 1.5 \cdot 11.2 = 58$
Third, calculate the effective combustible gases:
Y (%) $= CH_4$ (%) $+ 1.25\,H_2$ (%) $+ 6.4\,CO$ (%) $= 1.8 + 1.25 \cdot 3.3 + 0.4 \cdot 0.94$
$= 6.3$

Fourth, calculate the CH_4 ratio:

$$R = \frac{CH_4(\%)}{CH_4(\%) + CO(\%) + H_2(\%)} = \frac{1.8}{1.8 + 0.94 + 3.3} = 0.3$$

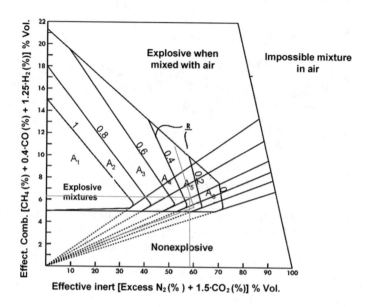

Although very close to the limit, the diagram indicates that the mixture is explosive.

(b)

First, calculate the percentage of N_2 excess:

N_2 excess $(\%) = N_2$ sample $(\%) - N_2$ normal $(\%) = 66.5 - 2.0 \cdot (79.04 \div 20.93) = 58.95$

Second, calculate the effective inert gases:

$X\ (\%) = N_2$ excess $(\%) + 1.5\ CO_2\ (\%) = 58.95 + 1.5 \cdot 26 = 97.95$

Third, calculate the effective combustible gases:

$Y\ (\%) = CH_4\ (\%) + 1.25\ H_2\ (\%) + 6.4\ CO\ (\%) = 3 + 1.25 \cdot 1.2 + 0.4 \cdot 1.3 = 5.02$

Fourth, calculate the CH_4 ratio:

$$R = \frac{CH_4}{CH_4 + CO + H_2} = \frac{3}{3 + 1.3 + 1.2} = 0.55$$

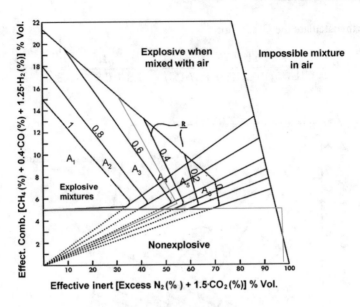

The point falls in the region of impossible mixtures, that is to say, it is impossible to form such a proportion of inert and combustible gases. Actually, the oblique line indicates the maximum effective combustibles for a given effective insert.

7.9   Flammable Dusts

For a dust explosion to take place, the presence of several factors is required (Abbasi and Abbasi 2007):

(a)   Combustible dust,
(b)   Adequate concentration,
(c)   Presence of an oxidant (oxygen),
(d)   Presence of an ignition source, and
(e)   A confined space.

Although one of the dust with the greatest risk of explosion is coal dust, other substances such as sulphide ores, which are less energy-rich, and oil shales (Weiss et al. 1995), must also be taken into account. Among sulphides, iron-containing powders such as pyrite, chalcopyrite and pyrrhotite are the most likely to trigger explosions. Further, disulphides tend to be more dangerous than mono-sulphides. The size of the dust influences the reactiveness of the sulphide mineral, so for dust larger than 180 μm the risk of explosion is reduced (Soundararajan et al. 1996). In addition to the above minerals, fine powders of the elements aluminium, magnesium and titanium are liable to cause explosions. It should be kept in mind that as the size of dust decreases, they can more easily be raised from the ground and carried on air currents.

Question 7.3 Indicate how the presence of a flammable dust affects the LEL of CH_4.

Answer

Since both are explosive, the presence of one of the two acts by lowering the LEL of the other.

7.10 Coefficients of Gases Indicating the Evolution of Fires

The presence of a fire inside a mine alters the composition of the mine gases. If other more sophisticated devices are absent, fires in coal seams can be detected at their earliest stages by the increase in temperature they cause, as well as by the rise in humidity in the return air. If this heating persists, fog will form, accompanied by exudation, or sweating, of walls. This will be followed by smells of oil or paraffin, culminating in a typical smoke smell (Heiss and Herbst 1945).

One of the main indicators of the existence of a fire is the presence of CO_2. However, high concentrations CO_2 may also be due to people breathing, wood decomposing, pyrites being oxidized, or the reaction of acidic water with carbonate rocks, among other causes. Hence, ratios based on CO_2 concentration should be taken with considerable caution. This is the reason that current indices consider not just carbon dioxide, but other gases such as CO, H_2, ethylene and propylene, which are released sequentially as coal heats up. A compilation of the most common ratios is given in Table 7.3 (Eqs. 7.29–7.33).

In the above, ΔO_2 is the oxygen deficiency (Eq. 7.34). In interpreting results, the ratio is usually expressed as a percentage.

$$\Delta O_2 = \left(\frac{20.93}{79.04}\right) N_2(\%) - O_2(\%) \qquad (7.34)$$

Table 7.3 Indicators of the state of evolution of fires in mines (see, for instance, McPherson 1993, p. 858)

Table of contents	Expression	
Graham's or carbon monoxide	$\frac{CO\,(\%)}{\Delta O_2}$	Eq. 7.29
Young	$\frac{CO_2(\%)}{\Delta O_2}$	Eq. 7.30
Willet	$\frac{CO(\%)}{\text{excess } N_2(\%) + CO_2(\%) + \text{combustibles }(\%)}$	Eq. 7.31
Jones-Trickett	$\frac{CO_2(\%) + 0.75\,CO(\%) - 0.25\,H_2(\%)}{\Delta O_2}$	Eq. 7.32
Carbon oxides	$\frac{CO}{CO_2}$	Eq. 7.33

As examples, the Graham and the Jones and Trickett formulae will be considered.

Graham's Ratio (1920)

This ratio works with the relationship between the carbon monoxide generated and the oxygen consumed. It is calculated by Eq. 7.35:

$$GR = \frac{100\ CO(\%)}{0.265(N_2 + Ar)(\%) - O_2(\%)} \tag{7.35}$$

Each mine must establish its own characteristic ranges, although the normal values for this ratio are below 0.5%. If the ratio is exceeded, it usually indicates heating of the coal. Values above 3% reliably reveal the presence of a fire.

The following is the set of average values normally used in European mines:

- 0.4%: Normal,
- 0.5%: Check,
- 1%: Coal heating,
- 2%: Severe heating, and
- 3%: Active fire.

This index is generally not considered to be very reliable if $\Delta O_2 < 0.3\%$.

Question 7.4 The following gas concentrations have been measured in a mine: $N_2 = 77.23\%$, $O_2 = 20.15\%$ and $CO = 18$ ppm. Estimate the state of the coal seam.

Answer

$$\Delta O_2 = \left(\frac{20.93}{79.04}\right) \cdot 77.23(\%) - 20.15(\%) = 0.3$$

Very little oxygen has been consumed, so the use of the Graham index does not seem advisable, and no further opinions can be offered.

Jones-Trickett Ratio (1954)

Is the ratio between the oxygen required for combustion and the amount consumed by burning. Its expression is the following (Eq. 7.36):

$$TR = \frac{CO_2(\%) + 0.75\ CO(\%) - 0.25\ H_2(\%)}{0.265(N_2 + Ar)(\%) - O_2(\%)} \tag{7.36}$$

As in the previous case, in accordance with the values recorded some conclusions can be reached. Among the commonest are the following:

- 0.5: CH_4 is burning,
- 0.7–0.9: Coal, lubricants or even a conveyor belt are burning,
- 1–1.6: Wood is burning, and
- Greater than 1.6: Such figures are not possible values for the ratio.

Exercise 7.7 Let there be a mine gas sample with the following characteristics[10]:

- O_2: 16,000 ppm,
- CO: 1.47%,
- CH_4: 2.3%,
- CO_2: 17%,
- H_2: 0, and
- N_2: 77.63%.

Calculate Graham's and Trickett's ratios and offer possible conclusions about the potential status of fires in the mine.

Solution

Calculation of the key figures:

First, it is determined whether Graham's expression is applicable:

$$\Delta O_2 = \left(\frac{20.93}{79.04}\right) N_2(\%) - O_2(\%) = \left(\frac{20.93}{79.04}\right) \cdot 77.63 - 1.6 = 18.9 > 0.3$$

This expression might thus give rise to errors. Nevertheless, it can be computed:

$$GR = \frac{100CO(\%)}{0.265(N_2 + Ar)(\%) - O_2(\%)} = \frac{100 \cdot 1.47}{0.265 \cdot 77.63 - 1.6} = 7.75$$

Thus suggesting the presence of an active fire.

The Jones-Trickett ratio is now calculated:

$$TR = \frac{CO_2(\%) + 0.75\, CO(\%) - 0.25\, H_2(\%)}{0.265(N_2 + Ar)(\%) - O_2(\%)} = \frac{17 + 0.75 \cdot 1.47 - 0}{0.265 \cdot 77.63 - 1.6} = 0.95$$

This suggests a fire in the coal, in fuel or in a conveyor belt. Since the ranges are not precise, it may even indicate the possibility of wood burning inside the mine.

7.11 Fire-Fighting Techniques

Fire-fighting methodologies include the displacement of O_2 and the reduction of heat. These can be achieved by isolating the affected portion of the mine by means of an adequate ventilation system, by creating inert atmospheres, by using foams and sludge, or even by controlled flooding. As they are directly related to ventilation, the

---

[10]Note that in the gas data the sum is 100%, without any account being taken of the concentration of argon, probably because of difficulties in obtaining a detector for this gas.

systems for creating inert atmospheres, *dynamic pressure balancing* and *variation of the ventilation regime* are listed below.

7.11.1   Formation of Inert Atmospheres

This approach seeks to create atmospheres that cannot sustain combustion, including ignition. This can be achieved in two ways: (a) by increasing the concentration of flammable gas above its UEL (for example, the injection of a large amount of CH_4 would displace O_2 and prevent combustion); (b) by the addition of an inert gas that cannot play a part in the combustion process, also displacing O_2.

Obviously, the second technique is safer, so inert gases are commonly used to displace O_2, as a measure to prevent explosions or to inhibit a combustion that has already begun (McPherson 1993, p. 853). Three gases are currently in use: CO_2, inert combustion gases and N_2.

Carbon Dioxide

As is well known, this gas is denser than air, making it particularly suitable for extinguishing fires in deep areas of mines. The gas helps to put a fire out by cooling and by displacing oxygen. However, its use suffers from certain disadvantages (McPherson 1993, p. 854):

- It can cause freezing of conduits.
- It is soluble in water, which causes part of it to be lost.
- It is corrosive once dissolved in water.
- It can act as an O_2 donor to fires.
- It does not follow the general circuit of ventilation, tending to accumulate in the lower zones of the workings.
- It is more expensive than N_2 and inert combustion gases.
- It interferes with the monitoring of the evolution of the fire by means of gas ratios.
- It tends to adsorb to carbonaceous surfaces.
- It can lead to the generation of CO.

Inert Combustion Gases

The group of techniques involving inert combustion gases includes the use of CO_2, H_2O, N_2 and small amounts of CO and H_2 from combustion. In the early days of this technique, inert coal combustion gases were used, but today they are obtained from the combustion gases of kerosene-fuelled jet engines. The gases are subjected to a previous cooling process and injected into the affected area. The most commonly used systems today are probably the GAG 3A, named for the Polish initial letters for gas-to-gas combustion (Mucho et al. 2005) and the Steamexfire, which adds into the gas stream a large amount of atomized water and steam. Both systems are preferred to the use of N_2 and CO_2, as they are able to supply higher airflow rates of inert gas. Despite this, the group of procedures has the following disadvantages:

- High operating costs (approximately twelve staff required for one twenty-four-hour period) and maintenance.
- A requirement for highly specialized equipment to implement the procedure.
- The gas stream contains O_2 concentrations of between 1 and 3%.
- Inert gases usually move through the mine at a temperature of up to 90 °C, making it harder to extinguish a fire.

Nitrogen

Initially, the use of N_2 was based on accumulated experience with CO_2. Nitrogen is generally preferred because (Ray and Singh 2007):

- It has a specific weight similar to, but slightly lower than, air.
- It is injected at low temperatures, which results in the cooling of the burned areas.
- It is easy to produce and its logistics are simple.
- It is not toxic.
- It is inert.
- It effectively displaces O_2.
- It lessens the chances that the atmosphere will be explosive.
- It can be used in areas that were sealed by the presence of fires.
- It helps to control spontaneous coal combustion.
- It is supplied as cold gas.

However, despite the above, there are multiple disadvantages to this approach:

- It cannot be stored for long periods, so it is sometimes not possible to obtain with any ease the large quantities needed at the time of a fire.
- It is more expensive than inert combustion gases, which can also be used for extinction, although definitely cheaper than CO.
- In many cases, there are inherent difficulties in transporting this gas into the mine from outside.

7.11.2 Dynamic Pressure Balancing

Areas of the mine that are affected by fires are isolated, for example, by means of explosion-resistant barriers. However, no matter how well constructed these may be, there will almost always be air leaks that can eventually lead to a revival of the fire. In order to avoid this phenomenon, additional walls are constructed, creating chambers in which the existing pressure is monitored by means of sensors that function by injecting N_2 if the pressure drops (Fig. 7.9). The procedure used to balance these pressures is called *dynamic balancing*. It is widely used in coal mining.

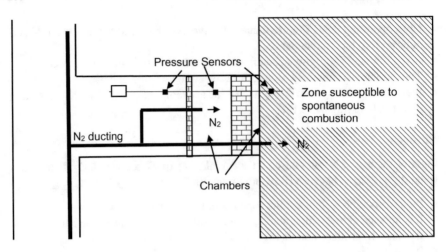

Fig. 7.9 Dynamic balancing by injection of N_2

7.11.3   Alteration of the Ventilation Mode

Another method of acting against fires involves varying the direction of the airflow inside the mine. This can be done locally or generally. With this approach, the aim is:

- To decrease the supply of oxygen.
- To control the direction and speed with which a fire is propagated.
- To drive combustion products to areas where they are less harmful.
- To mitigate fire-induced variations in the mine airflow circulation.
- To reduce fire-related landslides.

In all cases, the direction of ventilation must at first be such as to facilitate the evacuation of personnel. Once this has been ensured, it should be directed towards extinguishing the fire. If the mine does not release much coal gas, zero ventilation may be advisable, as a supply of oxygen-rich air could stoke the fire. However, this procedure faces limitations in the event the fire is located in an area of inclined workings, mainly because of the interference that the depression generated by the fire can exert on general ventilation. In this latter case, the recommendation would be to maintain the overall ventilation regime, but to isolate the workings affected by fire (Novitzky 1962, p. 514).

Once personnel have been evacuated, the main ventilation can be reversed if the fire originated in the downcast shaft or the surrounding areas, while reversal is not recommended if the fire takes place elsewhere. In any case, the reversal of the normal ventilation regime is a controversial decision, which has been put into practice only in a few rare instances. By way of summary it can be stated that ventilation reversal causes the following (McPherson 1993, pp. 831–833):

- A decrease ensues in the flow rates provided by the fan, because of the devices used to change the fan's direction.
- Areas that were initially expected to be free of gases may become filled with them, potentially affecting any miners who would use them as an escape route.
- Doors are prevented from having their closure assisted by ventilation, causing unexpected air movements.
- The pressure drop that is triggered can increase CH_4 emissions, heightening the risk of explosions.
- The ventilation network may become unstable for hours.
- The effects of the fire must be quantified, as they can cause local ventilation reversals.

In the case of centrifugal fans, reversal of ventilation is usually achieved with the help of a system of gates through which, for an exhaust system, the air inlet becomes the atmosphere, while the air outlet is led into the shaft. In the case of axial fans, the reversal of ventilation can be produced directly by varying the direction of rotation of the motor.

References

Abbasi, T., & Abbasi, S. A. (2007). Dust explosions–Cases, causes, consequences, and control. *Journal of Hazardous Materials, 140*(1–2), 7–44.

Budryk, W. (1956). *Pozary i wybuchy w kopalniach*, pp. 78–79, Katovice.

Geadah, M. (1985). *National inventory of natural and anthropogenic sources and emissions of ammonia (1980)*. Environmental Protection Programs Directorate, Environmental Protection Service, Environment Canada Report EPS5/IC/1.

Graham, J. I. (1920). The normal production of carbon monoxide in coal mines. *Transaction Institution of Mining Engineers, 60*, 222–234.

Hansen, R. (2010). Overview of fire and smoke spread in underground mines. In *Fourth International Symposium on Tunnel Safety and Security* (pp. 483–494). SP Fire Technology.

Heiss, F., & Herbst, F. (1945). Incendios, aparatos para la respiración y salvamento (Chap. 10). In *Tratado de laboreo de minas*. Madrid: Labor.

Jones, J. H., & Trickett, J. C. (1954). Some observations on the examination of gases resulting from explosions in collieries. *Mining Engineering* 114.

Kennedy, W. D. (1996). *Critical velocity: past, present and future, One Day seminar on smoke and critical velocity in tunnels*. ITC.

Kuenzer, C., Zhang, J., Tetzlaff, A., Van Dijk, P., Voigt, S., Mehl, H., et al. (2007). Uncontrolled coal fires and their environmental impacts: Investigating two arid mining regions in north-central China. *Applied Geography, 27*(1), 42–62.

Laboratorio Oficial Madariaga (LOM). (2013). *Técnicas de control y extinción de incendios en las obras subterráneas en la que se emplea técnica minera en su ejecución*. Madrid: Ministerio de Industria Comercio y Turismo.

Lee, C. K., Hwang, C. C., Singer, J. T., & Chaiken, R. F. (1979). *Influence of passageway fires on ventilation flows*. In Second International Mine Ventilation Congress, Reno, NV.

Luque, V. (1988). *Manual de ventilación de minas*. Asociación de Investigación Tecnológica de Equipos Mineros. Madrid: AITEMIN.

McPherson, M. J. (1993). Subsurface fires and explosions. In *Subsurface ventilation and environmental engineering*. Chapman & Hall.

Mucho, T. P., Houlison, I. R., Smith, A. C., & Trevits, M. A. (2005). *Coal mine inertisation by remote application*. Proceedings of the 2005 US National Coal Show, pp 7–9.

Ray, S. K., & Singh, R. P. (2007). Recent developments and practices to control fire in undergound coal mines. *Fire Technology, 43*(4), 285–300.

Schmidt, W., Grumbrecht, K., Bohm, H. J., & Blumel, H. (1973). On the mutual effect of open mine fires and ventilation design. *Gluckauf Forschungsheft, 34*(6), 213–220.

Simode, E. (1976). Stabilisation de l'aérage en cas d'incendie dans les travaux du fond: Théorie de Budryk. En: Aérage. Document SIM N3. Industrie Minérale. Mine, pp. 2–76.

Soundararajan, R., Amyotte, P. R., & Pegg, M. J. (1996). Explosibility hazard of iron sulphide dusts as a function of particle size. *Journal of Hazardous Materials, 51*(1–3), 225–239.

Stracher, G. B., & Taylor, T. P. (2004). Coal fires burning out of control around the world: thermodynamic recipe for environmental catastrophe. *International Journal of Coal Geology, 59*(1–2), 7–17.

Surkov, A. L. (1975). Determination of heat depression. In *Problems of safety in coal mines*. US Department of Interior.

Thomas, P. H. (1970). Movement of smoke in horizontal corridors against an air flow. *The Institution of Fire Engineers Quarterly, 30*(77), 45–53.

Trutwin, W. (1972). Estimation of the natural ventilating pressure caused by fire. In *International journal of rock mechanics and mining sciences & geomechanics abstracts* (vol. 9, no. 1, pp. 25–36). Pergamon.

Weiss, E. S., Cashdollar, K. L., Sapko, M. J., & Bazala, E. M. (1995). Secondary explosion hazards during blasting in oil shale and sulfide ore mines.

Zabetakis, M. G. (1965). *Flammability characteristics of combustible gases and vapors* (No. BULL-627). Washington DC: Bureau of Mines.

Zabetakis, M. G., Stahl, R. W., & Watson, H. A. (1959). *Determining the explosibility of mine atmospheres*. BuMines IC 790.

Zhang, J., Wagner, W., Prakash, A., Mehl, H., & Voigt, S. (2004a). Detecting coal fires using remote sensing techniques. *International Journal of Remote Sensing, 25*(16), 3193–3220.

Zhou, F., & Wang, D. (2005). Backdraft in descensionally ventilated mine fire. *Journal of Fire Sciences, 23*(3), 261–271.

Bibliography

Dougherty, J. J. (1969). Control of mine fires. Mining Extension Service, School of Mines, Appalachian Center, West Virginia University.

Francart, W. J., & Beiter, D. A. (1997). Barometric pressure influence in mine fire sealing. In *Proceedings of 6th international mine ventilation congress*, Pittsburgh, PA, May (pp. 17–22).

Froger, C., Jeger, C., & Pregermain, S. (1976). Feux de mine. In Aérage. Document SIM N3. *Industrie Minérale*. Mine 2–76.

Gillies, A. D. S., Wala, A. M., & Wu, H. W. (2004). Case studies from application of numerical simulation software to examining the effects of fires on mine ventilation systems. In *Proceedings of the 10th US mine ventilation symposium*, pp. 445–455.

Graham, J. I. (1914). Adsorption of oxygen by coal. *Transactions of the Institution of Mining Engineers, XLVIII*, 521.

Graham, J. I. (1917–1918). The origin of blackdamp. *Transactions of the Institution of Mining Engineers, LV*, pp. 294–312.

Jones, J. E., & Trickett J. C. (1954–1955). Some observations on the examination of gases resulting from explosions in colleries. *Transactions of the Institution of Mining Engineers, 114*, 768–790.

Justin, T. R., & Kim, A. G. (1988). Mine fire diagnostics to locate and monitor abandoned mine fires. *BuMines IC, 9184,* 348–355.

Kim, A. G. (2007). Greenhouse gases generated in underground coal-mine fires. En: Stracher, G. B. (Ed.). Geology of coal fires: case studies from around the world (Vol. 18). Geological Society of America.

Kissel, F. N., Diamond, W. P., Beiter, D. A., Taylor, C. D., Goodman, G. V., Cecala, A. B., & Volkwein, J. C. (2006). *Handbook for methane control in mining.* IC 9486. CDC Wokplace Safety and Health.

Koenning, T. H., & Bruce, W. E. (1987, October). Mine fire indicators. In *Proceedings of the 3rd US mine ventilation symposium,* University Park, PA (pp. 433–437).

León Marco, P. (1992). Fuegos en minas de carbón. Madrid: Instituto Geológico y Minero (IGME).

Litton, C. D. (1986). *Gas equilibrium in sealed coal mines.* BuMines RI 9031.

Mitchell, D. W. (1984). *Understanding a fire: Case studies.* Presented at workshop on combating mine fires, Sept. 20–21, Eighty-Four, PA, p. 14.

Mitchell, D. W. (1990). Interpreting the state of the fire. In: *Mine Fires.* Maclean Hunter Publishing, pp. 65–66.

NFPA. (2000). *NFPA 750 standard for the installation of water mist fire protection systems,* 2000 Edition. Quincy, MA.

Sánchez Arboledas, J. (1924). *Incendios y Fuegos Subterráneos.* Madrid: Artes de la Ilustración.

Timko, R. J., & Derick, R. L. (1995). Detection and control of spontaneous heating in coal mine pillars: A case study. BuMines RI 9553, p. 18.

U.S Code of Federal Regulations (2005). Title 30: Mineral Resources, Chapter I–Mine Safety and Health Administration, Department of Labor; Part 75-Mandatory Safety Standards-Underground Coal Mines, Subpart D-Ventilation, Sec. 75.321 Air Quality, Paragraph (a) (1).

Zabetaksi, M. G., Stahl, R. W., & Watson, H. A. (1959). *Determining the explosivity of mine atmospheres. U.S. Bureau of Mines.* Bulletin IC7901, p. 20.

Zhang, J., Wagner, W., Prakash, A., Mehl, H., & Voigt, S. (2004b). Detecting coal fires using remote sensing techniques. *International Journal of Remote Sensing, 25*(16), 3193–3220.

Chapter 8
Secondary Ventilation

8.1 Introduction

Secondary, auxiliary, or *ancillary ventilation* conveys air to those parts of the mine that are not directly served by the main ventilation circuit, such as development headings and certain stopes. The calculations we explore here recur in many engineering situations, not only in mining but also in civil works, for example, tunnel construction. The ultimate aim of these calculations is to generate a value for the necessary rate of air renewal within the mining works in order to dimension the appropriate fan. To this end, an extensive bibliography concerning the existing methodologies is compiled in this chapter.

8.2 Types of Secondary Ventilation

8.2.1 Forcing Ventilation

This type of ventilation pushes fresh air onto the gallery face. It has the following main characteristics:

- It favours the removal of the gases at the active face.
- It decreases the re-entry time after blasting.
- The air carried to the face is relatively fresh.
- The exhaust gases return through the gallery with the associated disadvantages.
- Dust is forced into suspension and dispersed by turbulence.

Depending on the number of fans and their arrangement, the *forcing system* can be *single* (Fig. 8.1a), *double* (Fig. 8.1b) or *multiple* (Fig. 8.1c).

The cone of air leaving the ventilation duct spreads out as it approaches the active face and as it does so its velocity decreases. Referring to Fig. 8.2, this generates a stationary zone, termed the *dead zone* (*a*) and a *ventilated zone* (*X*). In order to

C. Sierra, *Mine Ventilation*, https://doi.org/10.1007/978-3-030-49803-0_8

Fig. 8.1 Side view of a forcing ducted fan: **a** single, **b** double and **c** multiple

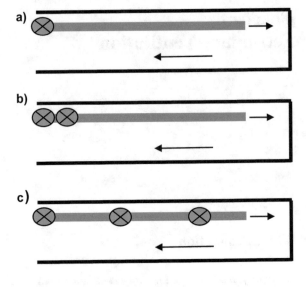

Fig. 8.2 Side view of the dead zone (*a*), ventilated zone (*X*) and recommended duct-to-face distance (*d*) in the forcing system for a development heading

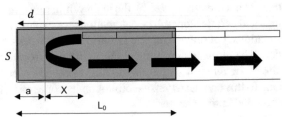

maximize the ventilated zone, the duct outlet must be placed at distances (d) $\leq$ 15 m. A common recommendation is that $d \leq 4\sqrt{S}$ or $d \leq 6\sqrt{S}$ where S is the face cross section in m^2 (e.g. Borisov et al. 1976, p. 381).[1] In Europe, most legislation requires that $d \leq 5\sqrt{S}$. Higher values of duct-to-face distance increase the role that natural diffusion plays at cleaning this zone. Certain foreign regulations use the diameter of the duct to calculate d setting it at 10–15 times this value (Kissell 2006).

There is an additional requirement in the case of coal mining for the airflow rate through the main gallery to be at least 1.3 times as big as the airflow rate through the forcing system (e.g. ASM-51, RGNBSM-IT.05.0.03).

---

[1]The smaller the duct-to-face distance the more likely it is that the duct will be damaged during blasting.

8.2.2 Exhausting Ventilation

In *exhausting ventilation* a fan is placed at one of the ends of a duct, extracting air from inside it and drawing out contaminated air causing clean air to sweep the face. Its most important features are:

- It provides a permanently clean zone between the entrance to the gallery and the beginning of the suction duct.
- It does not move contaminants directly through the gallery but confines them to the duct.
- It does not achieve an efficient sweep of the blasting gases at the face because the air that enters through the gallery does so across the whole available cross section, which decreases its speed and with it its turbulence. Moreover, air entering the gallery will not necessarily traverse the face before entering the duct, a factor which also diminishes the effectiveness of this system.
- Mixed exhaust ventilation scheme is not usually used unless the active face is being advanced by roadheaders or continuous miners.
- The scheme cannot be used in works with firedamp (e.g. ITC 05.0.03 ap.2 [MITC 2000]).
- Dust is directly evacuated meaning that this system is preferred in dusty conditions.

A diagram of this type of ventilation is presented in Fig. 8.3.

Fig. 8.3 Side view of an exhausting ventilation system at the development end

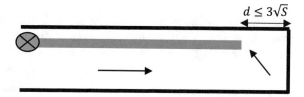

$$d \leq 3\sqrt{S}$$

In general $d \leq 3\sqrt{S}$ or $\leq 2\,\text{m}$ is usual. If works involve the use of coal cutting machines, it is recommended that $d \leq 1.5\sqrt{S}$. In international regulations, this value is limited to a maximum of 3 m, particularly in the presence of CH_4 emissions (Kissell 2006).

8.2.3 Mixed systems

Mixed systems comprise two fans both located in the main gallery one of which is in a forcing arrangement while the other is exhausting (Fig. 8.4). If the exhausted airflow rate is greater than the forced one, the air returns mainly through the exhaust duct. Conversely, if the forced airflow rate is higher than the exhausted one, the air

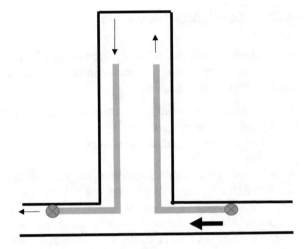

Fig. 8.4 Layout of the mixed ventilation scheme of development end

returns through the gallery. Although this system combines many of the advantages of the two previously mentioned systems it also has many of their disadvantages, and in addition, its installation is more complex and more expensive than either of them. The system uses either rigid ducts or steel reinforced canvas ducts.

8.2.4 Overlap System

In an *overlap ventilation system* two fans are used, with one of them placed very close to the active face. There are many possible configurations, but the most frequently used is shown in Fig. 8.5. Here, the forcing fan is closest to the face, while the

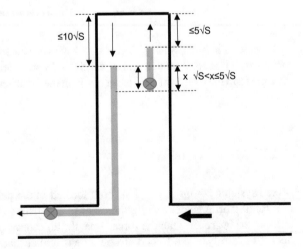

Fig. 8.5 Recommended layout of an overlap ventilation system for a development heading

exhausting fan, which must extract a greater flow to reduce recirculation, is placed furthest away. Normally the recommended configuration for the case of coal mines is (e.g. ASM-51):

- Distance between the face and air inlet of exhaust duct $<10\sqrt{S}$.
- Distance between the face and the forcing duct $<5\sqrt{S}$.
- The zone of overlap between the exhaust and forcing systems should be between $\sqrt{S}$ and $5\sqrt{S}$.
- With the additional requirement that the flow circulating through the main gallery must be at least 1.3 times greater than that circulating through the exhaust duct.

With regard to the height at which the ducts are to be laid, the exhausting duct is usually located on the roof of the gallery, as the gases in the gallery tend to be hot and will, therefore, rise.

8.3 Required Airflow Rate at the Face

8.3.1 Minimum Airflow Rate to Ensure Miners Can Breathe (Q_{op})

The minimum airflow rate required for each independent development heading must be a function of the number of persons present in the largest shift and assuming a consumption rate of 40 litres per person per second (e.g. ITC 04.7.01 [MITC 2000]). Therefore (Eq. 8.1):

$$Q_{op}[l\ s^{-1}] = 40n \qquad (8.1)$$

where n is the maximum number of workers that are in the face at a time.

8.3.2 Minimum Airflow Rate Required for Gases to Return at the Regulatory Minimum Speed (Q_r)

This is calculated as:

$$Q_r = A\,v$$

where A is the cross sectional area of the gallery (m^2).

For works using explosives, international regulations generally require that this minimum speed should be 0.2 m s^{-1}. Thus, the flow will be (e.g. ITC 05.0.01 [MITC 2000]) (Eq. 8.2):

$$Q_r \geq 0.2\,A \tag{8.2}$$

In coal mining galleries the requisite minimum speed is usually set at 0.30 m s^{-1} (ITC-04-07-01 [MITC 2000]), while in civil tunnels its range is 0.40–0.50 m s^{-1}.

8.3.3 Airflow Rate Needed to Dilute the Exhaust Gases from Diesel Engines (Q_m)

The relevant literature (e.g. Deniau 1976) suggests that for every horsepower, a diesel engine will emit exhaust gases at a rate (F), of about 0.07–0.09 l s^{-1}. The composition of these gases will be:

- CO: 100–400 ppm,
- NO$_x$: 200–600 ppm, and
- CO$_2$: 3–6%.

According to the approach of Bertard and Bodelle (1962), the airflow required to ventilate these gases can be calculated as the sum of the flows necessary to lower the concentrations of CO and NO$_x$ to admissible values as follows:

$$Q_m = F\left(\frac{CO}{C_{CO}} + \frac{NO_x}{C_{NOx}}\right) \tag{8.3}$$

where

- Q_m: Airflow rate required to dilute diesel equipment gases (m^3 s^{-1}).
- F: Exhaust gas flow rate (m^3 s^{-1}).
- CO: Concentration of CO in the exhaust gases.
- NO$_x$: Concentration of NO$_x$ in exhaust gases.
- C_{CO}: Maximum allowable concentration of CO. According to accepted values: 10^{-4} (fraction).
- C_{NOx}: Maximum permissible concentration of NO$_x$. According to accepted values: 0.25–10^{-4} (fraction).

Therefore, the above expression, if concentrations are expressed in terms of fractions is (Eq. 8.4):

$$Q_m = F(10{,}000\,CO + 40{,}000\,NO_x) \tag{8.4}$$

A number of national regulations express Q_m as a function of the power of the diesel engines. Gangal (2012) offers a compilation of these:

- Australia: 0.06 m^3 s^{-1}· Pw (kW),
- Canada: (0.045– 0.092) m^3 s^{-1}· Pw (kW),
- Chile: 0.063 m^3 s^{-1}· Pw (kW),

- South Africa: 0.063 m^3 s^{-1}· Pw (kW), and
- Spain: 0.066 m^3 s^{-1}· Pw (kW) (ITC 04.7.02 [MITC 2000]).

The main criticism that can be made of this approach is that, in the case of mines located at high altitude, gaseous emissions from diesel engines are much more dangerous. In any case, the above values should be taken as a minimum.

Overall, the Mine Safety and Health Administration (MSHA) states that a twofold increase in airflow sweeping a face will half *Diesel Particulate Matter* (DPM). The same authority publishes a *Particulate Index* (PI) for all diesel equipment approved for use in mines, which establishes the minimum air quantity required to lower the DPM concentration to 1000 μg m^{-3} (800 TC μg m^{-3}).

At the present time, there is a growing tendency to consider the *Exhaust Quality Index* (EQI). In this case, diesel engines are subjected to an approved test whereby sufficient air is supplied to reduce the value of EQI < 3. In this way, the amount of air required to properly ventilate the engine is established (Eq. 8.5)[2]:

$$EQI = \frac{CO}{50} + \frac{NO}{25} + \frac{DPM}{2} + 1.5\left[\frac{SO_2}{3} + \frac{DPM}{2}\right] + 1.2\left[\frac{NO_2}{3} + \frac{DPM}{2}\right] \quad (8.5)$$

where

- DPM: Diesel Particulate Matter (mg m^{-3}), and
- CO, NO, NO$_2$, SO$_2$: Gas concentrations (ppm).

8.3.4 ˙ Airflow Rate Required to Dilute Blasting Gases (Q$_e$)

The following calculations can be applied to one pollutant or several, although it is common to account only for the most critical. A gas can be considered as more problematic based on its toxicity (expressed as its TLV)[3] and on how difficult it is to dilute. Following de Souza and Katsabanis (1991), we can indicate the most problematic gas by means of a simple quotient of the initial concentration of the gas under consideration and its TLV. In this way, in general, the gas with the highest quotient is deemed to be more critical, since a greater fraction of it will have to be eliminated.

In general, the most problematic gases are usually the nitrogen oxides (NO$_x$), of which NO$_2$ is the most dangerous, not only because it is the most toxic (lowest TLV) and difficult to diffuse, but also because it is the most common given the great propensity of NO to oxidize rapidly to NO$_2$. This fact is quantified on the *Relative General Toxicity Index* (RGTI), used in some European countries (Zawadzka-Małota

---

[2]There are a number of parameters specific to industrial ventilation that should be used more frequently in the field of mining ventilation. These include *ventilation effectiveness, ventilation performance* or *contaminant removal effectiveness*. To obtain a deeper understanding of these issues a particularly relevant reference is Howard and Esko (2001).

[3]TLV is the *Threshold Limit Value*. This has already been discussed in previous chapters.

Table 8.1 Gases emitted according to the type of explosive. Taken from Greig (1982)

Explosive	Gas emitted $(m^3 kg^{-1}_{explosive})$			
	CO	NO + NO₂	CO₂	NH₃
Ammon dynamite	0.03	0.004	0.06	0.003
Ammon gelignites	0.05	0.006	0.07	0.003
ANFO	0.03	0.007	0.05	0.009
Dynagel	0.03	0.005	0.07	0.003
Watergel	0.09	0.002	0.05	–

2015). Using this criterion NO_2 has a toxicity 6.5 times greater than that of CO. This index is expressed as (Eq. 8.6):

$$L_{CO} = CO + 6.5\,NO_X \qquad (8.6)$$

where

- CO: Volume emitted per unit mass of explosive (l kg^{-1}), and
- NO$_X$: Volume emitted per unit mass of explosive (l kg^{-1}).

The composition of blasting fumes varies greatly depending on the composition of the explosive, so it is always best to refer to any data supplied by the manufacturer. However, these data are not always comprehensive, and so great disparities can be expected between these and practice. This is because tests on explosives are often carried out in special chambers with only small masses of explosives. This results in a reduction in the efficiency of the explosion reaction and means that some explosives are tested at sizes below the critical diameter thus no detonation reaction takes place. In this case, stoichiometric estimations are frequent.

In addition, many differences exist between the conditions of humidity and temperature in the test chamber compared to the locations where an explosive will be used. Not only this, unlike in a test chamber, within a mine, there is the possibility of the gases reacting with the rock and escaping through cracks. Here, the classic work of Greig (1982) (Table 8.1), in which values for different American explosives are compiled is still frequently referred to.

The García and Harpalani (1989) values, for the explosives Tovex 100, Tovex 220, Powermax 140, Iremite 42, are also widely accepted:

- CO: 0.0025–0.02 m^3 kg^{-1},
- CO$_2$: 0.042–0.097 m^3 kg^{-1},
- NO: 0.00062–0.0081 m^3 kg^{-1}, and
- NO$_2$: 0.00012–0.0068 m^3 kg^{-1}.

However, a more up to date reference is Zawadzka-Małota (2015) (Table 8.2).

If the combustion of the ANFO took place in an ideal stoichiometrically balanced reaction, then toxic gases would not be generated (Dick et al. 1982):

Table 8.2 Volume of gases generated in blasting for different explosives (Zawadzka-Małota 2015)

Explosive	Gas volume per unit mass of explosive ($l\ kg^{-1}$)					
	CO_2	CO	NO	NO_2	NO_x	$CO + 6.5\ NO_x$
Ammonite 1	145.15	5.87	3.58	1.15	4.74	36.65
Ammonite 2	110.22	2.63	0.77	0.56	1.22	10.56
Dynamite 1	167.45	5.93	0.92	0.07	0.99	12.38
Dynamite 2	181.60	4.37	2.61	4.50	0.64	8.53
Dynamite 3	185.12	4.58	1.89	0.11	2.00	17.59
Dynamite 4	171.53	1.56	5.49	0.46	5.96	40.31
MWE 1 (emulsion)	109.34	21.85	0.62	0.06	0.68	26.28
MWE 2 (emulsion)	123.72	21.43	1.09	0.06	1.15	28.90
MWE 3 (emulsion)	105.26	21.43	0.38	0.02	0.40	24.03
Methanite 1	91.85	9.29	3.69	0.16	3.86	34.35

Table 8.3 Mass and volume of gases generated in blasting per unit mass of ANFO

Gas	Mass of gas generated per unit mass of ANFO ($kg_{gas}\ kg^{-1}_{ANFO}$)	Gas Density ($kg\ m^{-3}$)	Volume of gas produced per unit mass of ANFO ($m^3_{gas}\ kg^{-1}_{ANFO}$)
CO	0.0163	1.25	0.01304
CO_2	0.1639	1.977	0.0829
NO_2	0.0035	1.36	0.0026

$$3NH_4NO_3 + CH_2 \rightarrow 7H_2O + CO_2 + 3N_2$$

However, incorrect fuel oil mixing, the presence of water and inadequate detonation velocities make it one of the most smoke-generating explosives.

This explosive is among the most widely employed, thus, as a precaution, many mining operators use it as the main reference for calculating their ventilation requirements. As an example, Australian blasting engineers generally use the values shown in Table 8.3.

Note that the disparities with respect to the above values and bibliography are mainly due to differences in ANFO composition resulting mainly from different percentages of fuel oil in the composition. The interested reader can consult the work of Rowland and Maineiro (2000), in which this issue is treated with respect to CO and NO_x emissions.

It is also interesting to recall the classical approach of Bertard and Bodelle (1962), for the estimation of the volume of CO generated in a blast. In it the following assumptions were made:

- 0.5–5 kg of explosives per m^2 of the cross section are consumed to advance any given gallery.

- The carbon content of explosives is a dimensionless value, α, which varies between 0.04 and 0.17. A table was provided for the various explosives in use at the time.[4]
- The gas throwback region (L_o) is about 50 m long.
- If p is the mass in kg of explosive used per m^2 of face section (S), then the product of these values gives the total mass of explosive needed for blasting. Multiplying this value by the carbon content of the explosive (α), gives the mass of carbon that reacts in the explosion. If this value is divided by the mass of a mole of carbon[5] and multiplied by 22.4 l (the volume occupied by a single mole of gas at STP), the volume of carbon gases generated is obtained. This, divided by the initial volume of the throwback region ($S\,L_o$) is the initial concentration. Therefore, the total carbon content ($CO + CO_2$) of the blasting fumes is close to $0.0375\alpha p$. The ratio of CO_2 to CO in the fumes is 2:1, thus the concentration of CO will be $0.0125\alpha p$.

Explosives are usually classified according to the volume of fumes they generate. The MSHA states that the maximum amount of gases produced by *Permissible Explosives* in underground coal mines or gas leaking mines must be 156 l kg^{-1}. Such explosives offer security in these conditions as they are designed such that their flame is of low volume, duration and temperature. These advantages are achieved through the addition of various salts to the explosive formula. In addition to CO, the classification considers gases such as CO_2, H_2S, NO, NO_2, NH_4 and SO_2, through conversion into their CO equivalent (CO_{eq}). This conversion is carried out by taking the quotient of the TLV-TWA[6] for CO and the TLV-TWA for a given toxic gas. The CO_{eq} for a mixture of gases is found by summing the CO_{eq} values for each gas in the mixture, thus:

$$CO_{eq} = 0.01CO_2 + 1CO + 5H_2S + 2NO + 17NO_2 + 2NH_4 + 25SO_2 \qquad (8.7)$$

After applying this expression, two categories are established:

- A: <78 l CO_{eq} kg^{-1} explosive, and
- B: 78–156 l CO_{eq} kg^{-1} explosive.

If the explosive generates more than 156 l CO_{eq} kg^{-1} then this explosive is deemed *not permissible* for underground use in coal mines. The explosive is then re-classified according to guidelines developed by the Institute of Makers of Explosives (IME),

---

[4]New generations of explosives have reduced the volume of the gases generated.

[5]Carbon reacts to form CO_2 or to form CO. One mol of C is required to make one mol of either of these gases.

[6]TWA stands for the Time-Weighted Average.

this time considering only the volumes of CO and H_2S produced. In this case, three categories are established:

- *Category* 1 (<22.5 l gases kg^{-1} explosive): Can be used in all types of underground works.
- *Category* 2 (22.5–47.2 l gases kg^{-1} explosive): Can be used in well-ventilated areas.
- *Category* 3 (47.2–94.85 l gases kg^{-1} explosive): Only suitable for surface blasting.

The European standard (EN 13631) does not set limits on the amount of gases that mining explosives can emit. This duty is left to national authorities. Polish regulations state that explosives used in mines must not produce more than 0.016 m^3 kg^{-1} of NO$_X$ and 0.027 m^3 kg^{-1} CO. In Belgium, Slovakia, France and the Czech Republic, a limit of 0.05 m^3 kg^{-1} has been established for the total volume of gases that can be emitted by a given explosive (Zawadzka-Małota 2015).

8.4 Gas Dilution Models

In this section, we present a number of methodologies used to approximate either the airflow rate required at the face or if the airflow rate is fixed, the time until an admissible gas concentration is reached (*re-entry time*). These approaches can be used to guague ventilation requirements for the dilution of engine fumes, mine gases (e.g. CH_4), fine dust, as well as blasting fumes. Large differences can be observed between the values derived from such methodologies and practical realities, particularly in the case of blasting. Therefore, they should be considered only as approximations to be used in the conceptual engineering phase, bearing in mind that the only valid approach to defining a safe time between blasting and allowing the re-entry of personnel is the measurement of gas concentrations *in situ*.

8.4.1 Dilution in Steady State

In general, the concentration of gases within a mining room (chamber) generated by either the processes of combustion or seam leaking increases exponentially, until, after sufficient time, it stabilizes (Fig. 8.6).

Once a steady state is reached (horizontal asymptote), the gases leave the room at the same speed at which they are generated. Applying the law of conservation of mass (Lomonósov-Lavoisier's law) as illustrated by Fig. 8.7 (Luo and Zhou 2013), the outward flow rate is balanced by the inward flow rate.

Thus:

$$Q_s = Q_v + Q_c$$

Fig. 8.6 Variation in gas concentration in a mining room over time showing how it reaches a steady state

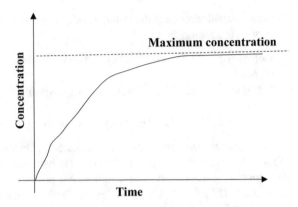

Fig. 8.7 Total volume flow rate conservation (incompressible fluid) in a steady state mining room

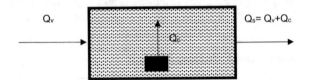

where

- Q_s: Airflow rate at the outlet (m^3 s^{-1}).
- Q_v: Ventilation airflow rate ("fresh air") (m^3 s^{-1}).
- Q_c: Pollutant flow rate (m^3 s^{-1}). Gases generated by diesel equipment or leaks (e.g. methane).

To work out the steady state conditions for a specific gas, its concentration in each of the flows has to be considered (Fig. 8.8), therefore:

$$Q_v B + Q_c n = Q_s s$$

where

- B: Concentration of the pollutant in the "fresh air",
- n: Concentration of the pollutant in the dirty stream, and
- s: Concentration of the pollutant in the output stream.

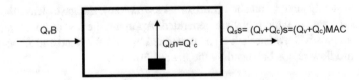

Fig. 8.8 Specific gas volume flow rate conservation (incompressible fluid) in a steady state mining room

In the above expression, it is common to refer to the pollutant flow rate as Q'_c under the assumption that the pollutant flow is composed of 100% pollutant with no other gases present.

Keeping in mind that since we are considering a steady state, B and Q_s will be constants.

Moreover, if the outlet concentration is assumed to be at the maximum permitted by law (Maximum Allowable Concentration, MAC), then the balance for the hazardous gas remains (Fig. 8.8):

$$Q_v B + Q_c n = (Q_v + Q_c)\, \text{MAC}$$

Therefore, solving for the airflow rate that the fan must provide, we have that (Eq. 8.8):

$$Q_V = \frac{Q_c\,(n - \text{MAC})}{\text{MAC} - B} \tag{8.8}$$

Exercise 8.1 During the excavation of a mine gallery, the machinery encounters a coal seam that releases 0.65 m^3 s^{-1} of gases, of which 95% is CH$_4$. For safety reasons, it is desirable that the concentration of CH$_4$ does not exceed a maximum level of 0.35%. Determine the airflow rate that must be delivered to the face if it is swept with: (a) clean air, (b) air from previous stopes with 0.1% CH$_4$ concentration. Assume that the concentration of CH$_4$ is stabilized in both cases.

Solution

(a) The contaminant concentration in the current Q_c is 0.95.

Maximum Allowable Concentration (MAC):

$$\text{MAC} = 0.35\% = 0.0035$$

As the air is clean, $B = 0.00$.
Substituting into Eq. 8.8 we have:

$$Q_v = \frac{Q_c\,(n - \text{MAC})}{\text{MAC} - B} = \frac{0.65 \frac{\text{m}^3}{\text{s}} \cdot \left(0.95 - \frac{0.35}{100}\right)}{\frac{0.35}{100} - 0.00} = 175.78 \ \frac{\text{m}^3}{\text{s}}$$

(b) For a certain concentration of CH$_4$ ($B = 0.001$) in the "fresh air" we have:

$$Q_v = \frac{Q_c\,(n - \text{MAC})}{\text{MAC} - B} = \frac{0.65 \frac{\text{m}^3}{\text{s}} \cdot \left(0.95 - \frac{0.35}{100}\right)}{\frac{0.35}{100} - \frac{0.10}{100}} = 246.09 \ \frac{\text{m}^3}{\text{s}}$$

Application of the Stationary Dilution Expression to Dust

The particles of the *respirable fraction* (<5 μm) sediment extraordinarily slowly, so for dilution calculations, they can be considered as a gas. This is so, because to reduce their concentration in air, what really matters is the airflow supply rate. Besides, coarse particles have a greater tendency to sediment, with air velocity being the most influential parameter for maintaining their concentration in the air, thus, the higher the speed, the greater the number of particles in suspension.[7]

In Fig. 8.9, it can be seen that as air velocity increases, the concentration of fine particles in suspension tends to decrease, while the concentration of coarse particles increases. The minimum total dust in the air is reached at a speed of about 2 m s^{-1}. Large concentrations of coarse dust cause discomfort to the miner but pose only a moderate risk to his respiratory system.

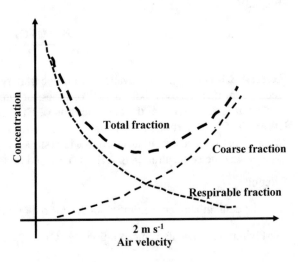

Fig. 8.9 Concentration of fine, coarse and total dust in the mine atmosphere at different air velocities

Accordingly, the expression seen in the previous section can be modified for the case of dust, leading to (Eq. 8.9):

$$Q_v = \frac{G}{MAC - B} \tag{8.9}$$

where

- G: Ratio of emission or emission factor (mg s^{-1}),
- MAC: Maximum allowable concentration (mg s^{-1}), and
- B: Dust concentration in clean air (mg s^{-1}).

By removing MAC from the numerator of Eq. 8.8 the new equation, Eq. 8.9, tends to give values that err on the safe side. This is so because MAC is a small value. Note

---

[7]Avoiding the lifting of particles from the ground is one of the reasons for limiting the maximum air speed in galleries and working faces. For general purposes, this limit is usually around 8 m s^{-1}.

that G represents the contaminant in the stream and not the airflow. That is to say, it is equivalent to Q_c.

8.4.2 Mixed Volume Model

The *Mixed Volume Model* is actually a first-order exponential decay model for the kinetic of mixing of dilution air with polluted air (Fig. 8.10). The model gives an

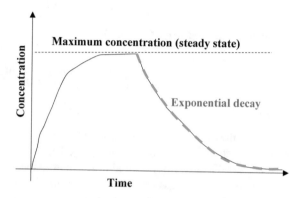

Fig. 8.10 Exponential decay observed for pollutant gas concentration after reaching steady state

idea of the time needed to reduce an initial concentration of an existing pollutant in a given space below a certain limit. This model assumes that:

- There is a uniform concentration of gas.
- The mass of harmful gas and dilution air is mixed perfectly. This requires a turbulent regime in which mixing is achieved rapidly.

The model is a classical approximation for estimating the dilution times for diesel engines and blasting fumes.

Derivation of the Exponential Decay Curve for Gas Concentration

If it is assumed that after a perfect mixing between fresh air and blast gases, the volume of contaminant gases will decrease at a constant rate with respect to time. In this way, we can start to define an equilibrium equation where the concentration of pollutant gases decreases by a fixed amount per unit volume (dx) of the pollutant gas in volume V. By the law of conservation of mass, this will be equal to the amount of pollutant swept away by the diluting air during a time dt. The amount of pollutant carried away by the fresh air will be equal to the dilution airflow rate (Q_d) multiplied by the concentration of the pollutant at a given instant (x), assuming that for a time dt this concentration x can be considered constant. Thus, the equation of equilibrium for the unit of time dt is given by the relation:

$$V \, dx = -Q_d \, x \, dt$$

where the negative sign indicates that the concentration of pollutants is decreasing, given that Q_d and x are positive values.

This is a separable first-order differential equation, and integrating between the limits $(t = 0, x = c_i)$ and $(t = t, x = c_f)$, where c_i and c_f are the initial and final values of pollutant concentration, respectively, we have:

$$-\frac{1}{Q_d} \int_{c_i}^{c_f} \frac{dx}{x} = \int_0^t \frac{dt}{V}$$

Thus, integrating we get Eq. 8.10:

$$t = -k \, \frac{V}{Q_d} \, \ln\left(\frac{c_f}{c_i}\right) = k \, \frac{V}{Q_d} \, \ln\left(\frac{c_i}{c_f}\right) \tag{8.10}$$

where

- Q_d: Airflow rate of fresh air supplied for dilution ($m^3 \, s^{-1}$).
- V: Volume of the gas-mixing zone (m^3). This is the cross section of the gallery, S times the distance from the end of the duct to the face, d ($V = S \, d$). In practice, the gallery has equipment working in it, piles of debris and explosion products that diminish this volume. Thus, the actual volume can be estimated at 80% of the total volume. However, in the interests of safety, this consideration is rarely taken into account.
- t: Time required to achieve a given concentration (s).
- c_i: Concentration of pollutant (volume of gas released in the explosion, divided by the volume, V) (in ppm or congruent units).
- c_f: Final concentration (legal limit) (in ppm or congruent units).
- k: Coefficient of turbulent regime. Varies from 1 (perfect mixture) to 10 (extremely bad mixture).
- x: Variable that represents the concentration of polluting gases in a given instant (in ppm or congruent units).

Exercise 8.2 A cul-de-sac, of transverse section $9 \, m^2$, is being ventilated by a forcing system at an airflow rate of $0.9 \, m^3 \, s^{-1}$. The duct end is located 11 m from the face. The explosive used has a carbon content of 0.08 (decimal) and the consumption requirement (p) is $4 \, kg \, m^{-2}$. Using the classical method of Bertard and Bodelle (1962), determine the time needed to reduce the concentration of CO to 25 ppm. You may suppose that the mixture of gases is perfect.

Solution

The classical method of Bertard and Bodelle (1962) states that:

$$C_i = 0.0125 \, \alpha \, p = 0.0125 \cdot 0.08 \cdot 4 = 0.004$$

Therefore, using Eq. 8.10:

$$t = k\frac{V}{Q_d}\ln\left(\frac{c_i}{c_f}\right) = 1 \cdot \frac{9\,\text{m}^2 \cdot 11\text{m}}{0.9\frac{\text{m}^3}{\text{s}}} \cdot \ln\left(\frac{0.004}{25 \times 10^{-6}}\right) = 558.27\,\text{s} = 9.3\,\text{min}.$$

8.4.3 Modelling the Contribution of Polluting Gases During Dilution

A model proposed by de Souza and Katsabanis (1991), takes into account the amount of gas remaining in a blasted area and quantifies the time taken and the precise airflow rate required to reduce the level of pollution below a given limit. It is very similar to the Mixed Volume Model discussed previously, but introduces new parameters such as the presence of a constant supply of gases during dilution. It has many similarities with the model proposed by Deniau (1976), for the ventilation of room-and-pillar mines.

In developing this model, it is assumed that, in the room or space to be ventilated, the concentration, x, of pollutant gases varies by an amount dx in a time dt. This variation is a consequence of the mixing between the toxic gases generated by mining activities and the gases existing in the chamber after dilution by incoming fresh air. In order to generalize the model, the approach considers that the fresh air supplied, at a flow rate Q, already contains some pollution, quantified as B. In this way, the variation in the concentration of a pollutant gas dx contained in volume V is equal to the volume of pollutant gas that enters in a given time $(Q_g + QB)dt$, minus the amount that is extracted in the same time $(Q + Q_g)x\,dt$, where Q_g is the flow rate of pollutant gases generated in the room. This gives us the differential equation[8]:

$$V\,dx = \left[(Q_g + Q\,B) - (Q + Q_g)x\right]dt$$

If the above expression is integrated, within the limits $(t = 0, x = X_0)$ and $(t = t, x = X)$ we have that:

$$\int_{X_0}^{X} \frac{dx}{(Q_g + Q\,B) - (Q + Q_g)x} = \int_{0}^{t}\frac{dt}{V}$$

---

[8]For there to be a diminishing quantity of pollutant gases in the volume V, the pollution entering the room $(Q_g + QB)$ must be smaller than the pollution exiting it $(Q + Q_g)$. In this way Vdx is negative, and so we obtain the expected exponential decay as explained in Sect. 8.4.2.

That is:

$$\frac{-1}{(Q+Q_g)} \ln\left[\frac{(Q_g+QB) - X(Q+Q_g)}{(Q_g+QB) - X_o(Q+Q_g)}\right] = \frac{t}{V}$$

Finally, solving for t (Eq. 8.11):

$$t = \frac{V}{(Q+Q_g)} \ln\left[\frac{(Q_g+QB) - X_0(Q+Q_g)}{(Q_g+QB) - X(Q+Q_g)}\right] \qquad (8.11)$$

where

- t: Time (s),
- Q_g: Pollutant gas flow rate (m^3 s^{-1}),
- Q: Fresh airflow rate (dilution airflow rate) (m^3 s^{-1}),
- V: Volume of the gas-mixing zone (sometimes approximated by the volume of the working zone) (m^3),
- B: Concentration of the pollutant in the clean air (as a fraction),
- X: Concentration of the pollutant gas in the mixture at time t (as a fraction), and
- X_0: Initial concentration of the polluting gas (as a fraction).

The following conditions apply to the model:

(1) B can only take positive numbers ($B \geq 0$),
(2)   In the case of effective dilution $X_o > X$, and
(3)   The concentration of pollutants in the supplied air must be lower than the desired concentration ($B < X$).

Analysis of the Solution Obtained

To obtain the desired exponential decay, the internal function of the natural logarithm in Eq. 8.11, must be >0.

$$\left[\frac{(Q_g+QB) - X_0(Q+Q_g)}{(Q_g+QB) - X(Q+Q_g)}\right] > 0$$

This condition is fulfilled if the numerator and denominator have the same sign, where the following two variants are possible:

(a)   The numerator and denominator are both positive.

In this assumption, the equations for numerator and denominator are given by:

- Numerator: $QB + Q_g - X_0(Q+Q_g) > 0$, and
- Denominator: $QB + Q_g - X(Q+Q_g) > 0$.

Rearranging the expression in the denominator we find that:

$$QB + Q_g > X\left(Q + Q_g\right)$$

Given that we previously established the condition that $B \geq 0$, it can be deduced that the flow rate of polluted fresh air (QB) plus the contribution from the generated gases (Q_g) is greater than the extraction capacity, given by $(Q + Q_g)X$. This condition, although mathematically correct, would not be technically desirable.

(b) The numerator and denominator are both negative.

Since in the case of effective dilution $X_o > X$, the most restrictive condition is expressed in that containing the value X, i.e. the denominator:

$$QB + Q_g - X\left(Q + Q_g\right) < 0$$

So, rearranging once more:

$$QB + Q_g < X\left(Q + Q_g\right)$$

According to this, the ventilation capacity is now greater than the incoming flow rate of contaminants, hence ventilation actually occurs.

Rearranging for Q, we get (Eq. 8.12), which is analogous to Eq. 8.8 and corresponds to the dilution in steady state.

$$Q > \frac{Q_g(1 - X)}{X - B} \tag{8.12}$$

This expression can also be obtained by solving the differential equation from which de Souza and Katsabanis's model has been obtained, for the steady state, that is, if $\frac{dx}{dt} = 0$.

If it is desired that the exhaust gases be allowed to reach the Maximum Allowable Concentration (MAC), we have Eq. 8.13:

$$Q = \frac{Q_c(1 - \text{MAC})}{\text{MAC} - B} \tag{8.13}$$

Dust Dilution

If it is assumed that dust occupies a negligible volume compared to the volume of air in circulation. Equation 8.11 becomes Eq. 8.14:

$$t = \frac{V}{Q} \ln\left(\frac{QB + G - X_0 Q}{QB + G - X Q}\right) \tag{8.14}$$

where

- t: Time (s),
- Q: Fresh airflow rate (m^3 s^{-1}),
- V: Volume of the working space (m^3),
- G: Dust generation rate (mg s^{-1}),
- B: Concentration of dust in clean air (mg m^{-3}),
- X: Dust concentration at time t (mg m^{-3}), and
- X_0: Initial dust concentration (mg m^{-3}).

In the case of the steady state:

$$Q\,B + G - X\,Q = 0$$

Therefore, the airflow rate necessary to dilute the dust will be (Eq. 8.15):

$$Q = \frac{G}{X - B} \qquad (8.15)$$

where X will normally be the MAC for dust.

Exercise 8.3 The airflow rate necessary to ventilate a stope measuring 75 m by 75 m by 1 m is 20 m^3 s^{-1}. If the concentration of CO immediately after blasting is 2500 ppm, estimate the precise time for this concentration to fall to 100 ppm.

Solution

In the absence of other information, we can consider $B = 0$ and $Q_g = 0$.
So Eq. 8.11, simplifies to the exponential decay equation (Eq. 8.10):

$$t = \frac{V}{(Q + Q_g)} \ln\left[\frac{(Q_g + QB) - X_0\,(Q + Q_g)}{(Q_g + QB) - X\,(Q + Q_g)}\right] = \frac{V}{Q}\ln\frac{X_0}{X} = \frac{75\,\text{m} \cdot 75\,\text{m} \cdot 1\,\text{m}}{20\,\frac{\text{m}^3}{\text{s}}} \cdot \ln\frac{2500}{100}$$

$$= 905.3\,\text{s} = 15\,\text{min}.$$

Exercise 8.4 The concentration of CH$_4$ inside a 3000 SCM stope is 7% and this needs to be reduced to non-explosive values of 1%. The face is swept with an airflow rate of 3 SCMS, and the CH$_4$ emanations from the coal seams are estimated at 0.2 SCMS. Determine the time necessary to achieve non-explosive concentrations of this gas. Consider the solution for the following three options:

(a) There are no CH$_4$ inputs to the stope and the supplied air is free of CH$_4$.
(b) There are CH$_4$ emanations of 0.2 SCMS, and in the stope and the sweeping air is clean.
(c) There are CH$_4$ emanations of 0.2 SCMS, and in the stope and the sweeping air has a concentration of 2500 ppm of CH$_4$.
(d) In the event that the ventilation airflow rate of 3 SCMS is not sufficient, indicate the minimum airflow rate to be supplied under conditions b and c.
(e) Resolve paragraphs a, b and c again assuming a fresh airflow rate of 27 SCMS.

(f) Plot the graph of the ventilation time versus fresh airflow rate if $Q_g = 0.2$ SCMS and $B = 0.0025$.

Note: SCM: Standard Cubic Metre; SCMS: Standard Cubic Metre Per Second.

Solution

(a) CH_4 is explosive for a concentration between 5 and 15%. The MITC (2000) sets a general limit of 1% CH_4 in mine air currents. To solve this question it is possible to use the expression for exponential decay contained in Eq. 8.10. Thus, assuming the turbulence parameter $k = 1$, we have:

$$t = k \frac{V}{Q_d} \ln\left(\frac{c_i}{c_f}\right) = 1 \cdot \frac{3000 \, \text{m}^3}{3 \frac{\text{m}^3}{\text{s}}} \cdot \ln\left(\frac{0.07}{0.01}\right) = 1945.9 \, \text{s} = 32.4 \, \text{min}$$

(b) As indicated above, the outlet flow must have a final concentration of 1% of CH_4 to comply with legal requirements. The stated clean air ventilation rate of 3 SCMS would be enough to exhaust 3 SCMS $\cdot$ 0.01 = 0.03 SCMS of CH_4. Therefore, given that the flow rate of CH_4 out of the stope is less than the flow rate of emanations into the stope, i.e. 0.03 SCMS of $CH_4 < 0.2$ SCMS of CH_4 ($Q\,X < Q_g$), the proposed conditions are not possible.

(c) As shown in (b) levels of CH_4 could not be reduced to safe values even using sweeping air that was totally free of CH_4. Thus, given that CH_4 inputs are the same here, the extra CH_4 contamination in the clean current makes it impossible to fulfil the conditions here either.

(d) The minimum ventilation airflow rate for the conditions described in paragraph (b) can be calculated as:

$$Q > \frac{Q_g(1 - X)}{X} = \frac{0.2 \, \text{SCMS}(1 - 0.01)}{0.01} = 19.8 \, \text{SCMS}$$

For the conditions outlined in (c), this limit is:

$$Q > \frac{Q_g(1 - X)}{X - B} = \frac{0.2 \, \text{SCMS}(1 - 0.01)}{0.01 - 0.0025} = 26.4 \, \text{SCMS}$$

(e) Solution of the above assumptions for a flow rate of 27 SCMS, noting that this value exceeds minimum flow rates found in the previous two scenarios, and will thus be adequate.

(e–a) Solution for 27 SCMS without any CH_4 leakage into the stope:

$$t = k \frac{V}{Q_d} \ln\left(\frac{c_i}{c_f}\right) = 1 \cdot \frac{3000 \, \text{m}^3}{27 \frac{\text{m}^3}{\text{s}}} \cdot \ln\left(\frac{0.07}{0.01}\right) = 216.2 \, \text{s} = 3.6 \, \text{min}$$

(e–b) There is a CH_4 leak of $Q_g = 0.2$ SCMS and the sweeping air is clean ($B = 0$), so:

$$t = \frac{3000 \, \text{m}^3}{(27 + 0.2) \frac{\text{m}^3}{\text{s}}} \cdot \ln\left[\frac{0.2 \frac{\text{m}^3}{\text{s}} - 0.07 \cdot (27 + 0.2) \frac{\text{m}^3}{\text{s}}}{0.2 \frac{\text{m}^3}{\text{s}} - 0.01 \cdot (27 + 0.2) \frac{\text{m}^3}{\text{s}}}\right] = 349.0 \, \text{s} = 5.8 \, \text{min}$$

(e–c) There is a CH_4 leak of $Q_g = 0.2$ SCMS and the sweeping air has a concentration $B = 0.0025$, then:

$$t = \frac{V}{(Q + Q_g)} \ln\left[\frac{Q \, B + Q_g - X_0 \, (Q + Q_g)}{Q \, B + Q_g - X \, (Q + Q_g)}\right]$$

$$t = \frac{3000 \, \text{m}^3}{(27 + 0.2) \frac{\text{m}^3}{\text{s}}} \cdot \ln\left[\frac{27 \cdot 0.0025 \frac{\text{m}^3}{\text{s}} + 0.2 \frac{\text{m}^3}{\text{s}} - 0.07 \cdot (27 + 0.2) \frac{\text{m}^3}{\text{s}}}{27 \frac{\text{m}^3}{\text{s}} \cdot 0.0025 + 0.2 \frac{\text{m}^3}{\text{s}} - 0.01 \cdot (27 + 0.2) \frac{\text{m}^3}{\text{s}}}\right] = 650.3 \, \text{s} = 10.8 \, \text{min}$$

(f) By substituting values in Eq. 8.11, the following table can be constructed:

Flow rate (SCMS)	t (s)	t (min)
27	650.3	10.8
32	358.7	6.0
38	259.8	4.3
45	200.3	3.3
54	155.9	2.6
64	125.5	2.1
76	101.9	1.7
91	82.5	1.4

Therefore, plotting it we have that:

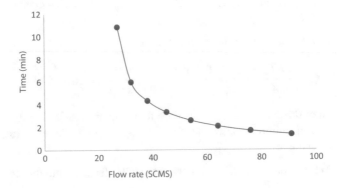

8.4.4 Classical Approaches of the Russian School

Novitzky (1962), pp. 247–249, provides a survey of the available Russian literature and compiles a set of equations for calculating ventilation airflow rates and re-entry times. These formulae are reproduced here only for historical interest as they are no longer of practical use.

Forcing Ventilation

According to Skochinsky and Komarov (1969), the time needed to ventilate the face can be estimated as:

$$t = \frac{7.8}{Q}(M_{exp} V^2)^{\frac{1}{3}}$$

where

- t: Required time (s),
- M_{exp}: Mass of explosive (kg),
- V: Volume of the drive prior to the explosion (m³), and
- Q: Ventilation airflow rate at the face (m³ s⁻¹).

Exhausting Ventilation

Here three expressions that were in regular use:

(a) Xenofontova's:

$$t = \frac{6}{Q}[M_{exp} S (75 + M_{exp})]^{\frac{1}{2}}$$

(b) Voronin's:

$$t = \frac{5.7}{Q}(M_{exp} V)^{\frac{1}{2}}$$

(c) Komarov & Kil'keev's:

$$t = \frac{18}{Q}[M_{exp} S(2.4 M_{exp} + 10)]^{\frac{1}{2}}$$

- t: Required time (s),
- S: Gallery cross section (m²),
- V: Volume of the drive prior to the explosion (m³),
- M_{exp}: Mass of explosive used the blast (kg), and
- Q: Ventilation airflow rate at the face (m³ s⁻¹).

8.4.5   Estimation as a Function of the Tunnel Transverse Section

By way of reference, in workings where no CH_4 is present, airflow rates from 0.25 to 0.3 m^3 s^{-1} per m^2 of gallery cross section can be used for the dilution of blasting gases in development ends (Expilly 1960). That is to say (Eq. 8.16):

$$Q = k\,S = 0.25\,S \qquad (8.16)$$

where

- Q: Airflow rate required at the face (m^3 s^{-1}),
- k: Wind speed ($0.25 - 0.3\,\frac{m^s \cdot s}{m^2}$), and
- S: Drive cross section (m^2).

Note that Eq. 8.16 can be compared with the definition of airflow rate ($Q = v\,S$), thus if $v = k$, we have wind speeds of 0.25–0.3 m s^{-1}. For tunnels in which works are not undertaken in conditions of positive pressure, BS 6164 requires an airflow rate of 9 m^3 per min per m^2 of gallery cross section ($0.15\,\frac{m^3 \cdot s}{m^2}$).

If it is assumed that the gases propagate to a distance of about 20 m from a blast site, Eq. 8.17 can be used (Vutukuri and Lama 1986, p. 101):

$$Q = \frac{20\,S\,N}{t} \qquad (8.17)$$

where

- Q: Required airflow rate at the face (m^3 s^{-1}),
- S: Tunnel cross section (m^2),
- t: Time in which the blast fumes are to be evacuated (s), and
- N: Number of air renovations (usually 5).

8.4.6   Dilution Factor Method

This expression (Eq. 8.18), is usually included in Chilean methodological guides (e.g. Andrade 2008):

$$t = \frac{100\,M_{exp}\,a}{d\,Q} \qquad (8.18)$$

where

- t: Time for the dilution of the gases (min).
- M_{exp}: Explosive mass detonated (kg).

- a: Volume of CO generated per unit of explosive mass (m^3 kg^{-1}). The value normally used for this parameter is 0.04 m^3 kg^{-1}.[9]
- d: Dilution factor (%). Maximum amount of CO allowed in the return air. Normally, a value of 0.008% is used.
- Q: Required airflow rate (m^3 min^{-1}).

A variant of this expression can be found in Russian literature (e.g. Borisov et al. 1976, p. 362) in the form (Eq. 8.19):

$$t = \frac{12.5\, M_{exp} B}{Q} \tag{8.19}$$

where

B: Volume in litres of CO generated in the explosion of 1 kg of explosive (approximately 40 l in general).

From the above expressions, the following rule of thumb can be deduced for a dilution time of 30 min:

$$Q = \frac{100\, M_{exp}\, 0.04\, m^3}{0.008 \cdot 30\, min}$$

Therefore:

$$Q[m^3\, min^{-1}] = 16.67\, M_{exp}$$

And in SI units (Eq. 8.20):

$$Q\left[\frac{m^3}{s}\right] = 0.28\, M_{exp} \tag{8.20}$$

8.5 Throwback Method

This method consists of the implementation of the following 5 steps (Howes 1994; WMC 2001; Stewart 2014):

1. Approximation of the initial volume of gases after detonation (Eq. 8.21):

$$L_i = \frac{K\, M}{F_a\, D\sqrt{A}} \tag{8.21}$$

---

[9]If you look at the work of Harris et al. (2003), you can see that only TNT flakes slightly exceed the total value of 0.04 m^3 kg^{-1}, so we are on the safe side. See Sect. 8.3.4, to note that these data are consistent with those for underground explosives.

where

- L_i: Fumes throwback length, i.e. the distance the fumes move backward, immediately after detonation (m);[10]
- K: 25,000;
- F_a: Face advance (distance between two consecutive faces) (m);
- D: Rock Density (kg m^{-3});
- A: Face transverse area (m^2); and
- M: Mass of explosive (kg).

2. Estimation of the volume of gases generated by the blasting. Generally, Table 8.3 is used, and calculations usually refer to NO$_2$ due to the fact that it is the most toxic (its TWA is 3 ppm) and difficult to dilute of the potential pollutants.
3. Estimation of the concentration of gases in the throwback region (Eq. 8.22):

$$C_i[\text{ppm}] = \frac{V_{\text{gas}}}{V_i} = \frac{V_{\text{NO}_2}}{A\,L_i} \times 10^6 \tag{8.22}$$

where

- A: Transverse section of the gallery (m^2), and
- V_i: Initial volume (m^2).

4. Calculation of the concentration of pollutants when the exhaust fumes retreat into the main airway. At this stage, the gases uniformly fill the length of the gallery (often termed the drive) but have not yet entered the main airway:

$$C_f[\text{ppm}] = \frac{V_{\text{gas}}}{V} = \frac{V_{\text{NO}_2}}{A\,L} \times 10^6$$

where

- V: Volume of the gallery (m^3), and
- L: Length of the gallery (drive) (m).

5. Calculation of the time it takes for diluted gases to reach the main airway (t_1) (Eq. 8.23):

$$t_1 = \frac{(L - L_i)A}{Q} \tag{8.23}$$

where

- Q: Ventilation airflow rate at the face (m^3 s^{-1}).

---

[10]In the Russian literature $L_i = 2.44M + 10$, is frequently quoted (e.g. Borisov et al. 1976, p. 381). This expression provides very different values from those found using Eq. 8.21, at least for ANFO.

6. Calculation of time needed to dilute gases (t_2) (Eq. 8.24):

$$t_2 = \left(\frac{V}{Q}\right) \ln\left(\frac{C_i}{C_f}\right)$$ (8.24)

7. Calculation of total ventilation time (t_t): $t_t = t_1 + t_2$

Exercise 8.5 A 5×5 m gallery, whose current length is 600 m is being excavated by drilling and blasting. The advance in it takes place at a rate of 3.5 m per cycle, in a rock of density 2.6 t m^{-3}. The specific consumption factor for ANFO is 2.05 kg m^{-3} of blasted rock, the ventilation airflow rate at the face is 25 m^3 s^{-1}. Calculate:

(a) Tonnes of rock blasted per cycle.
(b) Mass of ANFO used.
(c) Fumes throwback length.
(d) The time it takes the gases to reach the main gallery.
(e) Face fumes clearance time.
(f) Total ventilation time.

Note: Suppose you generate $0.0026\frac{\text{m}^3\text{NO}_2}{\text{kg of ANFO}}$

Solution

(a) Calculation of the mass of rock blasted:

$$5 \cdot 5 \cdot 3.5 = 87.5\,\text{m}^3$$
$$87.5\,\text{m}^3 \cdot 2.6\,\text{t m}^{-3} = 227.5\,\text{t per blast}$$

(b) Calculation of the mass of ANFO:

$$\frac{87.5\,\text{m}^3 \cdot 2.05\,\text{kg ANFO}}{\text{m}^3\text{ blasted rock}} = 179.38\,\text{kg of ANFO}$$

(c) Fume throwback length:

$$L_i = \frac{KM}{F_a D\sqrt{A}} = \frac{25\,000 \cdot 179.38\,\text{kg}}{3.5\,\text{m} \cdot 2\,600\frac{\text{kg}}{\text{m}^3} \cdot \sqrt{5\,\text{m} \cdot 5\,\text{m}}} = 98.56\,\text{m}$$

(d) To solve this problem the following parameters have to be calculated first:

$$\text{Throwback volume} = 98.56\,\text{m} \cdot 5\,\text{m} \cdot 5\,\text{m} = 2464\,\text{m}^3$$

Volume of gases immediately after blasting (V_{NO_2}):

$$V_{\text{NO}_2} = 0.0026\frac{\text{m}^3\,\text{NO}_2}{\text{kg ANFO}}179.38\,\text{kg ANFO} = 0.46638\,\text{m}^3\text{NO}_2$$

Concentration of gases immediately after blasting (C_i):

$$C_i = \frac{V_{NO_2}}{A\,L_i} \times 10^6\,\text{ppm} = \frac{0.466\,\text{m}^3}{2464\,\text{m}^3} \times 10^6\,\text{ppm} = 189\,\text{ppm}$$

Concentration of gases until reaching the principal airway (C_f):

$$C_f = \frac{V_{NO_2}}{A\,L} \times 10^6\,\text{ppm} = \frac{0.46638\,\text{m}^3}{600 \cdot 5 \cdot 5\,\text{m}^3} \times 10^6 = 31.09\,\text{ppm}$$

Time it takes for diluted gases to reach the principal airway (t_1):

$$t_1 = \frac{(L - L_i)A}{Q}$$

$$t_1 = \frac{(600 - 98.56) \cdot 5 \cdot 5\,\text{m}^3}{25\frac{\text{m}^3}{\text{s}}} = 501.4\,\text{s}$$

(e) Time required to dilute the gases (t_2):

$$t_2 = \left(\frac{V}{Q}\right) \ln\left(\frac{C_i}{C_f}\right) = \left(\frac{600 \cdot 5 \cdot 5\,\text{m}^3}{25\frac{\text{m}^3}{\text{s}}}\right) \ln\left(\frac{189}{31.09}\right) = 1083\,\text{s}$$

(f) Total ventilation time:

$$t_t = t_1 + t_2 = 501.4 + 1083 = 1584.4\,\text{s} \rightarrow 26.41\,\text{min}$$

8.6 Advection-Diffusion Method

Convection is a collective movement of molecules within a fluid. This phenomenon can be separated into two processes:

- *Advection*: Bulk motion of the fluid.
- *Diffusion*: Movement of particles from areas of higher to lower concentration.

The plume of fumes released in an explosion moves back through the main gallery by means of both processes. A gas detector located within the gallery would record concentrations of the gas under study rising suddenly shortly after detonation, then decaying over time. The exact nature of the curve recorded by the detector depends upon the distance between the detector and the gallery face (x) (Fig. 8.11).

The phenomenon can be modelled by means of the following expression (Taylor 1953, 1954; Widiatmojo et al. 2015) (Eq. 8.25):

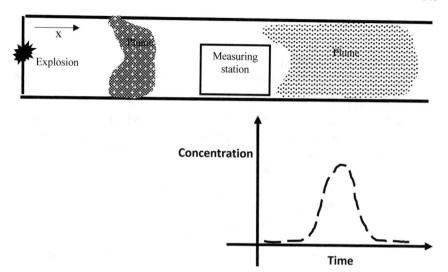

Fig. 8.11 Advection-diffusion of fumes in a cul-de-sac

$$C(x, t) = \left(\frac{V}{2A\sqrt{\pi Dt}} \right) e^{\frac{-(x-u_m \cdot t)^2}{4Dt}} \tag{8.25}$$

where

- C: Concentration of the gas at a distance x from the face at time t (fraction),
- V: Volume of gases after blasting (volume of gas in its initial state: $x = 0, t = 0$),
- L: Length of the gallery (m),
- A: Cross section of the gallery (m^2),
- t: Elapsed time after blasting (s),
- D: Diffusion coefficient (m^2 s^{-1}), and
- x: Distance from the detector to the point at which the gases are generated (m).

Scientific literature provides different approaches to the estimation of D. The oldest is Taylor's (1954), which we reproduce below due to its popularity and simplicity[11] (Eq. 8.26):

$$D = 5.05 \, du^* \tag{8.26}$$

where u^* is obtained from Eq. 8.27a:

$$u^* = u_m \sqrt{\frac{f}{8}} \tag{8.27a}$$

---

[11]This expression was obtained for saline transport along a pipe, thus its limitations are obvious for ventilation purposes.

where

- d: Duct diameter (m),
- u_m: Average air velocity (m s^{-1}), and
- f: Coefficient of friction (dimensionless) that can be obtained, for example, from the Colebrook equation.

Exercise 8.6 A gallery of cross section 9 m^2 whose current length is 330 m is under construction. The excavation is taking place by drilling and blasting, employing 150 kg of ANFO-type explosive in each cycle. The velocity of the throwback gases is estimated to be 0.6 m s^{-1}. Determine:

(a) The volume of gases formed at the face immediately after detonation.
(b) NO$_2$ concentration 3 min after blasting at a distance of 50 m from the face.
(c) The point in the gallery where the highest concentration of NO$_2$ is found 3 min after the blasting. What is this concentration?
(d) The function of the distribution of the NO$_2$ pollution along the gallery 3 min after the blasting.

You may take the turbulent diffusion coefficient for NO$_2$ in this gallery to be 17 m^2 s^{-1}.

Solution

(a) The volume of gases formed for $t = 0$ and $x = 0$.

Gas	Mass of gas generated per mass of ANFO ($kg_{gas}\ kg_{ANFO}^{-1}$)	Gas density ($kg_{gas}\ m_{gas}^{-3}$)	Volume of gas generated per mass of ANFO ($m_{gas}^3\ kg_{ANFO}^{-1}$)	Volume of gas formed (m_{gas}^3)
CO	0.0163	1.25	0.01304	1.956
CO$_2$	0.1639	1.977	0.0829	12.435
NO$_2$	0.0035	1.36	0.002574	0.3861
		Total		14.43

(b) The focus is on NO$_2$, which is expected to be the most toxic and the most difficult to disperse. Thus, since:

$$A = 9\,\text{m}^2$$
$$u_m = 0.6\,\text{m s}^{-1}$$
$$t = 3 \cdot 60\,\text{s} = 180\,\text{s}$$
$$x = 50\,\text{m}$$
$$D = 17\,\text{m}^2\,\text{s}^{-1}$$

$$C(x, t) = \left(\frac{V}{2A\sqrt{\pi Dt}} \right) e^{\frac{-(x - u_m \cdot t)^2}{4Dt}}$$

$$C(50, 180) = \left(\frac{0.3861 \text{ m}^3}{2 \cdot 9 \text{ m}^2 \cdot \sqrt{\pi \cdot 17 \frac{m^2}{s} \cdot 180 \text{ s}}} \right) e^{\frac{-(50 \text{ m} - 0.6 \frac{m}{s} \cdot 180 \text{ s})^2}{4 \cdot 17 \frac{m^2}{s} \cdot 180 \text{ s}}} = 1.66 \times 10^{-4}$$

This is equal to 166 ppm, which is greater than the acceptable safe limit for this gas of 10 ppm.

(c) The point (x) where the highest concentration of gas is found corresponds to the maximum value of Eq. 8.25. This value is obtained by taking the derivative of Eq. 8.25 and setting it to zero:

$$x - u_m t = 0; \quad x = u_m t$$

So, for $t = 180$ s (3 min), $x = 108$ m.

Then, the maximum concentration for $t = 180$ s is:

$$C(108, 180) = \left(\frac{0.3861 \text{ m}^3}{2 \cdot 9 \text{ m}^2 \cdot \sqrt{\pi \cdot 17 \frac{m^2}{s} \cdot 180 \text{ s}}} \right) e^{\frac{-(108 \text{ m} - 0.6 \frac{m}{s} \cdot 180 \text{ s})^2}{4 \cdot 17 \frac{m^2}{s} \cdot 180 \text{ s}}}$$

$$= 2.188 \times 10^{-4} = 218.8 \text{ ppm}$$

(e) Distribution of the contamination of NO_2 in the gallery after 180 s:

x (m)	$C(x, t = 180)$ (ppm)
0.0	84.4
5.0	92.0
6.3	93.9
8.3	97.2
12.5	103.8
25.0	124.6
33.3	138.7
50.0	**166.2**
75.0	200.1
100.0	217.6
150.0	189.4
200.0	109.6
250.0	42.1

(continued)

(continued)

x (m)	C(x, t = 180) (ppm)
300.0	10.8
330.0	3.9
330.0	3.9

8.7   Heat Dissipation with Ventilation

To determine the ventilation flow required to reduce the temperature to a certain value one can follow the procedure outlined by McPherson (1993, p. 292). The basis of this procedure relies on a calculation of the energy required to heat the air and water vapour (a function of the saturated vapour pressure) from a temperature t_{wi} at the inlet up to a temperature t_{wo} at the outlet, thus $t_{wo} > t_{wi}$. It consists of the following steps:

1. Calculate the saturated vapour pressure (P_{swi}) for the fresh air inlet temperature (t_{wi}).
2. Determine the sigma heat (S_i), for condition (1).
3. Specify an exhaust air temperature (t_{wo}). This temperature will be the maximum acceptable wet-bulb temperature.
4. Calculate the saturated vapour pressure (P_{swo}) for air at this maximum acceptable wet-bulb temperature (t_{wo}).
5. Calculate the sigma heat for condition 4 (S_o).
6. Calculate the required flow rate based on the total heat flow to the air (q_{eo}).

Necessary Equations

(a)   For the calculation of saturated vapour pressure (P_{sw}) (kPa) (Eq. 8.27b):

$$P_{sw} = 0.6106 \, e^{\left(\frac{17.27 t_w}{t_w + 237.3}\right)} \tag{8.27b}$$

where t_w is the wet-bulb temperature (°C). Note that this expression can often be used with the dry-bulb temperature in which case it will give P_{sd}, the dry-bulb saturated vapour pressure which is distinct from the wet-bulb saturated vapour pressure.

(b)   For the determination of sigma heat (S) (kJ kg^{-1}) (Eq. 8.28):

$$S = 0.622 \frac{P_{sw}}{P - P_{sw}} (2502.5 - 2.386 \, t_w) + 1.005 \, t_w \left[\frac{kg}{kg_{dry \, air}}\right] \tag{8.28}$$

where P is the barometric pressure (kPa) in the gallery.

(c) Calculation of airflow rate (Q) (m^3 s^{-1}) (Eq. 8.29):

$$Q = \frac{q_{eo}}{\rho(S_o - S_i)} \tag{8.29}$$

where

- ρ: Air density (kg m^{-3}),
- q_{eo}: Rate o heat flow to the air along the airway (kW),
- S_i: Air sigma heat at the inlet (kJ kg^{-1}), and
- S_o: Highest acceptable value of sigma heat in the air leaving the outlet (kJ kg^{-1}).

Exercise 8.7 The exhaust gases from diesel equipment located in a cul-de-sac are evacuated by means of a forcing ventilation system. The average wet-bulb temperature of the clean gases is 21 °C. The working power of the diesel equipment is 520 kW. The barometric pressure in the cul-de-sac is 111 kPa. Determine the airflow rate necessary for the gases to return to the main gallery at an average wet-bulb temperature of 29 °C.

Solution

Following the procedure outlined in the previous section:

1.   Calculate the saturated vapour pressure (P_{swi}) for the air at the wet-bulb temperature of the inlet (t_{wi}):

$$P_{swi} = 0.6106 \, e^{\left(\frac{17.27 \cdot 21}{21+237.3}\right)} = 2.49 \, \text{kPa}$$

2.   Determine sigma heat (S_i):

$$S_i = 0.622\frac{2.49}{111 - 2.49}(2502.5 - 2.386 \cdot 21) + 1.005 \cdot 21 = 56.11\frac{\text{kJ}}{\text{kg}_{\text{dry air}}}$$

3–4.   Calculate the saturated vapour pressure (P_{swo}) for gases at the maximum allowable temperature (t_{wo}):

$$P_{swo} = 0.6106 \, e^{\left(\frac{17.27 \cdot 29}{29+237.3}\right)} = 4.00 \, \text{kPa}$$

5.   Recalculate sigma heat for saturated pressure (S_o):

$$S_o = 0.622\frac{4.00}{111 - 4.00}(2502.5 - 2.386 \cdot 29) + 1.005 \cdot 29 = 85.72\frac{\text{kJ}}{\text{kg}_{\text{dry air}}}$$

6.   Calculate the required airflow rate based on the total heat flow to the air (q_{eo}):

It cannot be assumed that all the heat released by the diesel machinery will pass into the air in the short period of time it takes the air to travel from the inlet to the outlet. However, since no data are provided in this respect we will assume that 100% passes to the air as a first approximation. In addition, in the absence of other data, we will assume a value for air density of 1.2 kg m^{-3}. Therefore:

$$Q = \frac{520\,\text{kW}}{1.2\frac{\text{kg}}{\text{m}^3}\,(85.72 - 56.11)\frac{\text{kJ}}{\text{kg}_{\text{dry air}}}} = 14.63\,\frac{\text{m}^3}{\text{s}}$$

Any flow greater than this will be capable of lowering the temperature below the imposed threshold of 29 °C.

8.8 Duct Material Selection

When selecting the type of duct to be used in a ventilation system it is useful to know the characteristics of each type available:

Steel Ducts

- They are suitable for both forcing and exhaust systems.
- They present moderate airflow losses.
- They are suitable to be used for long length ducts.
- They are not reusable.

Fibre Ducts

- Like steel ducts, they are suitable for both forcing and exhaust systems.
- They are lighter than steel ducts.
- They have low coefficients of friction.
- They are reusable.

Spiral Ducts

- They can be elongated and shortened.
- It is possible to store them in a small space.
- They allow simple connections with ducts made of other materials.
- They adapt well to irregular geometries while still maintaining their cross section.
- They are lightweight, although they are heavier than the inflatable ducts due to the weight of the inner metal spiral.
- The installation quality can be poor, which will introduce abundant friction losses.

Inflatable (Layflat) Ducts

- They are low cost (20–80% less than the cost of fibre or metal).

- There will be less friction than in spiral ducts.
- They can only be used in forcing systems.
- They are very light.
- They can generally only be used for short lengths of ducting.

8.9 Determination of Duct Diameter

In principle, a duct with the largest possible diameter is always advisable, as this reduces the pressure that the fan must supply in order to overcome friction losses. This leads to the need for a less powerful fan and the lower pressure will, in turn, reduce leaks. However, the duct diameter is ultimately limited by the height of the mining machinery and the dimensions of the gallery where it will be situated. To aid selection, many manufacturers collate data on different types of ducting and their potential performance at different diameters using various airflow rates (e.g. Table 8.4).

Table 8.4 Minimum duct diameters for different ventilation airflow rates and recommended maximum velocities. Modified from de la Vergne (2008, p. 236)

Airflow rate	Plastics or fibre		Metal		Inflatable		Spiral	
Q $(m^3 \, s^{-1})$	D (cm)	v $(m \, s^{-1})$	D (cm)	v $(m \, s^{-1})$	D (cm)	v $(m \, s^{-1})$	D (cm)	v $(m \, s^{-1})$
2.4	38.1	21.1	40.6	18.5	43.2	16.4	50.8	11.8
4.7	53.3	21.1	55.9	19.2	58.4	17.5	71.1	11.8
7.1	66.0	20.8	68.6	19.2	73.7	16.6	86.4	12.1
9.4	76.2	20.6	78.7	19.3	83.8	17.0	101.6	11.6
18.9	109.2	20.2	111.8	19.3	119.4	16.9	142.2	11.9
23.6	121.9	20.2	124.5	19.4	132.1	17.2	157.5	12.1
35.4	149.9	20.1	154.9	18.8	162.6	17.0	198.1	11.5
47.2	172.7	20.1	177.8	19.0	188.0	17.0	228.6	11.5

8.10 Matching Duct Length and Fan Selection

The size of the fan is chosen bearing in mind that it will ultimately need to overcome the losses that will occur in the duct as the gallery is excavated, thus it is necessary to increase its length. In other words, the chosen fan must be capable of ventilating the final length of the cul-de-sac. In the initial stages of excavation, therefore, it will be oversized and its airflow rate must be regulated. Nowadays, this regulation tends to be carried by means of a variable-frequency drive that acts to moderate the rotation speed of the fan motor. The omission of such a frequency drive would entail an excess of airflow at the early stages of excavation that might be annoying for operators.

8.11   Air Velocity Inside the Duct

For plastic or fibre ducts, the maximum recommended air velocity is usually about 20 m s^{-1}. Other duct materials tend to have lower maximum air velocities: 19 m s^{-1} for metal ducts, 17 m s^{-1} for inflatable ducts and 12 m s^{-1} for spiral ducts (see Table 8.4).

8.12   Air Leaks in the Duct

Air leaks are generally a consequence of defects in the air ducts. If a duct is not sufficiently airtight, there will be a reduced airflow rate downstream of the fault. This results in a loss of air speed, and consequently, a reduction in dynamic pressure (Fig. 8.12) within the duct.

Fig. 8.12 Variation of the static (P_s), dynamic (P_v) and total (P_T) pressures along a permeable duct

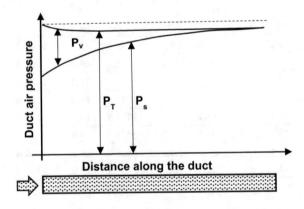

Intuitively, it can be deduced that an increase in static pressure within the duct would result in increased leakage. Similarly, the larger the area of the orifices through which air can escape, the greater the leakage. In this way, loss of airflow due to leaks can be approximated using what is termed the regulator or equivalent orifice formula (Eq. 8.30):

$$Q_p = \frac{A}{\rho_{st}} \left(\frac{\Delta P}{\rho_{real}} \right)^n = K \Delta P^n \tag{8.30}$$

where

- Q_p: Loss in airflow rate (m^3 s^{-1}).
- A: Orifice area (m^2).
- ρ_{st}: Standard air density (1.2 kg m^{-3}).
- ΔP: Static pressure difference between inside and outside the duct (Pa).
- n: A coefficient which approaches 1 for the laminar flow regime and varies between 0.5 and 1 for the turbulent flow regime. ASHRAE (2013) frequently uses $n = 0.65$.
- K: Dimensionless constant.

The above expression has recently been simplified by Auld (2004) (Eq. 8.31):

$$Q_p = \frac{\lambda \pi DL}{1000} \sqrt{\frac{\Delta P}{1000}} \qquad (8.31)$$

where

- Q_p: Loss in airflow rate (m^3 s^{-1}),
- λ: Leakage coefficient is calculated as the volume of air leaked per second per 1000 m^2 of conduit operating at 1000 Pa (m^3/s/1000 m^2/1000 Pa),
- D: Duct diameter (m),
- L: Conduit length (m), and
- ΔP: Static pressure difference between inside and outside the duct (Pa).

Gillies and Wu (1999), developed an expression to describe how pressure in the duct decreases in the direction of the outlet (Eq. 8.32):

$$\frac{Q_v}{Q_f} = \frac{P_v}{P_f} = 1 + 0.7 L_c \sqrt{R} \, l^{0.67} \qquad (8.32)$$

where

- Q_v: Airflow rate supplied by the fan (m^3 s^{-1}),
- Q_f: Loss in airflow rate supplied to the face (m^3 s^{-1}),
- P_v: Static pressure in the duct (kPa),
- P_f: Pressure required in (kPa) for the fan to deliver a flow Q_f (m^3 s^{-1}),
- L_c: Leakage coefficient (m^3 s^{-1} km^{-1} at 100 Pa),
- R: Duct resistance (Ns^2 m^{-8}), and
- l: Conduit length (m).

Additional approaches to the problem include that of Holdsworth, Prichard and Walton (1951) is (Eq. 8.33):

$$\frac{Q_v}{Q_f} = \left[1 + y^{(\mu+1)}\right]^{\frac{1}{1+\lambda}} \qquad (8.33)$$

where

- Q_v: Airflow rate supplied by the fan (m^3 s^{-1}),
- Q_f: Airflow supplied to the face (m^3 s^{-1}),
- y: A dimensionless parameter characteristic of the installation,

- λ: A dimensionless parameter depending on Re:

$$\lambda = 2 - \frac{29.7}{85 + \text{Re}}, \text{ and}$$

- μ: Constant depending on the nature of the orifices.

Voronin's expression (Novitzky 1962, p. 262) has been used widely in mining (Eq. 8.34):

$$\frac{Q_v}{Q_f} = \left(1 + \frac{KDL}{3l_t}\sqrt[2]{R}\right)^2 \tag{8.34}$$

where

- Q_v: Flow rate supplied by the fan (m^3 s^{-1}),
- Q_f: Flow supplied to the face (m^3 s^{-1}),
- D: Duct diameter (m),
- l_t: Length of each individual duct segment (m),
- L: Total length of the installation (m),
- R: Aerodynamic resistance of the installation (kmurgues), and
- K: Leakage coefficient. It was defined as the flow of air (m^3 s^{-1}) through the joint of an arbitrary duct 1 m long, when the pressure difference between the inside and outside is 1 mmwc. The values of this parameter for the previous units are between 10^3 and 3×10^3.

Robertson and Wharton (1980), developed one of the most widely used leakage coefficients in the mining industry. This coefficient was defined as (Eq. 8.35):

$$L_c = \frac{3(Q_1 - Q_2)(P_1 - P_2)}{2L\left(P_1^{1.5} - P_2^{1.5}\right)} 1000 \cdot 1000^{0.5} \tag{8.35}$$

where

- L_c: Leakage coefficient in m^3/s/100 m/1000 Pa. Le Roux (1972) offers the following ranges:

  - Excellent conditions: 0.03,
  - Good conditions: <0.25,
  - Inferior: 0.5, and
  - Bad: >1.0.

- Q_1: Flow rate at a point upstream of the leak (m^3 s^{-1}),
- Q_2: Flow rate at a point downstream of the leak (m^3 s^{-1}),
- P_1: Pressure at the upstream point (Pa),
- P_2: Pressure at the downstream point (Pa), and
- L: Distance between upstream and downstream points (m).

Table 8.5 Leakage coefficient θ as a function of conduit condition. Modified from Bertard and Bodelle (1962). These values can only be taken as an approximation and more accurate values should be obtained from the manufacturer

State	Type of conduit	θ $\left(\frac{m^2}{s\,mmwc^{0.5}}\right)$
Excellent	Compressed air conduit	10^{-6}
Very good	Conduit with flange and rubber gasket	0.5×10^{-5}
Good	Conduit with flange and rubber gasket	10^{-5}
Pretty good	Plastered joints (in the suction scheme[12])	0.5×10^{-4}
Mediocre	Band-type clamp and coupler (zip couplings)	10^{-4}
Bad	Band-type clamp and coupler (Velcro couplings)	0.5×10^{-3}
Very bad	Band-type clamp without coupler (eyelet joiners)	10^{-3}
Intolerable	Approximation of metallic tubes	$<10^{-3}$

It is very common in commercial catalogues to assume that the flow losses are concentrated in the joints between individual duct segments. These losses are generally quantified, thus (Audibert and Dressler 1946; Bertard and Bodelle 1962) (Eq. 8.36):

$$Q_i = \theta\, l\, \Delta h_i^{\gamma} \tag{8.36}$$

where

- Q_i: Loss in airflow rate at each joint ($m^3\ s^{-1}$).
- θ: Leakage coefficient measured in units: $\left(\frac{m^2}{s\,mmwc^{0.5}}\right)$. Its value depends upon numerous factors such as the size and shape of defects and values of this coefficient are provided in Table 8.5.
- l: Length of the conduit (m).
- Δh_i: Pressure difference between the inside and outside of the duct at the location of the joint (mmwc).
- γ: An exponent whose value is dependent upon the regime in which the leaks occur varying from 0.5 for the turbulent regime to 1 for the laminar regime. A common simplification often used in mine ventilation systems is to assume:

  - $\gamma = 1$ for very good quality rigid ductwork with leaks in the laminar regime;
  - $\gamma = 0.7$ for good installation (very low leakage), intermediate leakage rate, and a leak regimen between laminar and turbulent; and
  - $\gamma = 0.5$ for faulty installation with turbulent leakage.

Question 8.1 Taking Robertson and Wharton´s (1980) expression for leakage:

$$L_c = \frac{3(Q_1 - Q_2)(P_1 - P_2)}{2L\left(P_1^{1.5} - P_2^{1.5}\right)}$$

Demonstrate that this expression is dimensionally equivalent to the general air leakage law:

$$Q_p = K\sqrt{\Delta P}$$

where K is a constant.

Answer

The numerator can be simplified to give: $3\Delta Q\Delta P = K_1 \Delta Q\ \Delta P$.

The denominator, on the other hand, is: $K_2\ (\Delta P^{1.5})$, that is:

$$L_c = \frac{3\Delta Q\Delta P}{2L(\Delta P^{1.5})} = K_3\frac{\Delta Q}{\Delta P^{0.5}}$$

So, ignoring the Δ notation since this does not affect the dimensions of the expression and then rearranging:

$$\Delta P^{0.5} = \frac{K_3}{L_c}Q = K_4 Q$$

Solving for P, if $K_4^2 = R$, we get:

$$P = RQ^2$$

Or solving for Q

$$Q = K\sqrt{P}$$

Question 8.2 Transform the coefficients θ from Table 8.5, so that the expression 8.36, can be used with SI units. Assume that θ has been determined under turbulent conditions.

Answer

Taking Eq. 8.36, and solving for θ:

$$\theta = \frac{Q}{L\Delta h_i^\gamma}$$

Assuming turbulent conditions, then $\gamma = 0.5$:

$$\theta = \frac{Q}{L \Delta h_i^{0.5}}$$

Analysing the units in the expression:

$$\theta \left[\frac{\frac{m^3}{s}}{m \cdot mmwc^{0.5}} \right]$$

Bearing in mind that 1 mmwc = 9.81 Pa, then:

$$\frac{\frac{m^3}{s}}{m \cdot \cancel{mmwc^{0.5}}} \frac{1 \, \cancel{mmwc^{0.5}}}{(9.81 \, Pa)^{0.5}}$$

Therefore, if you divide the constant θ by $(9.81)^{0.5}$ you can convert it to SI units (θ_{SI}).

So, you finally get the table:

Quality of the installation	θ	θ_{SI}
Excellent	1×10^{-6}	3.2×10^{-7}
Very good	0.5×10^{-5}	1.6×10^{-6}
Good	1×10^{-5}	3.2×10^{-6}
Normal to regular	0.5×10^{-4}	1.6×10^{-5}
Bad to regular	1×10^{-4}	3.2×10^{-5}
Bad	0.5×10^{-3}	1.6×10^{-4}
Very bad	1×10^{-3}	3.2×10^{-4}
Intolerable	$<10^{-3}$	$<3.2 \times 10^{-4}$
Absolutely intolerable	0.5×10^{-3}	1.6×10^{-3}

8.13 Calculating Losses Along a Duct Using the Integral Method

8.13.1 General Relationship Between Airflow Rates and Pressures

Figure 8.13, below, shows a schematic diagram of a forcing system. The duct is permeable, and thus we assume that leaks are distributed evenly along the entire duct.

The system has the following characteristics:

- The ventilating fan supplies the pressure P_v with an airflow rate of Q_v at the coordinate origin ($x = 0$).
- At distance x from the fan, the pressure is $P(x) < P_v$ as a result of friction losses, and the flow rate $Q(x) < Q_v$ due to leakage along the distance x.
- Accordingly, at a distance of $x + dx$ from the fan, the general pressure conditions are: $P(x + dx) = P(x) - dP$, due to pressure losses in the segment dx; and the flow conditions: $Q(x + dx) = Q(x) - dQ$, due to air leaks in the segment dx.
- At the end of the duct ($x = L$), we have:

  - Pressure at the duct discharge (P_f): $P_f < P_v$
  - Flow rate at the duct discharge (Q_f): $Q_f < Q_v$

The value of the pressure drop along a dx length of duct is (Eq. 8.37):

$$-dP = r\, Q^\alpha\, dx \tag{8.37}$$

where

- α: Coefficient that in turbulent conditions takes the value $\alpha = 2$.
- r: Resistance per unit length ($k\mu\ m^{-1}$).

The leakage rate (dQ) for an element, dx, is (Eq. 8.38):

$$-dQ = \theta\, P^\gamma\, dx \tag{8.38}$$

where

- θ: Leakage coefficient.
- γ: Coefficient that for turbulent conditions is normally[13] $\gamma = 0.5$.

Taking the quotient of Eqs. 8.37 and 8.38, we obtain a relationship between pressure and airflow rate:

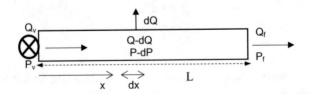

Fig. 8.13 A forcing ventilation system comprising a permeable duct of length L where the airflow and pressure vary from Q_v and P_v, at the fan, to Q_f and P_f at the duct output. Distance from the fan, along the length of the duct is denoted by x

---

[13]Some publications consider it appropriate to use the value of $\gamma = 0.65$.

$$\frac{-dP}{-dQ} = \frac{dP}{dQ} = \frac{r\,Q^\alpha\,dx}{\theta\,P^\gamma\,dx} = \frac{r\,Q^\alpha}{\theta\,P^\gamma}$$

Rearranging this gives (Eq. 8.39):

$$\theta\,P^\gamma\,dP = r\,Q^\alpha\,dQ \tag{8.39}$$

Integrating between P_v and P_f and Q_v and Q_f gives Eq. 8.40:

$$P_v^{\gamma+1} - P_f^{\gamma+1} = \left(\frac{\gamma+1}{\alpha+1}\right)\left(\frac{r}{\theta}\right)\left(Q_v^{\alpha+1} - Q_f^{\alpha+1}\right) \tag{8.40}$$

This expression allows us to see the interrelation between flow rates and pressures at the fan outlet and at the duct discharge point. Unfortunately, it is not practically useful since, in real-life applications too many of its parameters will be unknown. Furthermore, information about how these parameters vary with duct length has been lost (due to the elimination of this variable in the division process).

8.13.2 Simplified Solution Following Bertard and Bodelle (1962)

These authors place the origin, $x = 0$, where the duct discharges onto the gallery face, thus making all distances, x, along the duct negative. This then means that dP and dQ, in Eq. 8.37 and Eq. 8.38, respectively, take a positive sign. The particular considerations, in this case, are as follows:

- The static pressure at the duct outlet is zero ($P_f = 0$ for $x = 0$).
- The airflow rate provided by the fan, that is for $x = L$, is Q_o in the absence of leaking.
- If the duct is perfectly sealed, the leakage airflow rate (Q_l) is zero and the airflow rate at every point along the duct is then $Q_f = Q_o$.
- If P_o is the pressure provided by the fan in the absence of leakage. The pressure necessary to overcome friction losses is:

$$P_o = P(x = L) = r\,L\,Q_o^\alpha \text{ remembering that } P_f = P(x = 0) = 0$$

where α is the coefficient of regimen (2 for rigid ducts, 1.7 for flexible ones).

- In the absence of leaks and at a constant airflow rate is,[14] the pressure[15] $P(x)$ at a distance x, is:

$$dP(x) = r x Q_o^2$$

where r is the resistance per unit length.
   Then:

$$P_o = rLQ_o^2$$

So:

$$Q_o^2 = \frac{P_o}{rL}$$

Therefore, the pressure at distance x is:

$$P(x) = P_o \left(\frac{x}{L}\right)$$

If the duct is not airtight, there will be leaks, which can be calculated as before (Eq. 8.38). In this case, it is however, still appropriate to assume a direct proportionality between pressure and duct length, and knowing that the boundary condition $Q(x = 0) = Q_0 = Q_f$ holds, it follows that[16]:

$$dQ(x) = \theta \, P(x)^\gamma dx = \theta \left(P_o \frac{x}{L}\right)^\gamma dx$$

Integrating over the interval $(0, x)$:

$$Q(x) = Q(0) + \theta \left(\frac{P_o}{L}\right)^\gamma \left(\frac{x^{\gamma+1}}{\gamma+1}\right) \tag{8.41}$$

Given that $Q(0) = Q_f$ we have Eq. 8.42:

$$Q(x) = Q_f + \theta \left(\frac{P_o}{L}\right)^\gamma \frac{1}{\gamma+1} \left(\frac{x^{\gamma+1}}{1}\right) \tag{8.42}$$

If $x = L$ then $Q(x)$ is the airflow rate at the fan, Q_v, we have an equation to calculate the total leakage along the length of the duct:

$$Q_v = Q_f + \theta \left(\frac{P_o}{L}\right)^\gamma \frac{1}{\gamma+1} \left(\frac{L^{\gamma+1}}{1}\right) \tag{8.43a}$$

---

[14]Remember that r is the resistance per unit length.
[15] $\Delta P = P(x)$ because $P_f = 0$.
[16]Bertard and Bodelle (1962) assume that in modern ducts leaks are small.

And after some simplification we arrive at Eq. 8.43b:

$$Q_v - Q_f = \theta \frac{P_o^\gamma}{\gamma + 1} L \tag{8.43b}$$

Returning to Eq. 8.43a, we can rearrange terms such that:

$$\frac{Q_v - Q_f}{L^{(\gamma+1)}} = \theta \frac{1}{\gamma + 1} \left(\frac{P_o}{L}\right)^\gamma$$

Which can be substituted in Eq. 8.41, so obtaining Eq. 8.44:

$$Q(x) = Q_f + (Q_v - Q_f)\left(\frac{x}{L}\right)^{\gamma+1} \tag{8.44}$$

Taking the equation for friction losses (Eq. 8.37), at a point, x:

$$dP(x) = r\, Q(x)^\alpha\, dx$$

then substituting for $Q(x)$ from Eq. 8.44, we obtain Eq. 8.45:

$$dP(x) = r\left[Q_f + (Q_v - Q_f)\left(\frac{x}{L}\right)^{\gamma+1}\right]^\alpha dx \tag{8.45}$$

Which must be integrated to find the required fan pressure at point x. This can be done by first doing a Taylor expansion of Eq. 8.45. Taking the first two terms of this expansion, we have:

$$dP(x) \approx r\, Q_f^\alpha \left[1 + \frac{\alpha\,(Q_v - Q_f)}{Q_f}\left(\frac{x}{L}\right)^{\gamma+1}\right] dx$$

Which, when integrated for the interval $[0, L]$, with $P_f = 0$, gives us (Eq. 8.46):

$$P(x) = r\, Q_f^\alpha \left[x + \frac{\alpha\,(Q_v - Q_f)}{Q_f}\left(\frac{1}{L}\right)^{\gamma+1}\left(\frac{x^{\gamma+2}}{\gamma+2}\right)\right]$$

Therefore, the pressure provided by the fan (P_v) is as shown in Eq. 8.46:

$$P_v = P(x = L) = r\, Q_f^\alpha L\left[1 + \frac{(Q_v - Q_f)}{Q_f}\left(\frac{\alpha}{\gamma+2}\right)\right] \tag{8.46}$$

Since $P_0 = r\, Q_f^\alpha L$, this simplifies to give Eq. 8.47:

$$P_v = P_0 \left[1 + \left(\frac{\alpha}{\gamma + 2} \right) \frac{(Q_v - Q_f)}{Q_f} \right] \tag{8.47}$$

A more formal solution to Eq. 8.45, in the case of turbulent conditions, ($\alpha = 2$), is obtained if we expand the square:

$$dP(x) = r \left[Q_f^2 + 2 Q_f (Q_v - Q_f) \left(\frac{x}{L} \right)^{\gamma+1} + (Q_v - Q_f)^2 \left(\frac{x}{L} \right)^{2(\gamma+1)} \right] dx$$

Which when integrated for the range [0, L], with $P_f = 0$ as assumed previously, gives us Eqs. 8.48 and 8.49, as alternative expressions for the calculation of the fan pressure:

$$P_v = r \, Q_f^2 \, L \left[1 + \left(\frac{2}{\gamma + 2} \right) \frac{(Q_v - Q_f)}{Q_f} + \left(\frac{1}{2\gamma + 3} \right) \frac{(Q_v - Q_f)^2}{Q_f^2} \right] \tag{8.48}$$

$$P_v = P_0 \left[1 + \left(\frac{2}{\gamma + 2} \right) \frac{(Q_v - Q_f)}{Q_f} + \left(\frac{1}{2\gamma + 3} \right) \frac{(Q_v - Q_f)^2}{Q_f^2} \right] \tag{8.49}$$

The leak rate $(Q_v - Q_f)$ can be obtained from Eq. 8.43a. When leakage is high (poor installations), the value of $(Q_v - Q_f)^2$ becomes important, leading to discrepancies between results given by these expressions and the simplified ones.

In the above expressions, we usually work with the units proposed by Bertard and Bodelle (1962):

- r: kμ m^{-1},
- P_v: mmwc,
- Q_v: m^3 s^{-1},
- Q_f: m^3 s^{-1},
- L: m,
- α: Coefficient of regimen (2 for rigid ducts, 1.7 for flexible ones), and
- θ: Exponent of the leakage law. Values vary from 0.5 to 1.2 from higher to lower tightness of the conduction.

Exercise 8.8 For proper ventilation, the required airflow flow rate is 1.7 m^3 s^{-1} at the face of a cul-de-sac. To this end, a metal duct 400 m long of 0.4 m diameter and resistance of 0.2 kμ m^{-1} is used. For this type of rigid ducting θ is 0.5×10^{-5} (corresponding to a very good installation) and γ is 0.9. Determine:

(a) Airflow rate which must be provided by the fan to the duct.
(b) Static pressure supplied by the fan (calculation using the simplified expression).
(c) Static pressure supplied by the fan (calculation by means of the exact expression).

Solution

(a) We start by calculating the friction loss in the duct (P_o) when there is no leakage ($Q_o = Q_f$), with the expression:

$$P_0 = r L Q_f^\alpha$$

It is a rigid duct, so, as there are no data indicated in the statement, α can be approximated to 2 (turbulent flow). Therefore, substituting in our values we have:

$$P_0 = 0.2 \frac{k\mu}{m} \cdot 400\,m \cdot \left(1.7 \frac{m^3}{s}\right)^2 = 231.2\,mmwc$$

The airflow rate of the fan is given by:

$$Q_v = Q_f + L \frac{\theta\, P_0^\gamma}{\gamma + 1}$$

Therefore, substituting:

$$Q_v = 1.7 \frac{m^3}{s} + \left(400 \frac{0.5 \times 10^{-5} \cdot 231.2^{0.9}}{0.9 + 1}\right) \frac{m^3}{s} = 1.84 \frac{m^3}{s}$$

(b) The simplified expression for calculating the pressure at the fan is:

$$P_v = P_0 \left[1 + \frac{\alpha}{\gamma + 2}\left(\frac{Q_v - Q_f}{Q_f}\right)\right]$$

Substituting:

$$P_v = 231.2\,mmwc \left[1 + \frac{2}{0.9 + 2}\left(\frac{1.84 - 1.7}{1.7}\right)\right] = 244.4\,mmwc\ (2397.6\,Pa)$$

(c) A more accurate solution for the fan pressure can be obtained using Eq. 8.49:

$$P_v = P_0 \left[1 + \left(\frac{2}{\gamma + 2}\right)\frac{(Q_v - Q_f)}{Q_f} + \left(\frac{1}{2\gamma + 3}\right)\frac{(Q_v - Q_f)^2}{Q_f^2}\right]$$

Then, using the value obtained in (a) for $Q_v = 1.84\,m^3\,s^{-1}$, and substituting in the rest of the data from the question, we have:

$$P_v = 231.2\,mmwc \left[1 + \left(\frac{2}{0.9 + 2}\right)\left(\frac{1.84 - 1.7}{1.7}\right) + \left(\frac{1}{2 \cdot 0.9 + 3}\right)\frac{(1.94 - 1.7)^2}{1.7^2}\right]$$

$$= 245.29\,mmwc\ (2406.30\,Pa)$$

8.13.3 Special Solution for Turbulent Conditions

The primitive of Eq. 8.39 is:

$$\frac{\theta}{\gamma + 1} P^{\gamma+1} = \frac{r}{\alpha + 1} Q^{\alpha+1} + C_0$$

Under turbulent conditions, that is, $\alpha = 2; \gamma = 0.5$ this gives:

$$\frac{\theta}{1.5} P^{1.5} = \frac{r}{3} Q^3 + C_o \rightarrow \frac{2\theta}{3} P^{\frac{3}{2}} = \frac{r}{3} Q^3 + \frac{3}{3} C_o \rightarrow P^{\frac{3}{2}} = \frac{r}{2\theta} Q^3 + \frac{3C_o}{2\theta}$$

If:

$$C_1 = \frac{3C_o}{r}$$

Then:

$$P^{\frac{3}{2}} = \frac{r}{2\theta}(Q^3 + C_1) \tag{8.50}$$

We require Q^2 (Eq. 8.37), and this can be found by first solving Eq. 8.50 for Q^3:

$$Q^3 = \frac{2\theta}{r} P^{\frac{3}{2}} - C_1 = \frac{2\theta}{r}\left(P^{\frac{3}{2}} - \frac{C_1 r}{2\theta}\right) = \frac{2\theta}{r}\left(P^{\frac{3}{2}} + b\right)$$

where

$$b = -\frac{C_1 r}{2\theta}$$

Then, raising both sides of the equation to the power of 2/3, we obtain Eq. 8.51:

$$Q^2 = Q^{3 \cdot \frac{2}{3}} = \left(\frac{2\theta}{r}\right)^{\frac{2}{3}}\left(P^{\frac{3}{2}} + b\right)^{\frac{2}{3}} \tag{8.51}$$

Substituting this expression for Q^2 into Eq. 8.37, we have:

$$-dP = r\left(\frac{2\theta}{r}\right)^{\frac{2}{3}}\left(P^{\frac{3}{2}} + b\right)^{\frac{2}{3}} dx$$

Rearranging terms we obtain Eq. 8.52:

$$\left(P^{\frac{3}{2}} + b\right)^{-\frac{2}{3}} dP = -r\left(\frac{2\theta}{r}\right)^{\frac{2}{3}} dx \tag{8.52}$$

Whose integral form is Eq. 8.53:

$$\int \left(P^{\frac{3}{2}} + b \right)^{-\frac{2}{3}} dP = -r \left(\frac{2\theta}{r} \right)^{\frac{2}{3}} \int dx \qquad (8.53)$$

In this way we obtain two primitives:

$$\text{Primitive}: I(x) = -r \left(\frac{2\theta}{r} \right)^{\frac{2}{3}} \int dx = -r \left(\frac{2\theta}{r} \right)^{\frac{2}{3}} x$$

$$\text{Primitive}: I(P) = \int \left(P^{\frac{3}{2}} + b \right)^{-\frac{2}{3}} dP$$

Solving this second primitive is not straightforward and requires a number of steps. First, we have to make a substitution of variables:

$$t = P^{\frac{3}{2}}; P = t^{\frac{2}{3}}; dP = \frac{2}{3} t^{-\frac{1}{3}} dt$$

Then:

$$I(t) = \frac{2}{3} \int (t + b)^{-\frac{2}{3}} t^{-\frac{1}{3}} dt$$

Making a further substitution of variables:

$$\frac{(t + b)}{t} = z^3$$

Therefore:

$$(t + b) = t\, z^3; t = \frac{b}{z^3 - 1}; dt = \frac{-3bz^2}{(z^3 - 1)^2} dz$$

Thus the new primitive is:

$$I(z) = \frac{2}{3} \int \left(\frac{bz^3}{z^3 - 1} \right)^{-\frac{2}{3}} \left[\frac{b^{-\frac{1}{3}}}{(z^3 - 1)^{-\frac{1}{3}}} \right] \left[\frac{-3bz^2}{(z^3 - 1)^2} \right] dz$$

$$I(z) = -2 \int \frac{dz}{z^3 - 1}$$

To solve this, we can decompose the expression into partial fractions. Firstly, consider that:

$$(z^3 - 1) = (z - 1)(z^2 + z + 1)$$

Thus:

$$\frac{-2}{z^3 - 1} = \frac{A}{z - 1} + \frac{Bz + C}{z^2 + z + 1}$$

where[17]:

$$A = -\frac{2}{3}$$

$$B = \frac{2}{3}$$

$$C = \frac{4}{3}$$

In this way, the new integral is:

$$I(z) = -\frac{2}{3}\int \frac{dz}{z - 1} + \frac{2}{3}\int \frac{z + 2}{z^2 + z + 1}dz = I(z_1) + I(z_2)$$

Therefore:

$$I(z_1) = -\frac{2}{3}\int \frac{dz}{z - 1} = -\frac{2}{3}\ln(z - 1) = -\frac{1}{3}\ln(z - 1)^2$$

And[18]:

$$I(z_2) = \frac{2}{3}\int \frac{z + 2}{z^2 + z + 1}dz = \frac{1}{3}\int \frac{2z + 1}{z^2 + z + 1}dz + \frac{1}{3}\int \frac{3}{z^2 + z + 1}dz$$

Thus:

$$I(z_2) = \frac{1}{3}\ln(z^2 + z + 1) + \frac{2}{\sqrt{3}}\arctan\left(\frac{2z + 1}{\sqrt{3}}\right)$$

The solution to Eq. 8.53, changing the sign of both members, is:

$$I(z_1) + I(z_2) = \frac{1}{3}\ln(z - 1)^2 - \frac{1}{3}\ln(z^2 + z + 1) - \frac{2}{\sqrt{3}}\arctan\left(\frac{2z + 1}{\sqrt{3}}\right)$$

Therefore, we obtain Eq. 8.54:

[17]For the system to be solved: $A + B = 0$; $A - B + C = 0$; $A - C = -2$.
[18]The second term can be solved by expressing it in the form $1/(u^2 + 1)$ which when integrated gives arctan (u). Programmes such as Symbolab, Matlab and Mathematica, amongst others, will give an immediate result for the integral of $1/(x^2 + x + 1)$.

$$I(z) = \frac{1}{3} \ln \frac{(z-1)^2}{(z^2+z+1)} - \frac{2}{\sqrt{3}} \arctan \left(\frac{2z+1}{\sqrt{3}} \right) \tag{8.54}$$

Moreover:

$$I(z_1) + I(z_2) = r \left(\frac{2\theta}{r} \right)^{\frac{2}{3}} x + C_2$$

Therefore, we get Eq. 8.55:

$$I(z) = r \left(\frac{2\theta}{r} \right)^{\frac{2}{3}} x + C_2 \tag{8.55}$$

Returning to our true values[19]:

$$z = \left(1 - \frac{C_1 r}{2\theta} P^{-1.5} \right)^{\frac{1}{3}} = \left(1 - \frac{C_1 r}{2\theta} \frac{1}{P^{1.5}} \right)^{\frac{1}{3}}$$

If we now substitute the value of $P^{1.5}$, given by Eq. 8.50, we have:

$$z = \left[1 - \frac{C_1 r}{2\theta} \frac{2\theta}{r(Q^3 + C_1)} \right]^{\frac{1}{3}} = \left[1 - \frac{C_1}{(Q^3 + C_1)} \right]^{\frac{1}{3}} = \left[\frac{Q^3}{(Q^3 + C_1)} \right]^{\frac{1}{3}} = \frac{Q}{\sqrt[3]{Q^3 + C_1}}$$

Therefore, we get Eq. 8.56:

$$z = \sqrt[3]{1 - \frac{C_1 r}{2\theta} \frac{1}{P^{1.5}}} = \frac{Q}{\sqrt[3]{Q^3 + C_1}} \tag{8.56}$$

Which has the alternative forms shown in Eqs. 8.56a and 8.56b:

$$P = \left[\frac{C_1 r}{2\theta(1 - z^3)} \right]^{\frac{2}{3}} \tag{8.56a}$$

$$Q = \left(\frac{C_1 z^3}{1 - z^3} \right)^{\frac{1}{3}} \tag{8.56b}$$

---

[19]The changes made can be reverted with the expressions:

$\frac{(t+b)}{t} = z^3$; $z = \sqrt[3]{\frac{t+b}{t}}$; $t = P^{\frac{2}{3}}$; therefore: $z = \sqrt[3]{\frac{P^{\frac{3}{2}} + b}{P^{\frac{3}{2}}}} = \sqrt[3]{1 + \frac{b}{P^{\frac{3}{2}}}}$; $b = -\frac{C_1 \cdot r}{2\theta}$;

$z = \sqrt[3]{1 - \frac{C_1 r}{2\theta} P^{-\frac{3}{2}}}$.

Calculation Procedure

Assuming that conditions are such that pressure at the duct outlet (gallery face) is P_f $= 0$ and that the airflow rate, Q_f, at the face can be measured, the following procedure allows us to determine the operating conditions (Q_v, P_v) of the fan:

1st Calculation of the integration constant (C_1) can be done using Eq. 8.50, relating the values for pressure and airflow rate at the duct outlet:

$$P_f^{\frac{3}{2}} = \frac{r}{2\theta}\left(Q_f^3 + C_1\right)$$

Therefore:

$$C_1 = \left(\frac{2\theta}{r}\right)P_f^{\frac{3}{2}} - Q_f^3$$

Since $P_f = 0$, then:

$$C_1 = -Q_f^3$$

2nd Calculation of z at duct outlet.

$$z = \sqrt[3]{1 - \frac{C_1 r}{2\theta} \frac{1}{P_f^{1.5}}}$$

Note that, since $P_f = 0$, then $z \rightarrow \infty$.

3rd Calculation of the constant C_2 for the conditions at the duct outlet ($L = x$; P_f $= 0$ and $z \rightarrow \infty$).

First, using Eq. 8.54, we find $I(z)$:

$$I(z) = \frac{1}{3}\ln\frac{(z-1)^2}{(z^2+z+1)} - \frac{2}{\sqrt{3}}\arctan\left(\frac{2z+1}{\sqrt{3}}\right) =$$

Taking the limit $z \rightarrow \infty$:

$$\lim_{z \rightarrow \infty} \frac{1}{3}\ln\frac{(z-1)^2}{(z^2+z+1)} = \frac{1}{3}\ln(1) = 0$$

$$\lim_{z \rightarrow \infty} -\frac{2}{\sqrt{3}}\text{arctg}\left(\frac{2z+1}{\sqrt{3}}\right) = -\frac{2}{\sqrt{3}} \cdot \frac{\pi}{2} = -\frac{\pi}{\sqrt{3}}$$

Then, using Eq. 8.55, knowing that $x = L$: $I(z) = r\left(\frac{2\theta}{r}\right)^{\frac{2}{3}}L + C_2$

Therefore:

$$-\frac{\pi}{\sqrt{3}} = r\left(\frac{2\theta}{r}\right)^{\frac{2}{3}}L + C_2$$

$$C_2 = -\frac{\pi}{\sqrt{3}} - r\left(\frac{2\theta}{r}\right)^{\frac{2}{3}} L$$

4th Calculation of z at the location of the fan ($x = 0$). Having obtained the value of the integration constants, Eq. 8.54, which is only dependent on z, can be solved by numerical methods or using the Excel Solver function.

Table 8.6 and Fig. 8.14 shows $I(z)$ versus z. It can be observed that for $z \, \varepsilon(1.001, 20)$ $I(z)$ ranges from -6.181 to -1.816. Moreover, for $z = \infty$, $I(z)$ takes the value -1.838.

5th Calculation of the fan airflow rate and pressure. Once z is found, z the pressure and airflow rate are calculated with Eqs. 8.56a and 8.56b:

$$P = \left[\frac{C_1 r}{2\theta(1-z^3)}\right]^{\frac{2}{3}}$$

Table 8.6 Values of the I (z) function (Eq. 8.54)

z	I (z)	z	I (z)	z	I (z)	z	I (z)
1.001	−61.812	2.200	−20.287	1.23	−26.976	3.70	−18.874
1.01	−46.522	2.25	−20.187	1.24	−26.750	3.80	−18.836
1.02	−41.967	2.30	−20.094	1.25	−26.535	3.90	−18.800
1.03	−39.329	2.35	−20.007	1.30	−25.599	4.00	−18.767
1.04	−37.476	2.40	−19.927	1.35	−24.843	4.10	−18.736
1.05	−36.053	2.45	−19.851	1.40	−24.217	4.20	−18.708
1.06	−34.902	2.50	−19.781	1.45	−23.688	4.30	−18.682
1.07	−33.938	2.55	−19.714	1.50	−23.234	4.40	−18.657
1.08	−33.112	2.60	−19.652	1.55	−22.841	4.50	−18.634
1.09	−32.390	2.65	−19.594	1.60	−22.497	4.60	−18.613
1.10	−31.750	2.70	−19.539	1.65	−22.193	4.70	−18.592
1.11	−31.177	2.75	−19.487	1.70	−21.922	4.80	−18.574
1.12	−30.658	2.80	−19.437	1.75	−21.680	4.90	−18.556
1.13	−30.186	2.85	−19.391	1.80	−21.462	5.00	−18.539
1.14	−29.753	2.90	−19.347	1.85	−21.265	6.00	−18.416
1.15	−29.354	2.95	−19.305	1.90	−21.086	7.00	−18.342
1.16	−28.984	3.00	−19.266	1.95	−20.923	8.00	−18.294
1.17	−28.640	3.10	−19.193	2.00	−20.774	9.00	−18.262
1.18	−28.318	3.20	−19.127	2.05	−20.637	10.00	−18.238
1.19	−28.017	3.30	−19.067	2.10	−20.511	15.00	−18.182
1.20	−27.734	3.40	−19.012	2.15	−20.394	20.00	−18.163
1.21	−27.467	3.50	−18.962	2.20	−20.287	∞	−18.138
1.22	−27.215	3.60	−18.916				

$$Q = \left(\frac{C_1 z^3}{1 - z^3} \right)^{\frac{1}{3}}$$

Thus, for the particular case of $P_f = 0$:

$$P_v = Q_f^2 \left[\frac{r}{2\theta \left(1 - z^3 \right)} \right]^{\frac{2}{3}}$$

$$Q_v = Q_f \left(\frac{z^3}{1 - z^3} \right)^{\frac{1}{3}}$$

Appropriate units:
Any coherent units can be employed; however, it is common to use:
$Q_v, Q_f : m^3\ s^{-1}$,
$P_v, P_f : \mathrm{mmwc}$,
r: $\mathrm{k}\mu\ m^{-1}$,

Fig. 8.14 Graphical representation of I(z) function

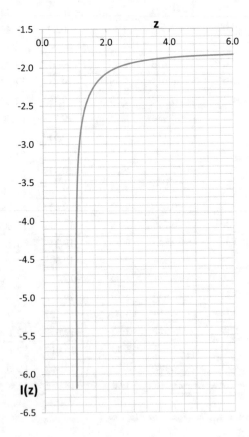

L: m, and
θ: No dimensions.

Exercise 8.9 For the conditions described in Exercise 8.8 determine:

(a) The airflow rate provided by the fan, and
(b) The pressure supplied by the fan.

Assume duct discharge pressure, $P_f = 0$.

Solution

(a)

Step 1. Calculate the constant C_1 with Eq. 8.50, using the assumption that $P_f = 0$:

$$C_1 = \left(\frac{2\theta}{r}\right)P_f^{\frac{3}{2}} - Q_f^3 = 0-1.7^3 = -4.913$$

Step 2. Calculate z at duct outlet.
Given that $P_f = 0$, then $z \to \infty$.
Step 3. Calculate the integration constant C_2:

$$C_2 = -\frac{\pi}{\sqrt{3}} - r\left(\frac{2\theta}{r}\right)^{\frac{2}{3}} L$$

$$I(z) = \frac{\pi}{\sqrt{3}} = 0.2 \cdot \left(\frac{2 \cdot 0.5 \times 10^{-5}}{0.2}\right)^{\frac{2}{3}} \cdot 400 + C_2$$

$$C_2 = -1.9224$$

Step 4. Calculate z at the fan location ($x = 0$) using Eq. 8.54:

$$I(z) = \frac{1}{3}\ln\frac{(z-1)^2}{(z^2+z+1)} - \frac{2}{\sqrt{3}}\arctan\left(\frac{2z+1}{\sqrt{3}}\right)$$

Given that for $x = 0$, Eq. 8.55 gives:

$$I(z) = r\left(\frac{2\theta}{r}\right)^{\frac{2}{3}} x + C_2$$

$$I(z) = C_2$$

Thus:

$$I(z) = -1.9224$$

Solved using Excel Solver gives $z = 3.0565$
Solving by means of Table 8.6, gives: $z = 3.00$
Step 5. Calculating the fan pressure and airflow rate by means of Eq. 8.56a:

$$P = \left[\frac{C_1 r}{2\theta (1 - z^3)} \right]^{\frac{2}{3}}$$

Substituting we have:

$$P_v = \left[\frac{-4.913 \cdot 0.2}{2 \cdot 0.5 \times 10^{-5} \cdot (1 - 3.0565^3)} \right]^{\frac{2}{3}} = 1.72 \frac{m^3}{s}$$

and Eq. 8.56b:

$$Q = \left(\frac{C_1 z^3}{1 - z^3} \right)^{\frac{1}{3}}$$

Substituting in all known values:

$$Q_v = \left(\frac{-4.913 \cdot 3.0565^3}{1 - 3.0565^3} \right)^{\frac{1}{3}} = 233.4 \, \text{mmwc} \, (2289.7 \, \text{Pa}).$$

8.14 Discrete Solution (Step by Step Method)

One calculation method that is frequently used assumes that a duct is composed of n segments each with the same aerodynamic characteristics. It is further assumed that leaks will occur at the joints between segments, and that these are always the same fraction (k) of the airflow rate at each joint (Flügge 1953).

In this way, the airflow rate provided by the fan (Q_v), must be equal to the one that arrives at the duct outlet (face) (Q_f), plus the losses (ΔQ) (Fig. 8.15):

$$Q_v = Q_f + \Delta Q$$

For segment $i = 0$ (a special segment which contains the fan and to which the remaining segments of the duct are connected):

$$\text{Airflow rate} : Q_v = Q_0$$

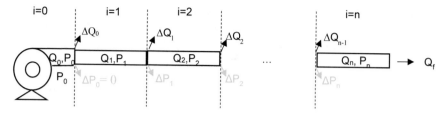

Fig. 8.15 Pressure, flow and friction losses in a segmented ventilation duct

At segment $i = 1$ (first duct segment):

$$\text{Airflow rate}: Q_1 = Q_0 - \Delta Q_0$$

where

ΔQ_0 is the leakage at the first joint between the fan and the duct.
At segment $i = 2$ (first duct segment):

$$\text{Airflow rate}: Q_2 = Q_1 - \Delta Q_1$$

Generalising for any segment i:

$$\text{Airflow rate}: Q_i = Q_{i-1} - \Delta Q_{i-1}$$

Assuming that each joint leaks the same fraction (k) of the airflow rate circulating in the duct, then:
Segment $i = 1$:

$$Q_1 = Q_0(1 - k)$$

Segment $i = 2$:

$$Q_2 = Q_0(1 - k)(1 - k)$$

Segment $i = 3$:

$$Q_3 = Q_0(1 - k)(1 - k)(1 - k)$$

Then, the flow rate (Q_i) for the ith segment of the geometric progression is (Eq. 8.57):

$$Q_i = Q_0 (1 - k)^i \qquad\qquad (8.57)$$

Toraño et al. (2002) used analogous reasoning to calculate pressures in a segmented duct, following Simode (1976). In this case, to provide a pressure (P_f), at the gallery face, the fan pressure (P_v) must exceed this value by a quantity sufficient to overcome pressure losses along the full length of the duct (ΔP), that is to say:

$$P_v = P_f + \Delta P$$

If we define the following variables:

- L_i: Length of an individual duct segment,
- $P_v = P_0$: Pressure provided by the fan at the duct entrance,
- $P_f = P_n$: Line outlet pressure (discharge pressure),
- R_i: Resistance per unit length of each duct segment,
- P_i: Internal pressure of each duct segment, and
- ΔP_i: Pressure loss at the end of a segment with a flow rate Q_i.

Thus, if we consider a duct formed of unitary segments each of length (L_i) and resistance (R_i), we have:
Segment $i = 0$:

$$\text{Pressure} : P_v = P_0$$

Segment $i = 1$:

$$\text{Pressure} : P_1 = P_0 - \Delta P_0$$

$$\text{Pressure loss in the segment} : \Delta P_1 = R_1 \, L_1 \, Q_1^2$$

Generalising for any segment i:

$$\text{Pressure} : P_i = P_{i-1} - \Delta P_{i-1}$$

$$\text{Pressure loss in the segment} : \Delta P_i = R_i \, L_i \, Q_i^2$$

The total friction pressure loss will be the sum of the partial losses in each segment:

$$\Delta P = \Delta P_1 + \cdots + \Delta P_i + \cdots + \Delta P_n$$

That is:

$$\Delta P = R_1 L_1 Q_1^2 + \cdots + R_i L_i Q_i^2 + \cdots + R_n L_n Q_n^2$$

And since, all the segments are equal of length (L_i) and resistance (R_i):

$$\Delta P = R_i L_i (Q_1^2 + \cdots + Q_i^2 + \cdots + Q_n^2)$$

Substituting in the expression for the airflow rate in the ith segment (Eq. 8.57):

$$\Delta P = R_i L_i Q_0^2 [(1 - k)^2 + (1 - k)^4 + (1 - k)^6 + \cdots + (1 - k)^{2n}]$$

The expression is a geometric progression[20] of ratio $(1 - k)^2$, therefore, the sum of the first n terms can be expressed as in Eq. 8.58:

$$\Delta P = R_i L_i Q_0^2 \frac{[(1 - k)^{2n+2} - (1 - k)^2]}{(1 - k)^2 - 1} \tag{8.58}$$

Now, as the airflow rate at the gallery face (Q_f) is:

$$Q_f = Q_n = Q_0 (1 - k)^n = Q_v (1 - k)^n$$

Therefore, the characteristic equation of the system as a function of the required airflow rate at the face is (Eq. 8.59):

$$\Delta P = R_i L_i \frac{Q_f^2}{(1 - k)^{2n}} \frac{[(1 - k)^{2n+2} - (1 - k)^2]}{(1 - k)^2 - 1} \tag{8.59}$$

Exercise 8.10 Given Eq. 8.59, you are asked to:

(a) Express it for a segment located before the beginning of the duct ($n = 0$).
(b) Express it for a number of segments (n) when leakage is very small (k tends to 0).

Solution

(a)

The conditions where $n = 0$ are equivalent to those pertaining to a fan output without any coupled segments, i.e. in free discharge. In this case, it is not possible to use the derived formula since it is only valid for cases where the fan is connected to a duct segment (n > 1).

---

[20]The sum of n terms of a r-ratio geometric progression is calculated by the formula: $S_n = \frac{a_n r - a_1}{r - 1}$.
Where a_1 and a_n are the first and last term of the progression, respectively.

(b)

Here we must use of the propierties of limits, thus, limit of a product of two functions is equivalent to the product of their limits. In this case, Eq. 8.59, can be expresed as the product of two limits, first:

$$\lim_{k \to 0}\left[\frac{1}{(1-k)^{2n}} \right] = 1$$

And second:

$$\lim_{k \to 0}\left\{ \frac{\left[(1-k)^{2n+2} - (1-k)^2 \right]}{\left[(1-k)^2 - 1 \right]} \right\}$$

Using the L'Hôpital's rule for the undeterminate form 0/0, we have:

- Numerator: $f'(k) = (-1)(2n+2)(1-k)^{2n+1} - 2(1-k)(-1)$
- Denominator: $g'(k) = 2(1-k)(-1)$

Hence:

$$\frac{f'(k)}{g'(k)} = \frac{2n}{2} = n$$

Therefore:

$$\Delta P = \lim_{k \to 0}\left\{ R_i L_i \frac{Q_f^2}{(1-k)^{2n}} \frac{\left[(1-k)^{2n+2} - (1-k)^2 \right]}{(1-k)^2 - 1} \right\} = R_i L_i Q_f^2 \cdot 1 \cdot n$$

Given that:

$$n R_i L_i = R_i L_T$$

We have:

$$\Delta P = R_i L_T Q_f^2$$

where L_T and R_T are the total length and total resistance of the duct, respectively.

In addition, if there are no leaks, the airflow rate of the fan is equal to the airflow rate at the duct outlet (gallery face), thus:

$$\Delta P = R_i L_T Q_0^2$$

Exercise 8.11 A secondary ventilation duct of length 400 m is formed of 50 m long segments. A fan blows to the system an airflow rate of 12 m³ s⁻¹ which reduces to 9.3 m³ s⁻¹ at the duct outlet due to leaks. Determine the k-leakage constant.

Solution

Solving for k from Eq. 8.57, we have:

$$k = 1 - 10^{\frac{1}{n}\log\left(\frac{Q_f}{Q_v}\right)} = 1 - 10^{\frac{1}{8}\log\left(\frac{9.3}{12}\right)} = 0.031$$

Exercise 8.12 The airflow requirements at a gallery face are $Q_v = 45.8$ m³ s⁻¹. The air is blown through a fibreglass duct of diameter $= 1.5$ m, length $= 660$ m and with an Atkinson constant, $K_{1.2} = 0.0028\frac{N\,s^2}{m^4}$. Duct segments are 12 m long and the leakage coefficient, k, is estimated to be 3.1×10^{-3}. You are asked to determine:

(a) The loss of airflow in the forcing system due to leaks,
(b) Pressure lost in the system, and
(c) The characteristic function of the system as a function of the airflow rate provided at the duct outlet (face).

Solution

(a)

The number of duct segments:

$$n = \frac{L_T}{L_i} = \frac{600}{12} = 55$$

Airflow rate to be provided by the fan:

$$Q_v = \frac{Q_f}{(1-k)^n} = \frac{45.8\frac{m^3}{s}}{\left(1 - 3.1 \times 10^{-3}\right)^{55}} = 54.33\frac{m^3}{s}$$

Leakage losses:

$$Q_l = (54.33 - 45.8)\frac{m^3}{s} = 8.53\frac{m^3}{s}$$

As a percentage of airflow at the face:

$$\%Q_r = \frac{8.53\frac{m^3}{s}}{45.8\frac{m^3}{s}} = 18.62$$

(b)

Since the resistance is the same for all the segments of the duct, it is possible to operate with the first one. Thus, the resistance per unit length is obtained as (Eq. 4.9):

$$R_i = K_i \frac{O_i}{A_i^3} = \frac{4K_i}{D}$$

As $K_{1.2} = 0.0028 \frac{Ns^2}{m^4}$, the resistance per unit length is:

$$R_i = K_i' \frac{O_i}{A_i^3} = \frac{64 K_i'}{\pi^2 D^5} = \frac{64 \cdot 0.0028 \frac{Ns^2}{m^4}}{\pi^2 (1.5\,m)^5} = 2.39 \times 10^{-3} \frac{Ns^2}{m^9}$$

Substituting in Eq. 8.59, we have:

$$\Delta P = \frac{2.39 \times 10^{-3} \frac{N\,s^2}{m^9} \cdot 12\,m \cdot \left(45.8 \frac{m^3}{s}\right)^2}{(1 - 3.1 \times 10^{-3})^{110}}$$

$$\frac{\left[(1 - 3.1 \times 10^{-3})^{110+2} - (1 - 3.1 \times 10^{-3})^2\right]}{(1 - 3.1 \times 10^{-3})^2 - 1}$$

$$= 3931.83\,Pa$$

(c)

From the expression above:

$$\Delta P = 1.87 \frac{N\,s^2}{m^8} Q_f^2$$

Exercise 8.13 An airflow rate of 7.5 m³ s⁻¹ needs to be delivered to the face of a cul-de-sac. For this purpose, a fan in a forcing configuration is connected to a duct 680 m long with a 700 mm diameter and resistance of 15 μ m⁻¹. Study how the different levels of airtightness of the installation (very well executed, well executed, poor and very poor) affect the fan pressure, fan airflow rate and fan power.

Use the models of Bertard and Bodelle (1962), the continuous model and the step by step methods. Assume that the exponent for friction loss $\alpha = 2$ and that the leakage exponent is 0.5.

Solution

As stated in the problem, $\alpha = 2$ and $\gamma = 0.5$ for all cases. Therefore:

Condition		Variable	Unit	Bertard and Bodelle (1962)	Continuous model		Step by step model	
					Solver	Table	Short	Long
Very well executed	θ: 0.000005	Number of segments (n)	–	–	–	–	4	16
		Fan pressure (P_v)	Pa	5889.0	5891.9	5995.9	5861.2	5857.7
		Fan airflow rate (Q_v)	m³ s⁻¹	7.55	7.56	7.56	7.51	7.5
		Fan power ($\eta = 1$)	kW	44	45	45	44	44
Well executed	θ: 0.00001	Fan pressure (P_v)	Pa	5921.6	5927.2	5991.9	5866.0	5858.8
		Fan airflow rate (Q_v)	m³ s⁻¹	7.61	7.61	7.62	7.51	7.5
		Fan power ($\eta = 1$)	kW	45	45	46	44	44
Poor	θ: 0.0001	Fan pressure (P_v)	Pa	6508.3	6611.9	6757.1	5952.7	5879.9
		Fan airflow rate (Q_v)	m³ s⁻¹	8.59	8.68	8.71	7.63	7.53
		Fan power ($\eta = 1$)	kW	56	57	59	45	44
Very poor	θ: 0.001	Fan pressure (P_v)	Pa	12,375.6	19,835.8	21,236.7	6925.6	6096.9
		Fan airflow rate (Q_v)	m³ s⁻¹	18.36	23.23	24.01	8.89	7.83
		Fan power ($\eta = 1$)	kW	227	461	510	62	48

The different calculation methods offer similar solutions for very well executed, well executed and even poor installations, however, they provide very different results for very poor installations. This can be explained bearing in mind that moderate leakage was assumed in the derivation of all formulae used in these methods.

8.15   Practical Rules for Estimating Leakage

In the case where exact operating conditions are unknown, as a rule of thumb, fan airflow rates can be estimated assuming losses of around 20–25% (Workplace Safety North 2010). This means that the volumetric flow supplied by the fan at the free outlet of the duct is reduced by 20–25%. This results in the loss of speed, and thus of dynamic pressure (often termed *velocity pressure* and sometimes expressed as *velocity head*).

Expressions of the type shown in Eq. 8.60, often appear in commercial catalogues:

$$Q_v = Q_f + \frac{Q_v\, LF}{100} \tag{8.60}$$

where

Q_v: Airflow rate at the fan ("ventilator", v), and
Q_f: Airflow rate at the face.

where LF is the % leak rate per 100 m of ducting. Thus, for a loss of 2% per 100 m, there would be a total loss of 20% at 1000 m. This expression is more accurate than the basic rule of thumb as it accounts for the total length of the duct.

If you have a leaky duct, the mean flow (Q_m) is usually approximated by the geometric mean of the flow rates at either end of the duct, i.e. (Eq. 8.61):

$$Q_m = \sqrt[2]{Q_f\, Q_v} \tag{8.61}$$

This can be approximated as shown in Eq. 8.62:

$$Q_m = Q_f + 0.5\, \Delta Q \tag{8.62}$$

where ΔQ is the airflow losses.

8.16   Friction Losses in Ducts

Where precise values for the Atkinson friction factor ($K_{1.2}$) of ducting is not available, the following ballpark figures can be used (de Souza 2004):

- Fibreglass: 0.0028 N s^2 m^{-4},
- Canvas: 0.0037 N s^2 m^{-4}, and
- Spiral: 0.00111 N s^2 m^{-4}.

In any case, manufacturers usually supply their own nomograms. The one in Fig. 8.16, corresponds to glass wool ducts from Flexadux.

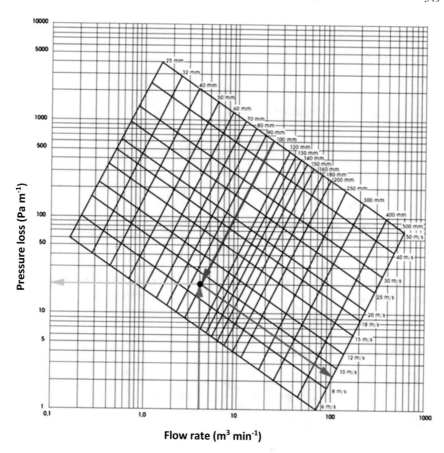

Fig. 8.16 Nomogram for the determination of friction losses as a function of airflow rate, speed and duct diameter. Modified from the Flexadux sales catalogue

Exercise 8.14 An airflow rate of 4 m³ min⁻¹ (0.067 m³ s⁻¹) traverses a 100 mm diameter duct. You are asked to:

(a) Estimate the pressure loss within the duct (Pa m⁻¹).
(b) Estimate the speed at which the air circulates under these conditions.
(c) Calculate the pressure loss by means of the Atkinson equation (with $K_{1.2} = 0.007 \frac{N\,s^2}{m^4}$).
(d) Calculate the speed at which the air moves in the duct.

Solution

(a)

Using the nomogram in Fig. 8.16, above: 20 Pa m⁻¹

(b)

Using the nomogram in Fig. 8.16: 9 m s^{-1}.

(c)

Applying the Atkinson equation:

$$\frac{\Delta P}{L} = K_{1.2} \frac{O\,Q^2}{A^3}$$

So:

$$\frac{\Delta P}{L} = 0.007 \frac{\text{N s}^2}{\text{m}^4} \cdot \frac{2 \cdot \pi \cdot 0.05\,\text{m} \cdot \left(0.067\frac{\text{m}^3}{\text{s}}\right)^2}{\left(\pi 0.05^2\,\text{m}^2\right)^3} = 20.37 \frac{\text{Pa}}{\text{m}}$$

It should be observed that since the nomogram (Fig. 8.15), has a logarithmic scale it does not give such accurate results as using Atkinson's expression.

(d)

The speed can be calculated as:

$$v = \frac{Q}{A} = \frac{0.067\frac{\text{m}^3}{\text{s}}}{\pi 0.05^2\,\text{m}^2} = 8.5 \frac{\text{m}}{\text{s}}$$

8.17 Shock Losses

To estimate shock losses, we must look at the dynamic pressure. This is calculated using the velocity, obtained as the quotient of the flow rate and the duct cross section. To estimate airflow velocity, it is generally acceptable to use either the flow rate at the point closest to the shock for which data is available or the most conservative estimated flow rate (depending on the designer's criteria). Thus, to estimate shock losses near the beginning of the duct, data from points close to the fan are usually used. For losses further along the duct, the average flow rate can be used.

It is important to note that, when constructing a conduit made of segments of ducting, it is advantageous to make these segments as long as possible. This reduces the number of joints, and thus the losses. Leakage between joints, tears in flexible ducts or damage due to machinery shocks should be repaired as soon as possible as these disturb airflow in the ventilation system. For the selection of the shock loss factors, Table 1.1 can be used.

8.18   Sizing the Auxiliary Fan

When selecting an auxiliary fan, the following must be considered, at least:

(a) The location of the fan with respect to the ducting (forcing, exhausting or booster).
(b) The type of secondary ventilation layout (forcing, exhausting, mixed or overlapped).
(c) The number of fans along the ducting (single, double, multiple).
(d) And whether static or total pressures are to be used for the sizing calculations.

Another fundamental consideration is the testing procedure used by the manufacturer to obtain characteristic fan curves (McPherson 1993, pp. 324–325). Thus, if the manufacturer places sensors in the test rig at points A and I (Fig. 8.17), shock losses will not be taken account of at all and must be included later by the design engineer. If sensors were positioned at points B and I, the manufacturer's data will include the shock losses incurred as air traverses the cone, however, those incurred entering the mouth of the duct will still be missed out. Finally, if pressure sensors were located at points C and I, then all losses have been accounted for by the manufacturer, and therefore, should not be added again during the sizing calculations.[21]

The case is similar for testing exhausting fans that discharge to the atmosphere. The sensor external to the duct must be placed well away from the cone outlet, where air velocity, and therefore, the dynamic pressure is zero, such that the static and total pressures are equal (McPherson 1993, pp. 324–325). In this case, measurements of fan operating characteristics will incorporate the effects of both friction losses in the cone and shock losses at the cone outlet, and so the design engineer can ignore them (Brake 2002). This is why manufacturing companies should provide information about how their tests were performed.

Another area where design engineers should take special care is in incorporating losses due to various accessories into calculations. Manufacturers do not usually take

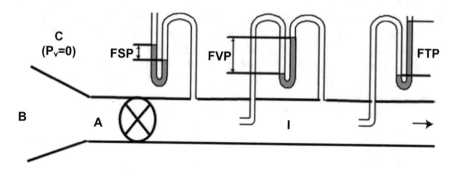

Fig. 8.17 Various sensor locations for the determination of a fan characteristic curve

---

[21]This fact is very important, because in some of our examples the losses at the duct entrance are added, and in others, they are ignored.

into account the effects of accessories when testing their fans, mainly because the end-user may require any number of different configurations.

In view of the above, when sizing an auxiliary fan, it is usual to add in losses from the following sources:

- Friction along the duct.
- Shock losses at the:

  - Fan inlet,
  - Grid,
  - Silencer,
  - Joints,
  - Elbows, and
  - Duct outlet.

Figure 8.18, is a diagram showing pressure losses in a booster fan. In this arrangement, a pressure measurement made just at the beginning of the inlet cone will give the smallest value of dynamic pressure (except for the outlet). At this same point, the static pressure (negative) will have the same absolute value as the dynamic pressure, to give a total pressure of zero. Static pressure and total pressure will both increase in terms of their absolute values until they reach the fan. As air traverses the fan, total pressure increases to a maximum value equal to the FTP (Fan Total Pressure). From that point on, total pressure and static pressure start to decrease due to the different losses, although there will be occasional gains in static pressure (static regain) at certain points along the duct, for example, where duct cross section increases.

At the system outlet, the fluid energy is all kinetic (i.e. due to air velocity), thus, the total and the dynamic pressures coincide. In the case of Fig. 8.18, there is a diffuser attached to the outlet, and the speed of air entering the diffuser will be higher than the speed of air leaving the diffuser. The kinetic energy of the air as it exits the system is sometimes referred to as the exit loss, which may generate some confusion.

The total exit loss for the system is calculated as an exit loss coefficient, X, multiplied by the dynamic pressure. This coefficient is dependent largely on the Reynolds number and the shape of the outlet. The shock loss factor may be greater than, less than, or equal to 1 depending on the Reynolds number and the configuration of the outlet (ASHRAE 2013; Carrier Corporation 1965, pp. 2–41). Values of $X_o = 1$ correspond to uniform speed profiles, larger than 1 (up to 3.67), to exponential, sinusoidal, asymmetric and parabolic profiles (ASHRAE 2013). In most circumstances under discussion here, exit losses are equal to one velocity head (i.e. $X = 1$).

Some authors (e.g. Brake 2002), believe that the problem can be addressed by considering it as a case where the conduit cross section expands abruptly as the flow enters the gallery. In our particular case, it is not exactly a loss of the system, as part of this energy will return through the gallery doing useful work, but a loss of the ventilation duct in which we focus our attention.

The Schauenburg Company provides the following rule of thumb for calculating the total static losses for flexible ducts:

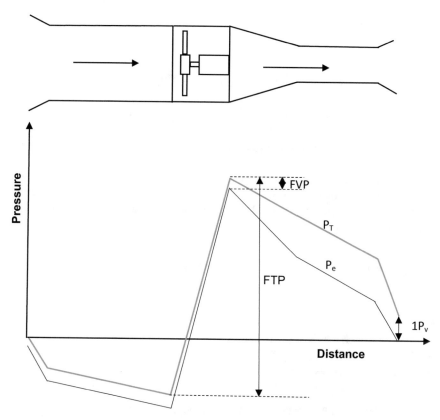

Fig. 8.18 Changes in total pressure (blue lines) and static pressure (black line) of air traversing a booster fan system. Modified from DuctSIM (n.d.), p. 31

$$C = AB$$

where C is the required pressure and A and B are defined below, as:

- A: Static friction losses in the duct per unit length:

While this can be calculated directly, the company provides graphs that enable its calculation per 100 feet of duct length.

- *B*: Equivalent duct length [22]:

The actual length of the duct

+

1.83 m (6 feet) equivalent length per coupling

+

30.5 m (100 feet) equivalent length for exit losses

The classic fluid mechanics approach to calculate pressure losses would be to apply Bernoulli's equation between two points: point 1 being at sufficient distance from the system intake and a point 2 being at sufficient distance from the duct outlet. In this way, we obtain:

$$\frac{1}{2}\rho v_1^2 + \rho g h_1 + P_1 - P_{f(i)} - P_{shock(i)} + P_{vent} = \frac{1}{2}\rho v_2^2 + \rho g h_2 + P_2 + P_{f(o)} + P_{shock(o)}$$

where

- $P_{f(i)}$: Pressure friction loss at the intake,
- $P_{shock(i)}$: Shock losses at the intake,
- P_{vent}: Pressure provided by the fan,
- $P_{f(o)}$: Friction losses at the outlet, and
- $P_{shock(o)}$: Shock losses at the outlet.

The following simplifications can be made here:

- $v_1 = v_2 = 0$ (still air),
- $h_1 = h_2$ (for horizontal driving),
- $P_1 = P_2 = P_{atm}$, and
- $P_{f(i)} = 0$ (if there is no upstream ducting and just a mouth coupled to the fan).

Later:

$$-P_{shock(i)} + P_{vent.} = P_{f(o)} + P_{shock(o)}$$

Hence:

$$P_{vent.} = FTP = P_{shock(i)} + P_{f(o)} + P_{shock(o)} \tag{8.63}$$

Once these losses have all been taken account of, we are, finally, in a position to calculate the size of the fan, or shaft power required to ventilate the system (Pw_{shaft}) using Eq. 8.64. ISO 5801 gives details on the different instalation types.

---

[22]Note that this manufacturer adds the fan inlet losses to the fan curve. Note that when this parameter is supplied it is no longer appropriate to work with static pressures alone.

$$Pw_{\text{shaft}} = \frac{\text{FTP}\, Q_v}{\eta_T} \qquad (8.64)$$

where

- FTP: Fan Total Pressure (Pa),
- Q_v: Airflow rate generated by the fan (m^3 s^{-1}), and
- η_T: Efficiency (fraction).

As a further consideration, when selecting the motor to drive the fan, it must be borne in mind that there are fluctuations between the air density in winter and summer. In this way, it is a common practice to over-size the fan by 20%.

In general, auxiliary fans are capable of delivering a flow rate of 14 m^3 s^{-1} at a pressure of up to 2.5 kPa over distances of up to about 760 m (ASHRAE, 2015). If ventilation needs are greater, additional booster fans are often used. In the most exceptional cases, the flow rates can be up to 28 m^3 s^{-1} with pressures of up to 3.7 kPa. The fan powers themselves are usually less than 75 kW, although the catalogues include engines of up to 185 kW (de Souza 2004).

The sizing calculations for secondary ventilation fans are similar whether considering a forcing or an exhausting system, except in the case of flexible ducts. If flexible ducts were used with an exhausting system, suctioning would cause cross section reduction, thereby increasing the resistance. As a result, flexible type ducts cannot be used for exhausting systems.

Exercise 8.15 In order to ventilate a cul-de-sac, a flow of 10 m^3 s^{-1} is taken from the main ventilation gallery by means of a duct of 1 m in diameter and 100 m long which contains a right-angle elbow as shown in the figure. It is known that:

- Atkinson's friction factor for the duct ($K_{1.2}$): 0.003 kg m^{-3},
- Air density: 1.18 kg m^{-3},
- Shock loss factors:

  - At the duct entrance: 1,
  - At the duct outlet: 1. This is because there is no grille, diffuser or contraction, it is assumed that the velocity profile is uniform, and the static pressure at the outlet is the atmospheric pressure, and
  - At the elbow: 0.8.

- Mechanical efficiency of the fan: 70%,

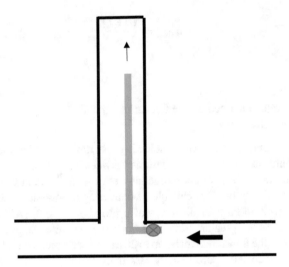

You are asked to calculate:

(a)   Pressure losses due to all resistances in the system,
(b)   Power loss due to all these resistances,
(c)   Velocity pressure at the fan outlet, and
(d)   Fan power.

You may simplify by assuming that there are no airflow leaks along the length of the system and that the fan and duct diameters are equivalent.

The following equations will be of use:

Shock loss resistance:

$$R_{sh} = \frac{X_{sh}\rho}{2A^2}$$

Friction resistance:

$$R_{fr} = \frac{K_{1.2}\, L\, O}{A^3}$$

where

O: Perimeter of the duct cross section.

Solution

(a)

1st Since there are no leaks, the flow rate at the inlet of the duct is equal to the flow rate at the outlet, i.e. the continuity equation can be applied. That is:

$$A_1 v_1 = A_2 v_2$$

Also, as $A_1 = A_2$, the air velocity along the duct remains constant.

The pressure gradient diagram will be similar to that shown in the figure below. For the resolution of this problem it is chosen to add to the total pressure losses on the blowing side the total losses at the fan inlet, that is, it is operated from the total pressure graph.

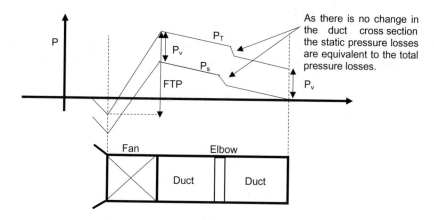

As there is no change in the duct cross section the static pressure losses are equivalent to the total pressure losses.

2nd We start by calculating the friction loss (P_{fr}) through the Atkinson expression:

$$P_{fr} = \frac{K_{1.2}L\,O}{A^3} Q^2 \left[\frac{\rho}{1.2\left(\frac{kg}{m^3}\right)} \right]$$

Substituting in the values provided gives:

$$P_{fr} = \frac{0.003\frac{N \cdot s^2}{m^4} \cdot \pi \cdot 1\,m \cdot 100\,m}{\left(\pi \cdot 0.5^2 m^2\right)^3} \left(10\frac{m^3}{s}\right)^2 \left[\frac{1.18\frac{kg}{m^3}}{1.2\left(\frac{kg}{m^3}\right)} \right] = 191.29\,Pa$$

3rd The shock losses at the inlet and discharge of the duct, as well as at the elbow, must now be calculated. The shock loss at the inlet is often included in the testing data provided by some manufacturers (Brake 2002), but since nothing is indicated here, it is calculated.

Concerning the outlet, the shock loss coefficient of $X = 1$ implies a loss of one dynamic pressure. Static pressure, P_s at the outlet will be equal to atmospheric pressure. Therefore, pressure variation as air flows out of the duct can be represented as shown in the figure (where P_T is the total pressure).

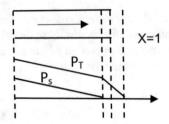

As the air velocity in the duct is constant, the calculation can be made by means of a unique total loss coefficient (X_T), which means:

$$X_T = 1 + 0.8 + 1 = 2.8$$

Thus, the total pressure loss due to shock (P_{TSh}) is:

$$P_{Tsh} = \frac{X_{Tsh}\rho}{2A^2}Q^2 = \frac{2.8 \cdot 1.18\frac{kg}{m^3}}{2\left(\frac{\pi \cdot 1^2}{4}m^2\right)^2}\left(10\frac{m^3}{s}\right)^2 = 267.81\,\text{Pa}$$

Hence, total losses $(P_{T_{loss}})$ can be calculated as:

$$P_{T_{loss}} = P_{fr} + P_{Tsh} = 459.10\,\text{Pa}$$

(b) So the power loss $(P_{w_{loss}})$ will be:

$$P_{w_{loss}} = P_{T_{loss}}\,Q = 4591\,\text{W}$$

(c) Using the values provided, the dynamic pressure is:

$$P_v = \frac{1}{2}\rho v^2 = \frac{1}{2}1.18\frac{kg}{m^3}\left(\frac{10}{\frac{\pi \cdot 1^2}{4}}\frac{m}{s}\right)^2 = 95.65\,\text{Pa}$$

Note that because we have assumed that there are no leaks (and therefore, no pressure loss due to leakage), the dynamic pressure at the fan outlet and the shock losses at the duct outlet are the same. That is:

$$P_{outlet} = \frac{X_T\rho}{2A^2}Q^2 = \frac{1 \cdot 1.18\frac{kg}{m^3}}{2\left(\frac{\pi \cdot 1^2}{4}m^2\right)^2}\left(10\frac{m^3}{s}\right)^2 = 95.65\,\text{Pa}$$

(d) So the shaft power of the fan:

$$Pw_{shaft} = \frac{FTP\,Q_v}{\eta_T} = \frac{459.10\,Pa \cdot 10\frac{m^3}{s}}{0.7} = 6558.57\,W = 6.6\,kW$$

Exercise 8.16 The construction, by drilling and blasting, of a main gallery is planned. This gallery will have a 30 m² cross section, 17 m perimeter and a final length of 200 m with $K_{1.2} = 0.015$ N s² m⁻⁴. The following machinery will be used during the works:

- 1 × 150 kW loader,
- 1 backhoe loader of 100 kW,
- 1 jumbo of 120 kW,
- 1 telescopic boom lift of 60 kW,
- 3 dumpers of 90 kW each,
- 1 gunite robot of 50 kW, and
- 1 concrete mixer truck of 100 kW.

The maximum number of people working on the face at the same is estimated to be 5, and an optimum air velocity in the gallery has been specified as 1 m s⁻¹. The ducts used in the ventilation system will be made of glass wool ($K_{1.2} = 0.0028$ Ns² m⁻⁴) with joints every 50 m. Each joint creates pressure losses equivalent to those produced by 2 m of duct length. The system extends 5 m into the existing main gallery by means of a 90° elbow (as in the case of the Exercise 8.15). The fan set up is shown in the figure below:

Tests performed by the manufacturer did not include shock losses in the bell-shaped inlet. The fan has a diameter of 1.27 m with a total efficiency of 83% (obtained for the conditions outlined in this exercise). The duct is 1.3 m in diameter.

Determine:

(a) Necessary airflow rate to supply adequate fresh air for operators to breathe.
(b) Minimum airflow rate to return the gases through the gallery at the specified minimum speed.
(c) Airflow rate required to dilute the engine exhaust gases assuming a combination of 1 loader, 1 backhoe and 3 dumpers working at the same time.
(d) Airflow rate necessary to dilute gases generated by the explosives used in blasting.
(e) The total airflow required at the gallery face.
(f) Airflow rate to be provided by the fan (estimate 25% leak rate at the joints).
(g) Air velocity at the duct outlet.
(h) Dynamic pressure at the face.

(i)   Air velocity at the fan outlet (duct inlet).
(j)   Dynamic pressure at the fan.
(k)   Average airflow rate in the duct.
(l)   Shock losses at the inlet cone of the fan (use $X = 0.5$).
(m)   Shock losses at the fan screen (the screen has 85% porosity).
(n)   Shock losses at the upstream silencer.
(o)   Shock losses at the downstream silencer.
(p)   Total friction losses in the duct.
(q)   Friction losses at the joints.
(r)   Shock losses at the elbow (use $X = 0.5$).
(s)   Losses at the duct outlet (discharge).
(t)   Return airflow pressure loss.
(u)   Total pressure loss.
(v)   Corrected total pressure loss.
(w)   Size of the fan required.
(x)   In this example, some parameters are above the expected normal values. Identify
       them and indicate the consequences this has on the design.

Formulae:

Shock loss coefficient at the screen, following Idelchik (1986):

$$X_r = 1.3\,(1 - FL) + \left(\frac{1}{FL} - 1\right)^2$$

where FL is the screen porosity.
        Shock losses of a component:

$$P_{component} = P_v\,X_{component}$$

where P_v is the dynamic pressure taken at a point sufficiently close to the component
in question.
        Required ventilation flow rate[23] (Q_m) per kW of diesel power.

$$Q_m = 0.06\frac{m^3}{s\,kW}\,Pw(kW)$$

        To solve this problem, it is advised that you follow the methodological outline
recommended by WSN (2013) for permeable ducts. Note differences with ISO 5801.

Solution

Fan flow calculations (items a to f)

---

[23] This flow rate is needed to dilute the combustion gases, not as a supply of air for combustion,
which would be much lower.

(a) Flow rate required to supply adequate fresh air for operators to breathe, (Q_{op}), is set as a standard 40 l/s. In the case of five operators:

$$Q_{op} = \frac{40\,l}{s \cdot operator} \cdot 5\,operators = 200\frac{l}{s} = 0.2\frac{m^3}{s}$$

(b) Minimum airflow rate (Q_r) to return gases through the gallery at the specified minimum speed of 1 m s^{-1}:

$$Q_r = A_{galllery}V_{gallery} = 30\,m^2 \cdot 1\frac{m}{s} = 30\frac{m^3}{s}$$

It should be noted that this velocity is more usually within the range 0.1 m s^{-1-6} m s^{-1}.

(c) Airflow rate required to dilute the engine exhaust gases (Q_m) is defined above and all that is required is to add up the powers of all the equipment that may be in operation in the gallery at the same time (in this case 1 loader, 1 backhoe, 3 dumpers):

$$Q_m = 0.06\frac{m^3}{s\,kW} \cdot (150 + 100 + 3 \cdot 90)kW = 31.2\frac{m^3}{s}$$

(d) Airflow rate required to dilute the gases generated by the explosives used in blasting (Q_e):

No data are provided on the quantities of explosive used in each cycle; however, an estimate can be made using a rule of thumb:

$$Q_e\left(\frac{m^3}{s}\right) = 0.28\frac{m}{s}\,A_{gallery}(m^2) = 8.4\frac{m^3}{s}$$

(e) Calculation of the total airflow required at the face (Q_f):

Selection of the highest flow rate between Q_e and ($Q_{op} + Q_m$):

$$Q_e = 8.4\frac{m^3}{s}$$

$$Q_{op} + Q_m = (0.2 + 31.2)\frac{m^3}{s} = 31.4\frac{m^3}{s}$$

The largest is 31.4 m^3 s^{-1}, which is in turn higher than 30 m^3 s^{-1} (airflow rate required for the gases to return at 1 m s^{-1}).

(f)   Airflow rate to be provided by the fan taking into account leakage (Q_v):

Leaks of 0.09 m³ s⁻¹ per joint are usually assumed. Another rule of thumb is to assume that 10–30% of the flow rate supplied by the fan will be lost (see Sect. 8.16). Here, the question states we are estimating a 25% loss rate, so, using Eq. 8.60:

$$Q_v = \frac{Q_f}{1 - LF} = \frac{31.4\frac{m^3}{s}}{1 - 0.25} = 41.87\frac{m^3}{s}$$

The first important parameter for sizing the fan has already been obtained: the flow rate. All that remains is to calculate operating pressures.

Dynamic pressure calculations and average aiflow rate (items g to k)

Note on duct diameter

To reduce friction in the duct, and thus energy costs, it follows from Atkinson's equation that, in principle, a duct of the largest possible diameter should be used. In practice choice of duct diameter is limited by the height of equipment and the dimensions of the gallery. Alternatively, some manufacturers propose tables for the selection of the appropriate diameter.

As it was indicated in the statement we will use 1.3 m.

(g)   Air velocity at the duct outlet onto the gallery face (v_f):

with $D = 1.3$ m

$$v_f = \frac{Q_f}{A_{duct}} = \frac{31.4\frac{m^3}{s}}{\pi\frac{1.3^2}{4}m^2} = 23.66\frac{m}{s}$$

(h)   Dynamic pressure at the face (P_{vf}):

$$P_{vf} = \frac{1}{2}\rho v_f^2 = \frac{1}{2}1.2\frac{kg}{m^3}\left(23.66\frac{m}{s}\right)^2 = 335.87\,Pa$$

(i)   Air velocity at the fan outlet (duct inlet) (v_v):

$$v_v = \frac{Q_v}{A_{duct}} = \frac{41.87\frac{m^3}{s}}{\pi\frac{1.3^2}{4}m^2} = 31.54\frac{m}{s}$$

(j)   The dynamic pressure at the fan (P_{vv}) can be approximated as:

$$P_{vv} = \frac{1}{2}\rho v_v^2 = \frac{1}{2}1.2\frac{kg}{m^3}\left(31.54\frac{m}{s}\right)^2 = 596.86\,Pa$$

(k) Average airflow in the duct:

As we have seen, leaks cause variation in the volumetric flow rate along a duct. This is because leakage reduces air velocity, and therefore, dynamic pressure. In the leaky duct, the average flow rate is usually calculated as the geometric mean of the flow rates at each end (Eq. 8.61):

$$Q = \sqrt[2]{Q_f \, Q_v}$$

$$Q = \sqrt[2]{31.4 \frac{m^3}{s} \cdot 41.86 \frac{m^3}{s}} = 36.25 \frac{m^3}{s}$$

Another way to approach this is by using Eq. 8.62:

$$Q = Q_f + 0.5 \, Q_{loss} = 31.4 \frac{m^3}{s} + 0.5(41.86 - 31.4) \frac{m^3}{s} = 36.63 \frac{m^3}{s}$$

Pressure losses in the system (items l to v)

In order to evaluate the shock losses in the system, we must estimate the airspeed at the points where these losses will be incurred, for example, close to obstructions, elbows or constrictions. If airflow rate data is available at these points, then these values can be used, however, for most cases, we will have to estimate the airflow rate. Thus, to estimate shock losses near the fan outlet (duct inlet), or close to the duct discharge, we can use the fan flow rate (Q_v), or the duct outlet flow rate (Q_f), respectively. For shock losses occurring along the remainder of the duct, the average flow rate can be used.

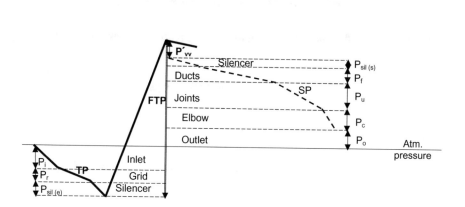

In this system, all shock losses take place in segments of ducting that have an equal cross section (ducts and elbow), which means that the static pressure losses (P_s) and the total losses (P_T) coincide. In this way, downstream of the fan we can look at the gradient of total pressure (TP) while upstream of the fan we look at the gradient of static pressure (SP). FTP is obtained by subtracting the total pressure at the fan outlet to the total pressure at the fan inlet (negative). It should be noted that the fan curves published by some manufacturers include losses incurred at the inlet. Therefore, these will not always have to be considered.

(l)   Shock losses at the inlet cone (P_i):

When calculating this loss, we should bear in mind that is may already be accounted for by the manufacturer in their published fan curve (Brake 2002). The dynamic pressure in the inlet is estimated to be equal to that existing at the fan outlet P_{dv} [calculated in (j)].[24] Then, using the shock loss coefficient, $X = 0.5$, which is given in the problem, we have [25]:

$$P_i = X\, P_{vv} = 0.5 \cdot 596.86\,\text{Pa} = 298.5\,\text{Pa}$$

(m)   Shock losses at the fan screen (P_r):

Assuming a porosity (FL) of 85%, as stated in the problem. This means that 85% of the air passage cross section is not occupied by the grid wires and can be expressed mathematically as:

$$FL = \frac{A_{\text{empty}}}{A_T} = 0.85$$

Then, the losses in the screen, using the expression given to us in the problem:

$$X_r = 1.3\,(1 - FL) + \left(\frac{1}{FL} - 1\right)^2 = 0.226$$

Then, using the dynamic pressure at the fan, the required shock loss will be:

$$P_r = X_r\, P_{vv}$$
$$P_r = 0.226 \cdot 596.86\,\text{Pa} = 135\,\text{Pa}$$

---

[24] A recalculation of the value will be carried out later, and it will be seen that this value is slightly higher.

[25] It should be noted that this is a somewhat high coefficient for a cone entry, although it is on the safety side of fan sizing.

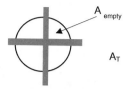

(n) Shock losses at the upstream silencer ($P_{sil\,(e)}$):

These losses vary greatly depending on the precise characteristics of fan configuration. For example, silencers can be located before and after the fan. In most cases, manufacturers will supply information about these losses which are generally in the region of 60–100 Pa.

(o) Shock losses at the downstream silencer ($P_{sil\,(s)}$):

As above: 60 Pa.

(p) Total friction losses in the ducts (P_f):

The duct is made from fibreglass for which there is a generally accepted value of $K_{1.2} = 0.0028\,\mathrm{Ns^2\,m^{-4}}$.
   Duct-to-face distance: $5\sqrt{S} = 5\sqrt{30} = 27\,\mathrm{m}$
   This is within the normal range of 20–30 m.

$$P_f = \frac{K_{1.2}\,O\,L\,Q^2}{A^3}$$

where: O is the perimeter (m) of the duct cross section, L is the duct length (m) and Q is the average airflow rate in the duct ($\mathrm{m^3\,s^{-1}}$) (item k).
   Considering that the elbow extends for 5 m into the main gallery, we have:

$$P_f = \frac{0.0028\frac{\mathrm{Ns^2}}{\mathrm{m^4}} \cdot (2 \cdot \pi \cdot 0.65)\mathrm{m}\,(200 + 5 - 27)\mathrm{m} \cdot \left(36.63\frac{\mathrm{m^3}}{\mathrm{s}}\right)^2}{\left(\pi \cdot 0.65^2\mathrm{m^2}\right)^3} = 1167.93\,\mathrm{Pa}$$

Another option is to use the manufacturer's nomograms.

(q) Friction losses at the joints (P_u):

This depends on the characteristics of the duct:

- layflat duct, 2 m of equivalent length can be added per joint.
- spiral duct, 2.5 m of equivalent length can be added per joint.

$$P_u = \frac{K\,O\,L_u\,Q^2}{A^3}$$

The problem states that 2 m should be added per joint. However, it does not indicate the number of joints, only that they exist every 50 m. Thus we can assume about 3 (duct) + 2 (elbow) +1 (fan) for the total number of joints.

$$P_u = \frac{0.0028\,\frac{N s^2}{m^4} \cdot (2 \cdot \pi \cdot 0.65)m\left[2 \cdot \left(\frac{200}{50} - 1 + 2 + 1\right)\right]m \cdot \left(36.63\frac{m^3}{s}\right)^2}{\left(\pi \cdot 0.65^2 m^2\right)^3} = 78.73\,\text{Pa}$$

It should be noted that there is an additional joint to connect the elbow to the 5 m of duct extending into the gallery.

(r) Shock losses at the elbow (P_c):

We have been given the shock loss coefficient of the elbow and, as the elbow is close—but far enough to consider that there are no interferences between them—to the fan, we will again use the dynamic pressure at the fan inlet [calculated in (j)]:

$$P_c = X_c\,P_{vv}$$
$$P_c = 0.5 \cdot 596.86\,\text{Pa} = 298.4\,\text{Pa}$$

(s) Static pressure loss at the discharge (P_o):

Pressure loss at the duct outlet is mostly of the total kind.[26] However, WSN (2013) establishes one velocity pressure ($X_o = 1$), thus:

$$P_o = X_o\,P_{vf}$$
$$P_o = 335.87\,\text{Pa}$$

(t) Return airflow pressure loss (P_{gr}):

This is the extra pressure that the fan must provide so that gases at the gallery face return through the cul-de-sac to the main ventilation circuit. This pressure can be estimated according to Atkinson's equation and the dimensions of the gallery:

$$P_{gr} = \frac{K\,O\,L\,Q^2}{A^3} = \frac{0.015\,\frac{N s^2}{m^4} \cdot 17\,\text{m} \cdot 200\,\text{m} \cdot \left(31.4\frac{m^3}{s}\right)^2}{\left(30\,\text{m}^2\right)^3} = 1.86\,\text{Pa}$$

---

[26]Luo and Zhou (2013), proposed the use of an equivalent length of 30.5 m. Moreover, sometimes flexible ducts can be constricted at the outlet so that air is forced closer into the face, thus acting as a nozzle. Despite this, static pressure loss at the duct outlet is considered 0 in many calculations.

This pressure is very low due to the fact that the cross-sectional area of the gallery (A) is very large in comparison with that of the duct. Thus, it can be neglected.

(u) Total losses (P_T):

This is the sum of all losses previously calculated:

$$P_T = P_i + P_{\text{sil}} + P_f + P_u + P_c + P_o + P_r + P_{gr}$$
$$= (298.5 + 120 + 1161.56 + 78.73$$
$$+ 298.4 + 335.78 + 135 + 0)\text{Pa} = 2427.97\,\text{Pa}$$

(v) Corrected total pressure loss (P_{Tc}):

The value obtained above should be corrected to take into account any possible deterioration of the installation. For this purpose, we introduce a quality factor for the installation (FC). In general:

- Bad quality: $FC = 1.4$
- Good quality: $FC = 1.2$

Assuming a good quality installation:

$$P_{Tc} = 1.2 \cdot 2427.97\,\text{Pa} = 2913.564\,\text{Pa}$$

(w) Size of fan required:

First, the total pressure to be generated by the fan (P_{Tv}) must be calculated. This value is the sum of the corrected total losses (P_{Tc}) plus the dynamic pressure at the fan outlet. As the diameter of the fan, resulting in area A', is different from that of the duct,[27] the previously calculated value of the dynamic pressure at the fan (P_{vv}) should be corrected to give P'_{vv}. So, firstly we must find the new air velocity at the fan:

$$v'_v = \frac{Q_v}{A'} = \frac{41.87\frac{\text{m}^3}{\text{s}}}{\pi \cdot \frac{1.27^2}{4}\text{m}^2} = 33.06\frac{\text{m}}{\text{s}}$$

Then the new pressure:

$$P'_{vv} = \frac{1}{2}\rho\,v^2_{v'} = \frac{1}{2}1.2\frac{\text{kg}}{\text{m}^3}\left(33.06\frac{\text{m}}{\text{s}}\right)^2 = 655.58\,\text{Pa}$$

---

[27]Note that the expansion from the fan diameter to the duct diameter will produce further losses. If these are small, they can be neglected.

From which we get the total pressure:

$$P_{Tv} = P_{Tc} + P'_{vv} = 2913.564\,\text{Pa} + 655.58\,\text{Pa} = 3569.14\,\text{Pa}$$

Fan selection

The operating point of the fan is, therefore:

$$P_{Tv} = 3553.4\,\text{Pa}$$

$Q_v = 41.87\,\text{m}^3\text{s}^{-1}$ [calculated in (f)].

If $\eta = 0.83$ (remember that η varies with P and Q and has to be obtained from graphs) as given in the statements:

$$Pw = \frac{P_{Tv}\,Q_v}{\eta} = \frac{3569.14\,\text{Pa} \cdot 41.87\frac{\text{m}^3}{\text{s}}}{0.83} = 180048.06\,\text{W}$$

$$Pw = 180\,\text{kW}$$

If the fan performance curve was provided, we would interpolate the curves and follow this scheme:

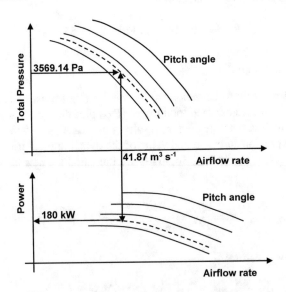

It should be noted that the installation might require such a large power that it cannot be supplied by a single fan. In this case, several fans are used in series. Another aspect to consider is that the calculations have been made for the maximum length of the gallery. However, during the advance, gallery length will be smaller than this maximum. As a result, the airflow rate must be regulated using a variable frequency drive to comply with regulations concerning maximum permitted air velocities in underground works.

(x) Before providing any result, students should check, at least, if their answers lie within the normal range. If we have a look at the last paragraph of Sect. 8.18, we can see that auxiliary fans usually provide up to 28 m s^{-1} at up to 3.7 kPa. Therefore, airflow requirements in the exercise are high. As the main reasons for this, we can highlight the high level of leakage and the fact that estimates for the machinery working in the tunnel at the same time could well be too large (the use of simultaneity factors would have been more appropriate). Moreover, the velocity at which air circulates inside the duct is above that generally recommended for this kind of duct (20 m s^{-1}). For this reason, it would be advisable to modify the design, taking into account these considerations.

References

Andrade, S. (2008). *Guía metodológica de seguridad para ventilación de minas.* Chile: Departamento de Seguridad Minera. SERNAGEOMIN.

Audibert, M. M., & Dressler, J. (1946). Etude expérimentale du fonctionnement des canalisation d´aérage secondarie. *Revue de l´Industrie Minérale,* 511 251–281.

ASM-51 (2010). Instrucción Técnica Complementaria. Explotación de capas de carbón por el método de sutiraje desde subniveles en fondo de saco. BOPA n 67.

ASHRAE (2013). *ASHRAE handbook-fundamentals. American Society of Heating.* Refrigerating and Air-Conditioning Engineers, Chapter 2. Atlanta, GA: Refrigerating and Air-Conditioning Engineers, Inc.

Auld, G. (2004). An estimation of fan performance for leaky ventilation ducts. *Tunnelling and Underground Space Technology, 19*(6), 539–549.

Bertard, C., & Bodelle, J. (1962). Chapitre IX. Aérage secondary. Aérage. *Revue de l'Industrie Minérale.* Document SIM N2.

Borisov, S., Klokov, M. Y., & Gornovoi, B. (1976). *Labores mineras.* Moscú: Editorial Mir.

Brake, D. J. (2002). Fan total pressure or fan static pressure: Which is correct when solving ventilation problems? *Mine Ventilation Society of South Africa, 55*(1), 6–11.

Carrier Corporation. Carrier Air Conditioning Company. (1965). *Handbook of air conditioning system design.* Table 10. McGraw-Hill Companies.

Deniau, R. (1976). Aérage des quartiers exploités par chambres et piliers. En: Aérage. Document SIM N3. Industrie Minérale. Mine 2–76.

de la Vergne, J. (2008). *Hard rock miner's handbook* (5th ed.). Alberta: Stantec Consulting Ltd.

de Souza, E. M., & Katsabanis, P. D. (1991). On the prediction of blasting toxic fumes and dilution ventilation. *Mining Science and Technology, 13,* 223–235.

de Souza, E. M. (2004). Auxiliary ventilation operation practices. In R. Ganguli & S. Bandopadhyay (Eds.), *Mine ventilation* (pp. 341–349). Taylor & Francis Routledge.

Dick, R. A., Fletcher, L. R., & D'Andrea, D. V. (1982). *Explosives and blasting procedures manual.* Washington, DC: US Department of the Interior, Bureau of Mines.

DuctSIM User´s Manual. (n.d.). Clovis, California: Mine Ventilation Services.

Expilly, P. (1960). *Ventilation des Souterrains en Construction.* Paris: Éditions Eyrrolles.

Flügge, G. (1953). Wirtschaftlichkeitsfragen der Sonderbewetterung. In *Conference at Technische Fachhochschule Georg Agricola.*

Gangal, M. (2012). Summary of worldwide underground mine diesel regulations. In *Proceedings of the 18th MDEC Conference*. Toronto, Ontario.

García, M. M., & Harpalani, S. (1989). Distribution and characterization of gases produced by detonation of explosives in an underground mine. *Mining Science and Technology, 8*(1), 49–58.

Gillies, A. D., & Wu, H. W. (1999). A comparison of air leakage prediction techniques for auxiliary ventilation ducting systems. In *Proceedings Eighth US Mine Ventilation Symposium, Society of Mining Engineers* (pp. 681–690).

Greig, J. (1982). Gases Encountered in Mines, Chapter 26. In J. Burrows (Ed.), *Environmental engineering in south African mines* (pp. 713–739).

Harris, M. L., Sapko, M. J., & Mainiero, R. J. (2003). *Toxic fume comparison of a few explosives used in trench blasting*. Mainiero National Institute for Occupational Safety and Health Pittsburgh Research Laboratory.

Holdsworth, J. F., Pritchard, F. W., & Walton, W. H. (1951). Fluid flow in ducts with a uniformly distributed leakage. *British Journal of Applied Physics, 2*(11), 321.

Howard, D. G., & Esko, T. (2001). *Industrial ventilation design guidebook*. Elsevier.

Howes, M. J. (1994). *Advanced ventilation course notes*.

Idelchik, I. E. (1986). *Handbook of hydraulic resistance*. Washington DC: Hemisphere Publishing Corp.

Kissell, F. N. (2006). *Handbook for methane control in mining*. National Institute for Occupational Safety and Health. Pittsburgh Research Laboratory: Pittsburgh.

Le Roux, W. L. (1972). *Mine ventilation notes for beginners*. The Mine Ventilation Society of South Africa.

Luo, Y. & Zhou, L. (2013). Coal mine ventilation. In C. J. Bise (Ed.), *Modern American coal mining: Methods and applications*. SME.

McPherson, M. J. (1993). *Subsurface ventilation and environmental engineering*. Chapman & Hall.

MITC (Ministerio de Industria Turismo y Comercio) (2000). SMI-Reglamento de General de Normas Básicas de Seguridad Minera e Instrucciones Técnicas Complementarias. (RGNBSM) Madrid.

Novitzky, A. (1962). *Ventilación de minas*. Buenos Aires: Yunque.

Robertson, R., & Wharton, P. B. (1980). Auxiliary ventilation systems—planning and application. In *Proceedings of CEC Information Symptoms* (p. 476). Luxemburg.

Rowland III, J. H., & Mainiero, R. J. (2000). Factors affecting ANFO fumes production. In *Proceedings of the 26th Annual Conference on Explosives and Blasting Technique* (Vol. I, pp. 163–174). Cleveland, OH: International Society of Explosives Engineers.

Simode (1976). Ventilation secondaire. *Industrie Minérale*. (Vol. 2–76, pp. 257–281).

Skochinsky, A., & Komarov, V. (1969). *Mine ventilation*. Mosco: Mir Publishers. Russia.

Stewart, C. M. (2014). Practical prediction of blast fume clearance and workplace re-entry times in development headings. In *10th International Mine Ventilation Congress*. The Mine Ventilation Society of South Africa.

Taylor, G. I. (1953). Dispersion of soluble matter in solvent flowing slowly through a tube. *Proceedings of the Royal Society of London, Series A: Mathematical and Physical Sciences, 219*(1137), 186–203.

Taylor, G. (1954). The dispersion of matter in turbulent flow through a pipe. *Proceedings of the Royal Society of London, Series A: Mathematical and Physical Sciences, 223*, 446–468.

Toraño, J., Menéndez, M., Rodríguez, R., & Cuesta, A. (2002). Non-iterative method of calculating ventilation requirements in tunnelling and mine roadway dead-ends. *Mining Technology, 111*(2), 115–122.

Vutukuri, V. S., & Lama, R. D. (1986). *Environmental Engineering in Mines*. Cambridge: Cambridge University Press.

Widiatmojo, A., Sasaki, K., Sugai, Y., Suzuki, Y., Tanaka, H., Uchida, K., et al. (2015). Assessment of air dispersion characteristic in underground mine ventilation: Field measurement and numerical evaluation. *Process Safety and Environmental Protection, 93*, 173–181.

WMC. (2001). *Underground ventilation major hazard standard.* WMC Environment, Health & Safety Management System.

Workplace Safety North (WSN). (2010). *Auxiliary mine ventilation manual.* Canada: Workplace Safety North.

Zawadzka-Małota, I. (2015). Testing of mining explosives with regard to the content of carbon oxides and nitrogen oxides in their detonation products. *Journal of Sustainable Mining, 14*(4), 173–178.

Correction to: Mine Ventilation

Correction to:
C. Sierra, *Mine Ventilation*,
https://doi.org/10.1007/978-3-030-49803-0

The original version of the book was inadvertently published without the Introduction section of each chapter, along with misplacement of some figures and missed typographical errors across chapters, which have now been incorporated and corrected accordingly.

The updated version of the book can be found at
https://doi.org/10.1007/978-3-030-49803-0

Printed in the United States
by Baker & Taylor Publisher Services